Python和Pygame
游戏开发指南

[美] Al Sweigart 著
李强 译

Making Games with Python and Pygame

人民邮电出版社

北京

图书在版编目（CIP）数据

Python和Pygame游戏开发指南 / （美）斯维加特 (Sweigart, A.) 著；李强译. -- 北京：人民邮电出版社，2015.12（2023.11重印）
ISBN 978-7-115-40735-1

Ⅰ. ①P… Ⅱ. ①斯… ②李… Ⅲ. ①游戏程序－程序设计－指南 Ⅳ. ①TP311.5-62

中国版本图书馆CIP数据核字（2015）第260784号

内容提要

Python 语言和 Pygame 都是开发图形化的计算机游戏的得力工具。Pygame 使得开发 2D 图形程序变得很容易，而且它可以免费下载和安装使用。

本书是一本中级编程图书，教读者如何用 Python 语言和 Pygame 库，来编写图形化的计算机游戏。本书共包括 10 章。本书首先从 Python 和 Pygame 基础知识开始，简要地介绍了 Pygame 库是如何工作的，以及它提供了哪些功能。然后，结合 7 款不同的游戏实例的开发过程，详细介绍了应用的技能和技巧。本书针对一些真实的游戏给出了完整的源代码，并且详细说明了这些代码如何工作，以便读者能够理解真正的游戏是如何使用 Pygame 的。

本书适合有一定 Python 基础知识的读者阅读和学习，进而掌握基本的游戏开发知识和技能，对于 Python 初学者来说，本书也可以作为通过游戏学习 Python 开发的实践教程。

◆ 著　　[美] Al Sweigart
　 译　　李强
　 责任编辑　陈冀康
　 责任印制　张佳莹　焦志炜

◆ 人民邮电出版社出版发行　北京市丰台区成寿寺路11号
　 邮编 100164　电子邮件 315@ptpress.com.cn
　 网址 http://www.ptpress.com.cn
　 北京七彩京通数码快印有限公司印刷

◆ 开本：800×1000　1/16
　 印张：21.5　　　　　　　　　　2015年12月第1版
　 字数：465千字　　　　　　　　2023年11月北京第25次印刷
　 著作权合同登记号图字：01-2015-5085 号

定价：69.90 元
读者服务热线：(010)81055410　印装质量热线：(010)81055316
反盗版热线：(010)81055315
广告经营许可证：京东市监广登字20170147号

版权声明

Simplified Chinese translation copyright ©2015 by Posts and Telecommunications Press

ALL RIGHTS RESERVED

Making Games with Python and Pygame ISBN-13:978-1469901732

By Al Sweigart

Copyright ©2012 by Al Sweigart

本书中文简体版由作者 Al Sweigart 授权人民邮电出版社出版。未经出版者书面许可,对本书的任何部分不得以任何方式或任何手段复制和传播。

版权所有,侵权必究。

作者简介

Al Sweigart（也可以称呼他 Al）是加利福尼亚州旧金山的一名软件开发者。他很喜欢骑自行车、当志愿者、泡咖啡吧以及制作有用的软件。《Making Games with Python & Pygame》是他编写的第 2 本书。他的第 1 本书是《Invent Your Own Computer Games with Python》。

他生于德克萨斯的休斯顿。他在德克萨斯大学 Austin 分校读完了计算机科学学位。当看到公园里的松鼠的时候，他会大笑，这使得人们认为他是傻瓜。

- Email: al@inventwithpython.com
- Blog: http://coffeeghost.net
- Twitter: @AlSweigart

前言

你好！本书将教你如何使用 Python 语言和 Pygame 框架（也叫作 Pygame 库）来开发图形化的计算机游戏。Pygame 使得开发 2D 图形程序变得很容易。Python 和 Pygame 框架都可以从 http://python.org 和 http://pygame.org 免费下载。你只需要有计算机和这本书，就可以开始开发自己的游戏了。

本书是一本中级编程图书。如果你完全是初次接触编程，你可能需要努力阅读源代码示例并搞清楚程序如何工作。然而，如果你此前学习过如何使用 Python 编程，这将会容易一些。《Invent Your Own Computer Games with Python》[1]这本书，将会教初学者如何编写非图形化的、基于文本的游戏，并且还有一些章节介绍如何使用 Pygame 库。

然而，如果你已经知道了如何用 Python 编程（或者甚至了解其他语言，因为 Python 是很容易掌握的语言）并且想要开始编写超越文本的游戏，那么，本书很适合你阅读。本书首先简短地介绍了 Pygame 库是如何工作的，以及它提供了哪些功能。然后，本书针对一些真实的游戏给出了完整的源代码，并且详细说明了这些代码如何工作，以便你能够理解真正的游戏是如何使用 Pygame 的。

本书内容安排

当你搞明白之后，编写视频游戏程序，只不过是点亮像素，让漂亮的图片出现在屏幕上以响应键盘和鼠标输入而已。

但很少有什么事情能像编写游戏程序这么有趣。

本书教你如何用 Python 语言和 Pygame 库来编写图形化的计算机游戏。本书假设你了解一些 Python 知识或一般的编程知识。如果你不知道如何编程，可以通过阅读《Invent Your Own Computer Games with Python》一书来学习；或者，你也可以直接开始坚持捧读本书。

本书包含了 7 款不同的游戏，它们都是你曾经玩过的流行游戏的翻版。和《Invent Your Own Computer Games with Python》一书中基于文本的游戏相比，这些游戏充满了乐趣和交互性，但仍然相当短小。所有的程序都在 600 行以内。当你考虑一下你所下载或者从商店购买的专业游戏往往有成百上千行的代码，这些游戏真的是很小了。而那些商业化的游戏，需要程序员和美工组成的整个团队一起工作数月，甚至是数年才能开发出来。

本书共包括 10 章内容和一个术语表。各章内容说明如下。

第 1 章介绍了 Python 和 Pygame 的基础知识，主要是在不同的平台下如何正确地安装

---

[1] 该书中文版将由人民邮电出版社出版。

前言

和配置 Python 和 Pygame。这是你使用 Python 和 Pygame 开发游戏的准备工作。这一章还介绍了如何使用本书中的代码以及本书的配套站点所提供的资料。

第 2 章介绍了 Pygame 的基础知识，包括 Pygame 是什么以及它能够提供哪些功能。本章还介绍了像素坐标、Surface 对象、颜色、动画、声音等和游戏编程相关的基础知识。

第 3 章完整地讲解了 Memory Puzzle 程序是如何工作的，并且通过介绍游戏程序，初次说明了嵌套的 for 循环、语法糖以及在同一程序中使用不同的坐标系统等概念。后续其他的游戏中，也会用到这些概念。

第 4 章详细说明了 Slide Puzzle 游戏的程序是如何工作的。这个游戏中所使用的众多方法和概念，在第 3 章中都介绍过。

第 5 章详细讲解了 Simulate 游戏的程序是如何工作的。

第 6 章详细介绍了 Wormy 游戏程序的实现和工作原理。本章初次介绍了碰撞检测的方法，这在后面的游戏程序中还会用到。

第 7 章讲解了 Tetromino 游戏的实现和工作原理。本章的游戏第一次引入了关卡的概念。它通过示例深入地揭示了如何把游戏思想变成流行的、真正的程序。

第 8 章介绍了 Squirrel Eat Squirrel 游戏程序的原理和实现。这是第一款一次拥有多个敌人在游戏板上移动的游戏。本章引入了相机的概念，介绍了如何通过相机视图将游戏世界的一部分显示在屏幕上。此外，本章引入了数学正弦函数来处理真实的松鼠跳动。你会发现，利用一些数学知识进行游戏编程，可以使得游戏效果更为生动逼真。

第 9 章介绍了一个更为复杂的 Star Pusher 游戏的原理和实现。该游戏拥有使用贴片图形独特设计的关卡。我们学习了如何修改关卡文件，从而改变众多游戏对象在游戏世界中的位置。

第 10 章给出了 4 款其他的游戏的源代码。这一章并没有详细介绍 4 款游戏的工作原理。但是，这些不同类型的游戏程序，能够帮助读者思考自己想要制作什么样的游戏以及如何编写其代码。本章最后还列出了进一步学习 Python 和 Pygame 编程的更多资源。

术语表列出了本书中介绍的一些术语及其概念说明，以方便读者在阅读过程中查阅参考。

本书的目标读者

这本书针对中级程序员，他们已经学习过什么是变量和循环，但是现在想要知道真正的游戏程序看起来是什么样子。在我初次学习编程之后，还并不是真正地知道如何使用这些技能来开发一些很酷的东西，这之间还有很长的距离。我希望本书中的游戏能够使你对于程序如何工作有足够的了解，从而为你实现自己的游戏打下坚实的基础。

配套 Web 站点

本书的 Web 站点是 http://inventwithpython.com/pygame。本书所提及的所有程序和文件，都可以从这个 Web 下载。如果你对于本书以及其中的程序有任何的问题，请通过 al@inventwithpython.com 给我写邮件。

目录

第 1 章　安装 Python 和 Pygame ·············· 1
 1.1　预备知识 ··································· 1
 1.2　下载和安装 Python ················· 1
 1.3　Windows 下的安装说明 ········ 1
 1.4　Mac OS X 上的安装说明 ······ 2
 1.5　Ubuntu 和 Linux 上的安装
   说明 ··· 2
 1.6　启动 Python ···························· 2
 1.7　安装 Pygame ·························· 3
 1.8　如何阅读本书 ·························· 4
 1.9　特色的程序 ······························ 4
 1.10　下载图形文件和声音文件 ···· 4
 1.11　行号和空格 ···························· 4
 1.12　图书中的文本折行 ················ 5
 1.13　在线检查代码 ························ 5
 1.14　配套网站上的更多信息 ········ 6

第 2 章　Pygame 基础知识 ··············· 7
 2.1　GUI vs. CLI ····························· 7
 2.2　使用 Pygame 的 Hello World 程序
   源代码 ···································· 7
 2.3　建立一个 Pygame 程序 ········· 8
 2.4　游戏循环和游戏状态 ············ 10
 2.5　pygame.event.Event 对象 ···· 11
 2.6　QUIT 事件和 pygame.quit()
   函数 ······································ 12
 2.7　像素坐标 ································ 13
 2.8　关于函数、方法、构造函数和
   模块中的函数（及其差别）的
   一些提示 ······························ 13
 2.9　Surface 对象和窗口 ············· 14
 2.10　颜色 ······································ 15
 2.11　颜色的透明度 ······················ 16
 2.12　pygame.Color 对象 ············· 17
 2.13　Rect 对象 ····························· 17
 2.14　基本的绘制函数 ·················· 19
 2.15　pygame.PixelArray 对象 ···· 22
 2.16　pygame.display.update()函数 ··22
 2.17　动画 ······································ 22
 2.18　帧速率和 pygame.time.Clock
    对象 ····································· 25
 2.19　用 pygame.image.load()和 blit()
    绘制图像 ···························· 26
 2.20　字体 ······································ 26
 2.21　抗锯齿 ·································· 28
 2.22　播放声音 ······························ 28
 2.23　本章小结 ······························ 29

第 3 章　Memory Puzzle 游戏 ················· 31
 3.1　如何玩 Memory Puzzle 游戏 ········ 31
 3.2　嵌套的 for 循环 ···················· 31
 3.3　Memory Puzzle 的源代码 ············ 33
 3.4　声明和导入 ···························· 40
 3.5　幻数很糟糕 ···························· 40
 3.6　使用 assert 语句全面检查 ············ 41
 3.7　判断一个数字是偶数还是奇数 ····· 42
 3.8　较早崩溃和经常崩溃 ················· 42
 3.9　让源代码更好看一些 ·············· 43
 3.10　使用常量变量而不是字符串 ········ 44
 3.11　确保有足够的图标 ·············· 44
 3.12　元组 vs.列表，不可变 vs.可变 ····· 45
 3.13　单项元组需要一个结尾的逗号 ···· 46

i

3.14	在列表和元组之间转换……46	3.42	为何要那么麻烦地使用 main() 函数……66
3.15	global 语句以及为什么全局变量是罪恶的……47	3.43	为什么要为可读性操心……67
3.16	数据结构和 2D 列表……48	3.44	本章小结……71
3.17	"开始游戏"动画……49		
3.18	游戏循环……50	**第 4 章**	**Slide Puzzle……72**
3.19	事件处理循环……50	4.1	如何玩 Slide Puzzle……72
3.20	检查鼠标光标在哪一个方块之上……51	4.2	Slide Puzzle 的源代码……72
3.21	处理第一次点击的方块……52	4.3	第二款游戏和第一款相同……80
3.22	处理不一致的一对图标……53	4.4	设置按钮……81
3.23	处理玩家获胜……53	4.5	使用愚笨的代码变聪明……83
3.24	将游戏状态绘制到屏幕……54	4.6	主游戏循环……83
3.25	创建"揭开的方块"数据结构……55	4.7	点击按钮……84
3.26	创建游戏板数据结构：第 1 步——获取所有可能的图标……55	4.8	用鼠标滑动贴片……85
		4.9	用键盘滑动贴片……85
3.27	第 2 步——打乱并截取所有图标的列表……56	4.10	使用 in 操作符实现"等于其中之一"的技巧……86
3.28	第 3 步——将图标放置到游戏板上……56	4.11	WASD 和箭头按键……86
		4.12	实际执行贴片滑动……87
3.29	将一个列表分割为列表的列表……57	4.13	IDLE 和终止 Pygame 程序……87
		4.14	检查特定的事件并且将事件添加到 Pygame 的事件队列……88
3.30	不同的坐标系……58		
3.31	从像素坐标转换为方块坐标……59	4.15	创建游戏板数据结构……89
3.32	绘制图标以及语法糖……59	4.16	不记录空白的位置……89
3.33	获取游戏板控件的图标的形状和颜色的语法糖……61	4.17	通过更新游戏板数据结构来移动……90
		4.18	何时不使用断言……90
3.34	绘制盖住的方块……61	4.19	获取一次并不是那么随机的移动……91
3.35	处理揭开和覆盖动画……62		
3.36	绘制整个游戏板……63	4.20	将贴片坐标转换为像素坐标……92
3.37	绘制高亮边框……63	4.21	将像素坐标转换为游戏板坐标……92
3.38	"开始游戏"动画……64	4.22	绘制一个贴片……93
3.39	揭开和盖住成组的方块……64	4.23	让文本显示在屏幕上……93
3.40	"游戏获胜"动画……65	4.24	绘制游戏板……94
3.41	判断玩家是否已经获胜……65	4.25	绘制游戏板的边框……94

4.26	绘制按钮 ·················· 95	6.1	Wormy游戏的玩法 ············ 125
4.27	实现贴片滑动动画 ············ 95	6.2	Wormy的源代码 ·············· 125
4.28	Surface的copy()方法 ········· 96	6.3	栅格 ························ 131
4.29	创建新的谜题 ················ 98	6.4	设置代码 ···················· 131
4.30	实现游戏板重置动画 ·········· 99	6.5	main()函数 ·················· 132
4.31	时间vs.内存的权衡 ·········· 100	6.6	单独的runGame()函数 ········ 133
4.32	没人在乎几个字节 ············ 101	6.7	事件处理循环 ················ 134
4.33	没人在乎几百万个纳秒 ········ 101	6.8	碰撞检测 ···················· 134
4.34	本章小结 ···················· 101	6.9	检测和苹果的碰撞 ············ 135
		6.10	移动虫子 ···················· 136
第5章	Simulate ···················· 102	6.11	insert()列表方法 ············· 136
5.1	如何玩Simulate游戏 ·········· 102	6.12	绘制屏幕 ···················· 137
5.2	Simulate的源代码 ············ 102	6.13	在屏幕上绘制"Press a key"文本 ······················· 137
5.3	常用初始内容 ················ 108	6.14	checkForKeyPress()函数 ······ 137
5.4	设置按钮 ···················· 109	6.15	初始屏幕 ···················· 138
5.5	main()函数 ·················· 110	6.16	旋转初始屏幕文本 ············ 139
5.6	程序中用到的一些局部变量 ···· 110	6.17	旋转并不完美 ················ 140
5.7	绘制游戏板并处理输入 ········ 111	6.18	决定苹果出现在哪里 ·········· 141
5.8	检查鼠标点击 ················ 112	6.19	游戏结束屏幕 ················ 141
5.9	检查键盘按下 ················ 112	6.20	绘制函数 ···················· 142
5.10	游戏循环的两种状态 ·········· 113	6.21	不要复用变量名 ·············· 144
5.11	搞清楚玩家是否按下了正确的按钮 ······················ 113		
5.12	新纪元时间 ·················· 115	第7章	Tetromino ··················· 146
5.13	将游戏板绘制到屏幕 ·········· 116	7.1	一些Tetromino术语 ··········· 146
5.14	相同的旧的terminate()函数 ···· 116	7.2	Tetromino的源代码 ··········· 147
5.15	复用常量变量 ················ 117	7.3	常用设置代码 ················ 159
5.16	实现按钮闪烁动画 ············ 117	7.4	设置按下键的定时常量 ········ 159
5.17	绘制按钮 ···················· 120	7.5	更多的设置代码 ·············· 160
5.18	实现背景颜色改变的动画 ······ 120	7.6	设置砖块模式 ················ 161
5.19	游戏结束动画 ················ 121	7.7	将"一行代码"分隔到多行 ······ 164
5.20	将像素坐标转换为按钮 ········ 122	7.8	main()函数 ·················· 165
5.21	显式比隐式好 ················ 123	7.9	开始新的游戏 ················ 166
		7.10	游戏循环 ···················· 167
第6章	Wormy ······················ 125	7.11	事件处理循环 ················ 167

7.12	暂停游戏 ·········· 167		第 8 章	Squirrel Eat Squirrel ·········· 191
7.13	使用移动变量来处理用户输入 ·· 168		8.1	如何玩 Squirrel Eat Squirrel ······· 191
7.14	检查移动或旋转是否有效 ······· 168		8.2	Squirrel Eat Squirrel 的设计 ······ 191
7.15	找到底部 ·········· 171		8.3	Squirrel Eat Squirrel 的源代码 ·· 192
7.16	通过按下按键来移动 ·········· 172		8.4	常用设置代码 ·········· 202
7.17	让砖块 "自然" 落下 ·········· 174		8.5	描述数据结构 ·········· 203
7.18	将所有内容绘制到屏幕上 ······· 175		8.6	main()函数 ·········· 204
7.19	制作文本的快捷函数 makeTextObjs() ·········· 176		8.7	pygame.transform.flip()函数 ····· 205
7.20	相同的旧的 terminate()函数 ····· 176		8.8	更为详细的游戏状态 ·········· 205
7.21	使用 checkForKeyPress()函数等待按键事件 ·········· 176		8.9	常用的文本创建代码 ·········· 206
			8.10	相机 ·········· 206
7.22	通用文本屏幕函数 showTextScreen() ·········· 177		8.11	"活动区域" ·········· 208
			8.12	记录游戏世界中的物体的位置 ·· 208
7.23	checkForQuit()函数 ·········· 178		8.13	从一些草开始 ·········· 209
7.24	calculateLevelAndFallFreq() 函数 ·········· 178		8.14	游戏循环 ·········· 209
			8.15	检查去掉保护状态 ·········· 209
7.25	用函数 getNewPiece()产生新的砖块 ·········· 180		8.16	移动敌人松鼠 ·········· 210
			8.17	删除较远的草对象和松鼠对象 ·········· 211
7.26	给游戏板数据结构添加砖块 ······ 181			
7.27	创建一个新的游戏板数据结构 ·········· 181		8.18	当从列表中删除项的时候，反向遍历列表 ·········· 211
			8.19	添加新的草对象和松鼠对象 ···· 213
7.28	isOnBoard()和 isValidPosition()函数 ·········· 182		8.20	相机延迟以及移动相机视图 ···· 213
			8.21	绘制背景、草、松鼠和生命值指示 ·········· 214
7.29	检查、删除和填满一行 ·········· 184			
7.30	将游戏板坐标转换为像素坐标 ·· 186		8.22	事件处理循环 ·········· 216
7.31	在游戏板上或屏幕上的其他位置绘制方块 ·········· 187		8.23	移动玩家并考虑跳动 ·········· 218
			8.24	碰撞检测：吃或被吃 ·········· 219
7.32	将所有内容绘制到屏幕上 ······· 187		8.25	游戏结束屏幕 ·········· 221
7.33	绘制得分和关卡文本 ·········· 188		8.26	获胜 ·········· 221
7.34	在游戏板上或屏幕的其他位置绘制一个砖块 ·········· 188		8.27	绘制图形化的生命值指标 ······· 221
			8.28	相同的旧的 terminate()函数 ····· 222
7.35	绘制 "Next" 砖块 ·········· 189		8.29	正弦函数 ·········· 222
7.36	本章小结 ·········· 189		8.30	对 Python 2 的向后兼容 ·········· 225
			8.31	getRandomVelocity()函数 ·········· 226

8.32	找到一个地方添加新的松鼠和草 ····· 226		9.10	绘制地图 ················· 273
8.33	创建敌人松鼠数据结构 ····· 228		9.11	检查关卡是否完成 ······ 275
8.34	翻转松鼠图像 ·············· 228		9.12	本章小结 ················· 276

第 10 章 4 款其他游戏 ············ 277

- 8.35 创建草数据结构 ·········· 229
- 8.36 检查是否在活动区域之外 ······ 229
- 8.37 本章小结 ····················· 230

- 10.1 Flippy，Othello 的翻版 ····· 277
- 10.2 Flippy 的源代码 ············ 279
- 10.3 Ink Spill，Flood It 游戏的翻版 ················ 291
- 10.4 Ink Spill 的源代码 ·········· 292
- 10.5 Four-In-A-Row，Connect Four 的翻版 ············· 303
- 10.6 Four-In-A-Row 的源代码 ···· 304
- 10.7 Gemgem，Bejeweled 的翻版 ··· 312
- 10.8 Gemgem 的源代码 ·········· 313
- 10.9 本章小结 ···················· 326

第 9 章 Star Pusher ·············· 231

- 9.1 如何玩 Star Pusher ········· 231
- 9.2 Star Pusher 的源代码 ······· 232
- 9.3 初始化设置 ················· 245
- 9.4 Star Pusher 中的数据结构 ···· 259
- 9.5 读取和写入文本文件 ······· 260
- 9.6 递归函数 ···················· 268
- 9.7 栈溢出 ······················ 269
- 9.8 使用基本条件防止栈溢出 ··· 271
- 9.9 漫水填充算法 ·············· 271

术语表 ·································· 328

第 1 章 安装 Python 和 Pygame

1.1 预备知识

　　如果在阅读本书之前，你了解一点 Python 编程的话（或者知道如何使用 Python 以外的另一种语言编程），可能会有所帮助。然而，即便你还不了解，仍然可以阅读本书。编程并不像人们所想象得那么难。如果你遇到一些困难，可以读我的另一本书《Invent Your Own Computer Games with Python》，查找令你感到混淆的某一个主题。

　　在阅读本书之前，你不需要知道如何使用 Pygame 库。本书第 2 章是介绍 Pygame 的主要功能和函数的一个简短教程。考虑到你有可能没有读过《Invent Your Own Computer Games with Python》这本书，并且可能还没有在自己的计算机上安装 Python 和 Pygame，本章将介绍安装过程。如果你自己已经安装了 Python 和 Pygame，那么，你可以直接跳过本章。

1.2 下载和安装 Python

　　在开始编写程序之前，需要先在计算机上安装叫作 Python 解释器的软件（你可能需要请别人帮忙）。解释器（interpreter）是能够理解你用 Python 编写（或者说录入）的指令的一个程序。没有解释器，计算机就不能够运行 Python 程序。从现在开始，我们将"Python 解释器"称为"Python"。

　　可以从 Python 编程语言的官方站点 http://www.python.org 下载 Python 解释器软件。你可能需要某些人的帮助才能够下载和安装 Python 软件。根据你的计算机使用的是 Windows、Mac OS X 或 Ubuntu 这样的 Linux OS，安装过程会有些不同。你可以从 http://invpy.com/installing 找到人们在自己的计算机上安装 Python 的在线视频。

1.3 Windows 下的安装说明

　　当你访问 http://python.org 的时候，应该会在左边看到链接的一个列表（如"About"、"News"、"Documentation"、"Download"等）。在 Download 链接上点击，以打开下载页面，然后，找到名为"Python 3.2 Windows Installer(Windows binary -- does not include source)"的一个文件，并且点击该链接，以下载针对 Windows 的 Python。

　　在刚刚下载得到的 python-3.2.msi 文件上双击，以启动 Python 安装程序（如果它没有启动，尝试用鼠标右键点击该文件并且选择 Install）。一旦安装程序启动，只要不断地点击 Next 按钮并随着步骤接受安装程序中的选项（不需要做任何修改）就可以了。当安装完成之后，点击 Finish 按钮。

1.4 Mac OS X 上的安装说明

Mac OS X 10.5 带有 Apple 预安装的 Python 2.5.1。在编写本书的时候，Pygame 只支持 Python 2，而不支持 Python 3。然而，本书中的程序在 Python 2 和 Python 3 下都能工作。

Python Web 站点还有一些关于在 Mac 上使用 Python 的额外信息：http://docs.python.org/dev/using/mac.html。

1.5 Ubuntu 和 Linux 上的安装说明

Pygame for Linux 只支持 Python2，而不支持 Python 3。如果你的操作系统是 Ubuntu，可以这样来安装 Python：打开一个终端窗口（从桌面点击 Applications > Accessories > Terminal），并输入"sudo apt-get install python2.7"，然后按下回车键。你将需要输入根密码才能安装 Python，如果你不知道这个密码的话，向计算机的所有者询问并输入它。

你还需要安装 IDLE 软件。从终端中录入"sudo apt-get install idle"。要安装 IDLE，也需要根密码（请计算机的所有者为你输入该密码）。

1.6 启动 Python

我们将使用 IDLE 软件来输入程序并运行它。IDLE 英文全称为 Interactive DeveLopment Environment，中文含义为交互式开发环境。这个开发环境是使得编写 Python 程序更为容易的一个软件，就像是字处理器软件使得写书很容易一样。如果你的操作系统是 Windows XP，你应该能够通过点击 Start 按钮，然后选择 Programs，Python 3.1，IDLE(Python GUI) 来运行 Python。对于 Windows Vista 或 Window 7 来说，只要点击左下角的 Windows 按钮，输入"IDLE"并选择"IDLE(Python GUI)"就可以了。

如果你的操作系统是 Max OS X，通过打开 Finder 窗口并在 Applications 上点击，然后点击 Python 3.2，点击 IDLE 图标，以启动 IDLE。

如果你的操作系统是 Ubuntu 或 Linux，通过打开一个终端窗口，然后输入"idle3"并按下回车键，来启动 IDLE。也可以点击屏幕顶部的 Applications，然后选择 Programming，然后选择 IDLE 3。

当你初次运行 IDLE 时，会出现叫作交互式 shell（interactive shell）的窗口。shell 就是一个程序，它允许你向计算机输入指令。Python shell 允许你输入 Python 指令，然后，shell

会将这些指令发送给 Python 解释器去执行，如图 1-1 所示。

图 1-1

1.7 安装 Pygame

　　Pygame 不是 Python 所附带的。和 Python 一样，Pygame 也是可以免费使用的。你必须下载并安装 Pygame，这与下载和安装 Python 解释器一样容易。在 Web 浏览器中，访问 http://pygame.org 并且在该 Web 站点左边的"Downloads"上点击。本书假设你使用的是 Windows 操作系统，但是 Pygame 在每种操作系统上的工作方式都是相同的。你需要针对自己的操作系统以及所安装的 Python 版本，下载相应的 Pygame 安装程序。

　　你不需要下载 Pygame 的"源代码"，而是要下载针对你的操作系统的"二进制程序"。对于 Windows，下载 pygame-1.9.1.win32-py3.2.msi 文件（这是针对 Windows 上的 Python 3.2 的 Pygame）。如果你安装的是不同的 Python 版本（例如，Python 2.7 或 Python 2.6），那么，要下载针对你的 Python 版本的.msi 文件。在编写本书的时候，Pygame 当前的版本是 1.9.1。如果你在 Web 站点上看到了更新的一个版本，请下载并安装较新的 Pygame 版本。

　　对于 Mac OS X，针对你所拥有的 Python 版本，下载.zip 文件或.dmg 文件并运行它们。对于 Linux，打开一个终端并运行"sudo apt-get install python-pygame"。

　　在 Windows 上，双击所下载的文件以安装 Pygame。要检查 Pygame 是否安装正确，在交互式 shell 中输入如下内容：

```
>>> import pygame
```

　　如果按下回车键后什么也没有出现，那么我们知道已经成功地安装了 Pygame。如果出现了 ImportError: No module named pygame 错误，那么，尝试再次安装 Pygame，并且确保正确地输入 import pygame。

　　本章有 5 个小程序，它们展示了如何使用 Pygame 所提供的不同功能。在本书的最后一章中，我们将使用这些不同的功能，用 Python 和 Pygame 编写一个完整游戏。本书的 Web

站点上提供了安装 Pygame 的一个视频教程：http://invpy.com/videos。

1.8　如何阅读本书

这本书和其他的编程图书不同，因为它专注于几个游戏程序的完整的源代码。本书不会教授编程概念，然后让你自己去搞清楚如何使用这些概念来编写程序，而是向你展示一些程序，然后，说明它们是如何组织起来的。通常，你应该按照顺序阅读本书中的各章。这些游戏中还有很多概念会重复地用到，并且只是会在它们所出现的第一个游戏中加以介绍。但是，如果你认为这个游戏很有趣，可以向前跳到那一章阅读。如果你觉得有些概念不了解，稍后再回头来阅读之前的几章。

1.9　特色的程序

每一章都关注一个游戏程序，并且说明代码的不同部分是如何工作的。一行一行地录入本书中的代码，将有助于复制这些游戏。然而，你也可以从本书的 Web 站点下载源代码。在 Web 浏览器中，访问 http://invpy.com/ 并且按照说明来下载源文件。但是，自己输入代码真地能够帮助你更好地了解代码。

1.10　下载图形文件和声音文件

尽管可以在阅读本书的时候输入代码，但你还是需要从 http://invpy.com/downloads 下载书中的游戏所用到的图形文件和声音文件。确保将这些图像文件和声音文件放在与 .py Python 文件相同的目录中，否则的话，你的 Python 程序可能会无法找到这些文件。

1.11　行号和空格

当你自己编写代码的时候，不要输入在每一行开头处出现的行号。例如，如果你在本书中看到如下代码：

```
1. number = random.randint(1, 20)
2. spam = 42
3. print('Hello world!')
```

不需要输入左边的"1"以及紧接其后的空格。只要输入如下内容就可以了。

```
number = random.randint(1, 20)
spam = 42
print('Hello world!')
```

这些编号只是为了便于在书中引用特定的代码行。它们并非实际程序的一部分。

除了行号之外，确保按照代码的样子进行输入。注意，其中一些代码行从页面的最左端开始，但是却缩进了 4 个或 8 个空格。确保在每一行的开始处放置合适数目的空格（因为 IDLE 中的每个字符具有相同的宽度，你可以通过统计所看到的代码行之上或之下的字符数，来计算应该使用的空格数）。

例如，在如下的代码中可以看到，第二行缩进了 4 个空格，因为上面一行的 4 个字符（"whil"）超出了缩进位置。第 3 行又缩进了 4 个空格（4 个字符，第三行之上的一行的"if n"超出了缩进位置）。

```
while spam < 10:
    if number == 42:
        print('Hello')
```

1.12 图书中的文本折行

一些代码行太长了，无法在图书页面中放到一行之中，并且，代码的文本会换到新的一行中。当你在文本编辑器中输入这些代码行的时候，将所有代码输入到同一行而不要按下回车键。

你可以查看代码左边的行号，从而分辨出新的一行开始了。例如，如下的代码只有两行，即便第一行换行了。

```
1. print('This is the first line! xxxxxxxxxxxxxxxxxxxxxxxxxxxxx
xxxxxxxxxxxxxxx')
2. print('This is the second line, not the third line.')
```

1.13 在线检查代码

本书中的一些程序有点长，尽管通过录入这些程序的源代码对于学习 Python 是很有帮助的，但你可能偶尔会有些录入错误，从而导致程序崩溃。而这些录入错误可能并不是很明显。

你可以将自己的源代码的文本复制并粘贴到本书 Web 站点上的在线比较工具中。这个比

第 1 章　安装 Python 和 Pygame

较工具将会显示出你输入的源代码和书中的源代码的任何不同之处。这是找出程序中的录入错误的一种很容易的方法。

复制和粘贴文本是很有用的计算机技能，特别是对于计算机编程来说。本书的 Web 站点上有一个视频，介绍如何进行复制和粘贴：http://invpy.com/copypaste。

这个在线比较工具的 Web 页面是 http://invpy.com/diff/pygame。本书的 Web 站点上也有一个视频介绍如何使用该工具。

1.14　配套网站上的更多信息

有很多编程知识需要学习。但是，你现在还不需要学习所有这些内容。在本书中有几处地方，你可能会想要学习这些额外的内容，但是，如果我将这些内容都包含在本书中的话，本书的篇幅会增加不少。这样一来，这本又大又厚的图书会令你崩溃，会烦死你。相反，我把这些内容放在本书的 Web 站点上一个叫作"more info"的链接之中。要理解本书中的内容，你并不需要阅读这些额外信息，但是，如果你感到好奇的话，可以去看看。这些链接都很短，并且以 http://invpy.com 打头。

这个"more info"中的所有信息，也都可以通过 http://invpy.com/pygamemoreinfo 来下载。

第2章 Pygame 基础知识

Python 带有诸如 random、math 或 time 这样的几个模块，它们为程序提供额外的功能，同样，Pygame 框架也包含了几个模块，其功能包括绘制图形、播放声音、处理鼠标输入等。

本章将介绍 Pygame 所提供的基本模块和功能，并且假设你已经了解 Python 编程的基础知识。如果你对理解一些编程概念感到困难，可以阅读我的《Invent Your Own Computer Games with Python》一书，这本书的目标读者是程序设计的完全初学者。

《Invent Your Own Computer Games with Python》这本书中有几章会介绍 Pygame。一旦你了解了 Pygame，可以通过在线文档 http://pygame.org/docs 看看 Pygame 所提供的其他模块。

2.1 GUI vs. CLI

使用 Python 的内建函数编写的 Python 程序，只能够通过 print() 和 input() 函数来处理文本。程序可以在屏幕上显示文本，并且让用户通过键盘来输入文本。这类程序有一个命令行界面（command line interface，CLI）。这些程序多少有些局限性，因为它们不能显示图形，没有颜色，并且不能使用鼠标。这种 CLI 程序只是使用 input() 函数从键盘获取输入，甚至用户必须按下回车键，然后程序才能够响应输入。这意味着不可能制作实时（也就是说，持续运行代码而不需要等待用户）动作的游戏。

Pygame 提供了使用图形化用户界面（graphical user interface，GUI）来创建游戏的功能。使用基于图形的 GUI 的程序可以显示带有图像和颜色的窗口，而不再是一个基于文本的 CLI。

2.2 使用 Pygame 的 Hello World 程序源代码

我们用 Pygame 开发的第一个程序，是在屏幕上显示一个带有"Hello World"的窗口的小程序。通过点击 IDLE 的 File 菜单，然后选择 New Window，打开一个新的文件编辑器。在 IDLE 的文件编辑器中，输入如下的代码并将其保存为 *blankpygame.py*。然后，按下 F5 键或者从文件编辑器顶部的菜单选择 Run > Run Module，运行该程序。

记住，不要输入每一行开始处的行号和句点（那些只是为了方便在本书中引用）。

```
1. import pygame, sys
2. from pygame.locals import *
3.
4. pygame.init()
5. DISPLAYSURF = pygame.display.set_mode((400, 300))
6. pygame.display.set_caption('Hello World!')
7. while True: # main game loop
8.     for event in pygame.event.get():
9.         if event.type == QUIT:
10.             pygame.quit()
11.             sys.exit()
12.     pygame.display.update()
```

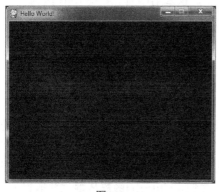

图 2-1

当运行这个程序的时候,将会出现一个黑色的窗口,如图 2-1 所示。

是的,你刚刚创建了世界上最无趣的视屏游戏。它只是一个空白的窗口,在窗口的顶部显示了一个"Hello World!"(在所谓的窗口的标题栏中,标题栏会保存标题文本)。

但是,创建一个窗口只是制作图形化游戏的第一步。当你点击窗口右上角的 X 按钮的时候,程序会终止并且窗口会消失。

调用 print()函数来让文本出现在窗口中的方法无效,因为 print()是一个用于 CLI 程序的函数。对于使用 input()获取来自用户的键盘输入,也是一样的。Pygame 使用其他的函数进行输入和输出,我们将在本章稍后介绍它们。现在,我们来详细看一下"Hello World"程序中的每一行代码。

2.3 建立一个 Pygame 程序

Hello World 的前几行,几乎在你使用 Pygame 编写的每一个程序中都会用作开头的几行。

```
1. import pygame, sys
```

第 1 行是一条简单的 import 语句,它导入 pygame 和 sys 模块,以便我们可以在程序中使用这些模块中的函数。Pygame 所提供的所有那些处理图形、声音以及其他功能的 Pygame 函数,都位于 pygame 模块中。

当导入 pygame 模块的时候要注意,你也会自动地导入位于 pygame 模块之中的所有模块,如 pygame.images 和 pygame.mixer.music。不需要再用其他的 import 语句来导入

2.3 建立一个 Pygame 程序

这些位于该模块之中的模块。

```
2. from pygame.locals import *
```

第 2 行也是一条 import 语句。然而，它使用了 from modulename import *的格式，而不是 import modulename 的格式。通常，如果你想要调用模块中的一个函数，必须在导入该模块之后，使用 modulename.functionname()的格式。然而，通过使用 from modulename import *，你可以省略掉 modulename.部分，而直接使用 functionname()来调用（就像是调用 Python 的内建函数一样）。

针对 pygame.locals 使用这种形式的 import 语句，是因为 pygame.locals 包含了几个常量变量，它们前面不需要 pygame.locals，也可以很容易地识别出是 pygame.locals 模块中的变量。对于所有其他的模块，通常会使用常规的 import modulename 格式（http://invpy.com/namespaces 更为详细地介绍我们想要这么做的原因）。

```
4. pygame.init()
```

第 4 行是 pygame.init()函数调用，在导入了 pygame 之后并且在调用任何其他的 Pygame 函数之前，总是需要调用该函数。现在不需要知道这个函数到底做些什么，只需要知道，要让众多的 Pygame 函数能够工作，我们需要先调用这个函数。如果你看到诸如 pygame.error: font not initialized 的一个错误，检查看看是否在程序的开始处忘记调用 pygame.init()了。

```
5. DISPLAYSURF = pygame.display.set_mode((400, 300))
```

第 5 行调用了 pygame.display.set_mode()函数，它返回了用于该窗口的 pygame. Surface 对象（本章后面将会介绍 Surface 对象）。注意，我们给该函数传入了两个整数的一个元组值：(400, 300)。这个元组告诉 set_mode()函数创建一个宽度和高度分别为多少个像素的窗口。(400, 300)将会创建一个宽 400 像素、高 300 像素的窗口。

记住给 set_mode()传递两个整数的一个元组，而不是两个整数自身。调用该函数的正确方式是这样的: pygame.display.set_mode((400, 300))。诸如 pygame.display. set_mode(400, 300)的一个函数调用，将会导致 TypeError: argument 1 must be 2-item sequence, not int 这样的一个错误。

返回的 pygame.Surface 对象（为了简便起见，我们将其称为 Surface 对象），存储在一个名为 DISPLAYSURF 的变量中。

```
6. pygame.display.set_caption('Hello World!')
```

第 6 行通过调用 pygame.display.set_caption()函数，设置了将要在窗口的顶部显示的标题文本。在这个函数调用中，传入了字符串值'Hello World!'，以使得该文本作为标题出现，如图 2-2 所示。

图 2-2

2.4 游戏循环和游戏状态

```
7. while True: # main game loop
8.     for event in pygame.event.get():
```

第 7 行是一个 while 循环，它有一个直接为 True 值的条件。这意味着它不会因为该条件求得 False 而退出。程序执行退出的唯一方式是执行一条 break 语句（该语句将执行移动到循环之后的第一行代码）或者 sys.exit()（它会终止程序）。如果像这样的一个循环位于一个函数中，一条 return 语句也可以使得执行退出循环（同时退出函数的执行）。

本书中的游戏，其中都带有这样的一些 while True 循环，并且带有一条将该循环称为"main game loop"的注释。游戏循环（game loop，也叫作主循环，main loop）中的代码做如下 3 件事情。

1．处理事件。
2．更新游戏状态。
3．在屏幕上绘制游戏状态。

游戏状态（game state）只不过是针对游戏程序中用到的所有变量的一组值的一种叫法。在很多游戏中，游戏状态包括了记录玩家的生命值和位置、敌人的生命值和位置、在游戏板做出了什么标记、分数值或者轮到谁在玩等信息的变量的值。任何时候，如玩家受到伤害（这会减少其生命值），敌人移动到某个地方，或者游戏世界中发生某些事情的时候，我们就说，游戏的状态发生了变化。

如果你已经玩过那种允许你存盘的游戏，"保存状态（save state）"就是你在某个时刻所存储的游戏状态。在大多数游戏中，暂停游戏将会阻止游戏状态发生变化。

由于游戏状态通常是作为对事件的响应而更新（例如，点击鼠标或者按下键盘），或者是随着时间的流逝而更新，游戏循环在一秒钟之内会不断地、多次检查和重复检查是否发生了任何新的事件。在主循环中，就有用来查看创建了哪个事件的代码（使用 Pygame，这通过调用 pygame.event.get()函数来做到）。在主循环中，还有根据创建了哪个事件而更新游戏状态的代码。这通常叫作事件处理（event handling），如图 2-3 所示。

2.5 pygame.event.Event 对象

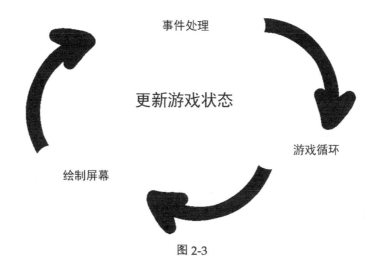

图 2-3

2.5 pygame.event.Event 对象

任何时候，当用户做了诸如按下一个按键或者把鼠标移动到程序的窗口之上等几个动作之一（在本章后面会列出这些动作），Pygame 库就会创建一个 pygame.event.Event 对象来记录这个动作，也就是"事件"（这种叫作 Event 的对象，存在于 event 模块中，该模块本身位于 pygame 模块之中）。我们可以调用 pygame.event.get() 函数来搞清楚发生了什么事件，该函数返回 pygame.event.Event 对象（为了简单起见，我们直接称之为 Event 对象）的一个列表。

这个 Event 对象的列表，包含了自上次调用 pygame.event.get() 函数之后所发生的所有事件（或者，如果从来没有调用过 pygame.event.get()，会包括自程序启动以来所发生的所有事件）。

```
7. while True: # main game loop
8.     for event in pygame.event.get():
```

第 8 行是一个 for 循环，它会遍历 pygame.event.get() 所返回的 Event 对象的列表。在这个 for 循环的每一次迭代中，一个名为 event 的变量将会被赋值为列表中的下一个事件对象。pygame.event.get() 所返回的 Event 对象的列表，将会按照事件发生的顺序来排序。如果用户点击鼠标并按下键盘按键，鼠标点击的 Event 对象将会是列表的第一项，键盘按键的 Event 对象将会是第二项。如果没有事件发生，那么 pygame.event.get() 将返回一个空白的列表。

2.6　QUIT 事件和 pygame.quit()函数

```
 9.        if event.type == QUIT:
10.            pygame.quit()
11.            sys.exit()
```

Event 对象有一个名为 type 的成员变量（member variable，也叫作属性，attributes 或 properties），它告诉我们对象表示何种事件。针对 pygame.locals 模块中的每一种可能的类型，Pygame 都有一个常量变量。第 9 行检查 Event 对象的 type 是否等于常量 QUIT。记住，由于我们使用了 from pygame.locals import *形式的 import 语句，主要输入 QUIT 就可以了，而不必输入 pygame.locals.QUIT。

如果 Event 对象是一个停止事件，就会调用 pygame.quit()和 sys.exit()函数。pygame.quit()是 pygame.init()函数的一种相反的函数，它运行的代码会使得 Pygame 库停止工作。在调用 sys.exit()终止程序之前，总是应该先调用 pygame.quit()。通常，由于程序退出之前，Python 总是会关闭 pygame，这不会真的有什么问题。但是，在 IDLE 中有一个 bug，如果一个 Pygame 程序在调用 pygame.quit()之前就终止了，将会导致 IDLE 挂起。

我们没有 if 语句来针对其他的 Event 对象类型运行代码，因此，当用户点击鼠标、按下键盘按键或者导致创建任何其他类型的 Event 对象的时候，没有事件处理代码。用户可能会做一些事情来创建这些 Event 对象，但是，这并不会对程序有任何改变，因为程序不会有任何针对这些类型的 Event 对象的事件处理代码。在第 8 行中的 for 循环执行完后，就处理完了 pygame.event.get()所返回的所有 Event 对象，程序继续从第 12 行开始执行。

```
12.    pygame.display.update()
```

第 12 行调用了 pygame.display.update()函数，它把 pygame.display.set_mode()所返回的 Surface 对象绘制到屏幕上（记住，我们将这个对象存储在了 DISPLAYSURF 变量中）。由于 Surface 对象没有变化（例如，没有被本章稍后将会介绍的某些绘制函数修改），每次调用 pygame.display.update()的时候，将会重新绘制相同的黑色图像。

这就是整个程序。在第 12 行代码执行之后，无限的 while 循环再次从头开始。这个程序只是让一个黑色的窗口出现在屏幕上，不断地检查 QUIT 事件，然后重复地将这个没有变化的黑色窗口重新绘制到屏幕上，除此之外，什么也不做。接下来，我们来学习像素、Surface 对象、Color 对象、Rect 对象和 Pygame 绘制函数，以了解如何让一些有趣的内容出现在这个窗口中，而不只是一片黑压压的颜色。

2.7 像素坐标

"Hello World"程序所创建的窗口,只不过是屏幕上叫作像素(pixel)的小方点的组合。每个像素最初都是黑色的,但是可以设置为一种不同的颜色。假设我们只有一个 8 像素×8 像素的 Surface 对象,而不是一个 400 像素宽和 300 像素高的 Surface 对象,并且,为 X 轴和 Y 轴添加了数字,然后,就可以很好地将其表示为如图 2-4 所示的样子。

我们可以使用一个笛卡尔坐标系统(Cartesian Coordinate system)来表示一个特定的点。X 轴上的每一列和 Y 轴的每一行都有一个地址,也就说从 0~7 的一个整数,我们可以通过指定 X 轴和 Y 轴的整数来定位任何的像素。

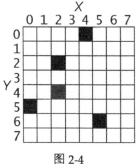

图 2-4

例如,在上面的 8×8 图像中,我们可以看到 XY 坐标为(4, 0)、(2, 2)、(0, 5)和(5, 6)的像素显示为黑色,而坐标为(2, 4)的像素显示为灰色。XY 坐标也叫作点(point)。如果你已经上过数学课或者学习过笛卡尔坐标,你可能会注意到,Y 坐标从最顶部的 0 开始,然后向下增加,而不是向上增加。这就是为什么在 Pygame 中(以及几乎每一种编程语言中),笛卡尔坐标是有效的。

Pygame 框架通常将笛卡尔坐标表示为两个整数的一个元组,例如,(4, 0)或(2, 2)。第 1 个整数是 X 坐标,而第 2 个整数是 Y 坐标(我在《Invent Your Own Computer Games with Python》一书的第 12 章详细地介绍了笛卡尔坐标)。

2.8 关于函数、方法、构造函数和模块中的函数(及其差别)的一些提示

函数和方法几乎是相同的东西。二者都可以通过调用来执行其中的代码。一个函数和一个方法之间的区别在于方法总是要附加给一个对象。通常,方法修改和特定对象相关的某些内容(你可以将附加的对象当作传递给方法的一种永久性的参数)。

如下是调用名为 foo()的一个函数。

```
foo()
```

如下是对同样名为 foo()的一个方法的调用,该方法附加给了一个对象,这个对象存储在一个名为 duckie 的变量中。

```
duckie.foo()
```

对于模块中的一个函数的调用,看上去可能像是一个方法调用。为了对二者加以区分,你需要先看一下第一个名称,看看它是一个模块的名称,还是包含了对象的一个变量的名称。你可以分清楚,sys.exit()是对于模块中的一个函数的调用,因为程序的开始处有一条类似import sys 的 import 语句。

构造函数(constructor function)是可以和常规函数一样调用的函数,只不过其返回值是一个新的对象。单看源代码,函数和构造函数是一样的。构造函数(也可以简单地称为constructor,或者有时候称为 ctor)只是对于那些返回一个新的对象的函数的叫法。通常,构造函数是以一个大写字母开头的。这就是为什么当你编写自己的程序的时候,函数名应该总是以小写字母开头。

例如,pygame.Rect()和 pygame.Surface()都是 pygame 模块中的构造函数,它们分别返回一个新的 Rect 对象和 Surface 对象(稍后将会介绍这些对象)。

如下是函数调用、方法调用以及调用模块中的一个函数的示例。

```
import whammy
fizzy()
egg = Wombat()
egg.bluhbluh()
whammy.spam()
```

即便这些名称都放到一起,你也可以分辨出哪一个是函数调用,哪一个是方法调用,而哪一个是对方法中的一个函数的调用。名称 whammy 引用一个模块,因为你可以看到,在第一行已经导入了它。名称 fizzy 的前面以及其后面的圆括号中都没有任何内容,因此,我们知道这是一个函数调用。

Wombat()也是一个函数调用,只不过它是一个构造函数(开头的大写字母并不能保证它是一个构造函数,而不是一个常规的函数,但这是一个额外的标志),会返回一个对象。该对象存储到名为 egg 的一个变量中。egg.bluhbluh()调用是一个方法调用,之所以能够分辨出来,是因为 bluhbluh 附加给了一个变量,而该变量中存储了一个对象。

同时,whammy.spam()是一个函数调用,而不是一个方法调用。你可以看出它不是一个方法,因为名称 whammy 在前面是作为一个模块导入的。

2.9　Surface 对象和窗口

Surface 对象是表示一个矩形的 2D 图像的对象。可以通过调用 Pygame 绘制函数,来改变 Surface 对象的像素,然后再显示到屏幕上。窗口的边框、标题栏和按钮并不是 Surface 对象的一部分。

特别是 pygame.display.set_mode()返回的 Surface 对象叫作显示 Surface(display Surface)。绘制到显示 Surface 对象上的任何内容,当调用 pygame.display.update()函数的

时候，都会显示到窗口上。在一个 Surface 对象上绘制（该对象只存在于计算机内存之中），比把一个 Surface 对象绘制到计算机屏幕上要快很多。这是因为修改计算机内存比修改显示器上的像素要快很多。

程序经常要把几个不同的内容绘制到一个 Surface 对象中。在游戏循环的本次迭代中，一旦将一个 Surface 对象上的所有内容都绘制到了显示 Surface 对象上（这叫作一帧，就像是暂停的 DVD 上的一幅静止的画面），这个显示 Surface 对象就会绘制到屏幕上。计算机可以很快地绘制帧，并且我们的程序通常会每秒运行 30 帧，即 30 FPS。这叫作帧速率（frame rate），我们将在本章后面介绍它。

本章后面的第 2.14 节和 2.19 节将会介绍在 Surface 对象上绘制。

2.10 颜色

光线有 3 种主要的颜色：红色、绿色和蓝色（红色、蓝色和黄色是绘画和颜料的主要颜色，但是计算机显示器使用光，而不是颜料）。通过将这 3 种颜色的不同的量组合起来，可以形成任何其他的颜色。在 Pygame 中，我们使用 3 个整数的元组来表示颜色。元组中的第 1 个值，表示颜色中有多少红色。为 0 的整数值表示该颜色中没有红色，而 255 表示该颜色中的红色达到最大值。第 2 个值表示绿色，而第 3 个值表示蓝色。这些用来表示一种颜色的 3 个整数的元组，通常称为 RGB 值（RGB value）。

由于我们可以针对 3 种主要的颜色使用 0～255 的任何组合，这就意味着 Pygame 可以绘制 16 777 216 种不同的颜色，即 256×256×256 种颜色。然而，如果试图使用大于 255 的值或者负值，将会得到类似"ValueError: invalid color argument"的一个错误。

例如，我们创建元组(0, 0, 0)并且将其存储到一个名为 BLACK 的变量中。没有红色、绿色和蓝色的颜色量，最终的颜色是完全的黑色。黑色就是任何颜色都没有。元组(255, 255, 255)表示红色、绿色和蓝色都达到最大量，这最终得到白色。白色是红色、绿色和蓝色的完全的组合。元组(255, 0, 0)表示红色达到最大量，而没有绿色和蓝色，因此，最终的颜色是红色。类似的，(0, 255, 0)是绿色，而(0, 0, 255)是蓝色。

可以组合红色、绿色和蓝色的量来形成其他的颜色。表 2-1 列出了几种常见的颜色的 RGB 值。

表 2-1

颜　　色	RGB 值	颜　　色	RGB 值
湖绿色	(0, 255, 255)	海军蓝	(0, 0, 128)
黑色	(0, 0, 0)	橄榄色	(128, 128, 0)
蓝色	(0, 0, 255)	紫色	(128, 0, 128)
紫红色	(255, 0, 255)	红色	(255, 0, 0)

续表

颜　色	RGB 值	颜　色	RGB 值
灰色	(128, 128, 128)	银色	(192, 192, 192)
绿色	(　0, 128,　0)	青色	(　0, 128, 128)
浅绿色	(　0, 255,　0)	白色	(255, 255, 255)
褐红色	(128,　0,　0)	黄色	(255, 255,　0)

2.11　颜色的透明度

当你通过一个带有红色色调的玻璃窗口看过去，其背后的所有颜色都会增加一个红色的阴影。你可以通过给颜色值添加第 4 个 0~255 的整数值来模仿这种效果。这个值叫作 alpha 值（alpha value）。这是表示一种颜色有多么不透明的一个度量值。通常，当你在一个 Surface 对象上绘制一个像素的时候，新的颜色完全替代了那里已经存在的任何颜色。但是，使用带有一个 alpha 值的颜色，可以只是给已经存在的颜色之上添加一个带有颜色的色调。

例如，表示绿色的 3 个整数值的一个元组是(0, 255, 0)。但是，如果添加了第 4 个整数作为 alpha 值，我们可以使其成为一个半透明的绿色(0, 255, 0, 128)。255 的 alpha 值是完全不透明的（也就是说，根本没有透明度）。颜色(0, 255, 0)和(0, 255, 0, 255)看上去完全相同。alpha 值为 0，表示该颜色是完全透明的。如果将 alpha 值为 0 的任何一个颜色绘制到一个 Surface 对象上，它没有任何效果，因为这个颜色完全是透明的，且不可见。

为了使用透明颜色来进行绘制，必须使用 convert_alpha()方法创建一个 Surface 对象。例如，如下代码创建了一个可以在其上绘制透明颜色的 Surface 对象。

```
anotherSurface = DISPLAYSURF.convert_alpha()
```

一旦在该 Surface 对象上绘制的内容存储到了 anotherSurface 中，随后 another Surface 可以"复制"（blit，也就是 copy）到 DISPLAYSURF 中，以便它可以显示在屏幕上（参见本章稍后的 2.19 节）。在那些并非从一个 convert_alpha()返回的 Surface 对象上，你不能够使用透明颜色，这也包括从 pygame.display. set_mode()返回的显示 Surface，注意这一点是很重要的。

如果我们创建了一个颜色元组来绘制著名的隐形粉红独角兽，应该会使用(255, 192, 192, 0)，它最终看上去是完全不可见的，就像是 alpha 值为 0 的任何其他颜色一样。毕竟，它是隐形的。图 2-5

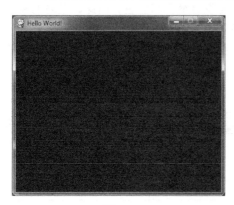

图 2-5

是绘制隐形粉红独角兽的屏幕截图。

2.12 pygame.Color 对象

你需要知道如何表示一种颜色，因为 Pygame 的绘制函数需要一种方式来知道你想要使用何种颜色进行绘制。3 个整数或 4 个整数的一个元组是一种方式。另一种方式是一个 pygame.Color 对象。你可以调用 pygame.Color()构造函数，并且传入 3 个整数或 4 个整数，来创建 Color 对象。可以将这个 Color 对象存储到变量中，就像可以将元组存储到变量中一样。尝试在交互式 shell 中输入如下的内容。

```
>>> import pygame
>>> pygame.Color(255, 0, 0)
(255, 0, 0, 255)
>>> myColor = pygame.Color(255, 0, 0, 128)
>>> myColor == (255, 0, 0, 128)
True
>>>
```

Pygame 中的任何绘制函数（我们将会学习一些），如果有一个针对颜色的参数的话，那么可以为其传递元组形式或者 Color 对象形式的颜色。即便二者是不同的数据类型，但如果它们都表示相同的颜色的话，一个 Color 对象等同于 4 个整数的一个元组（就像 42 == 42.0 将会为 True 一样）。

既然你知道了如何表示颜色（作为 pygame.Color 对象，或者是 3 个整数或 4 个整数的一个元组，3 个整数的话，分别表示红色、绿色和蓝色，4 个整数的话，还包括一个可选的 alpha 值）和坐标（作为表示 X 和 Y 的两个整数的一个元组），让我们来了解一下 pygame.Rect 对象，以便可以开始使用 Pygame 的绘制函数。

2.13 Rect 对象

Pygame 有两种方法来表示矩形区域（就像有两种方法表示颜色一样）。第一种是 4 个整数的元组。

1．左上角的 X 坐标。
2．左上角的 Y 坐标。
3．矩形的宽度（以像素为单位）。
4．矩形的高度（以像素为单位）。

第二种方法是作为一个 pygame.Rect 对象，我们后面将其简称为 Rect 对象。例如，如下的代码创建了一个 Rect 对象，它的左上角位于(10, 20)，宽度为 200 像素，高度为 300 像素。

```
>>> import pygame
>>> spamRect = pygame.Rect(10, 20, 200, 300)
>>> spamRect == (10, 20, 200, 300)
True
```

这种表示的方便之处在于 Rect 对象自动计算矩形的其他部分的坐标。例如，如果你需要知道变量 spamRect 中所存储的 pygame.Rect 对象的右边的 X 坐标，只需要访问 Rect 对象的 right 属性。

```
>>> spamRect.right
210
```

如果左边的 X 坐标为 10 并且矩形的宽度为 200 像素，Rect 对象的 Pygame 代码会自动计算出矩形的右边的 X 坐标必须位于 210。如果重新设置 right 属性，所有其他的属性也会自动计算求得。

```
>>> spamRect.right = 350
>>> spamRect.left
150
```

表 2-2 列出了 pygame.Rect 对象所提供的所有属性（在我们的示例中，Rect 对象存储在名为 myRect 的一个变量中）。

表 2-2

属 性 名 称	说　　明
myRect.left	矩形的左边的 X 坐标的 int 值
myRect.right	矩形的右边的 X 坐标的 int 值
myRect.top	矩形的顶部的 Y 坐标的 int 值
myRect.bottom	矩形的底部的 Y 坐标的 int 值
myRect.centerx	矩形的中央的 X 坐标的 int 值
myRect.centery	矩形的中央的 Y 坐标的 int 值
myRect.width	矩形的宽度的 int 值
myRect.height	矩形的高度的 int 值
myRect.size	两个整数的一个元组：(width, height)
myRect.topleft	两个整数的一个元组：(left, top)
myRect.topright	两个整数的一个元组：(right, top)
myRect.bottomleft	两个整数的一个元组：(left, bottom)
myRect.bottomright	两个整数的一个元组：(right, bottom)
myRect.midleft	两个整数的一个元组：(left, centery)

续表

属 性 名 称	说 明
myRect.midright	两个整数的一个元组：(right, centery)
myRect.midtop	两个整数的一个元组：(centerx, top)
myRect.midbottom	两个整数的一个元组：(centerx, bottom)

2.14 基本的绘制函数

Pygame 提供了几个不同的函数，用于在一个 Surface 对象上绘制不同的形状。这些形状包括矩形、圆形、椭圆形、线条或单个的像素，通常都称为绘制图元（drawing primitives）。打开 IDLE 的文件编辑器并且输入如下的程序，将其保存为 *drawing.py*。

```
 1. import pygame, sys
 2. from pygame.locals import *
 3.
 4. pygame.init()
 5.
 6. # set up the window
 7. DISPLAYSURF = pygame.display.set_mode((500, 400), 0, 32)
 8. pygame.display.set_caption('Drawing')
 9.
10. # set up the colors
11. BLACK = (  0,   0,   0)
12. WHITE = (255, 255, 255)
13. RED   = (255,   0,   0)
14. GREEN = (  0, 255,   0)
15. BLUE  = (  0,   0, 255)
16.
17. # draw on the surface object
18. DISPLAYSURF.fill(WHITE)
19. pygame.draw.polygon(DISPLAYSURF, GREEN, ((146, 0), (291, 106), (236, 277), (56, 277), (0, 106)))
20. pygame.draw.line(DISPLAYSURF, BLUE, (60, 60), (120, 60), 4)
21. pygame.draw.line(DISPLAYSURF, BLUE, (120, 60), (60, 120))
22. pygame.draw.line(DISPLAYSURF, BLUE, (60, 120), (120, 120), 4)
23. pygame.draw.circle(DISPLAYSURF, BLUE, (300, 50), 20, 0)
24. pygame.draw.ellipse(DISPLAYSURF, RED, (300, 250, 40, 80), 1)
25. pygame.draw.rect(DISPLAYSURF, RED, (200, 150, 100, 50))
26.
27. pixObj = pygame.PixelArray(DISPLAYSURF)
28. pixObj[480][380] = BLACK
29. pixObj[482][382] = BLACK
```

```
30.    pixObj[484][384] = BLACK
31.    pixObj[486][386] = BLACK
32.    pixObj[488][388] = BLACK
33.    del pixObj
34.
35.    # run the game loop
36.    while True:
37.        for event in pygame.event.get():
38.            if event.type == QUIT:
39.                pygame.quit()
40.                sys.exit()
41.        pygame.display.update()
```

当这个程序运行的时候，会显示如图 2-6 所示的窗口，直到用户关闭该窗口。

注意我们是如何为每种颜色设置常量变量的。这么做使得代码更具有可读性，因为在源代码中看到 GREEN，很容易理解这是表示绿色，比使用(0, 255, 0)要清晰很多。

绘制函数是根据它们所绘制的形状来命名的。传递给这些函数的参数，则告诉它们在哪一个 Surface 对象上绘制，将形状绘制到何处（以及形状的大小是多少），用什么颜色绘制，以及使用的线条的宽度是多少。在 drawing.py 程序中，可以看到这些函数是如何调用的，如下是针对每个函数的一个简短介绍。

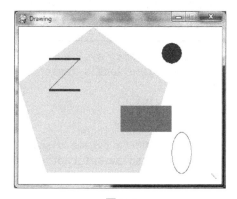

图 2-6

- fill(color) ——fill()方法并不是一个函数，而是 pygame.Surface 对象的一个方法。它将会使用传递给 color 参数的任何颜色来填充整个 Surface 对象。
- pygame.draw.polygon(surface, color, pointlist, width) ——多边形是由多个扁平的边所组成的形状。Surface 和 color 参数告诉函数，将多边形绘制到哪一个 Surface 上，以及用什么颜色绘制。

  pointlist 参数是一个元组或者点的列表（也就是说，用于 XY 坐标的元组，或者两个整数的元组的列表）。多边形是通过这样的方式来绘制的，即在每个点以及元组中其后续的点之间绘制线条，然后，从最后的点到第一个点绘制一个线条。你也可以传递点的列表，而不是点的元组。

  width 参数是可选的。如果你漏掉了这个参数，多边形将会绘制为填充的，就像是屏幕上的绿色多边形那样，会用颜色来填充它。如果确实给 width 参数传递了一个整数值，则只是绘制出多边形的边框。这个整数表示多边形的边框会有多少个像素那么宽。给 width 参数传递 1，将会绘制一个边框细瘦的多边形，而传递 4、10 或

2.14 基本的绘制函数

者 20，将会绘制一个边框较粗一些的多边形。如果给 width 参数传入 0，多边形将会是填充的（这和完全漏掉了 width 参数的效果一样）。

所有的 pygame.draw 绘制函数的最后都有可选的 width 参数，并且，它们的工作方式和 pygame.draw.polygon()的 width 参数相同。可能 width 参数的一个更好的名称应该是 thickness，因为这个参数控制了你绘制的线条有多粗。

- pygame.draw.line(surface, color, start_point, end_point, width) ——这个函数在 start_point 和 end_point 参数之间绘制一条直线。
- pygame.draw.lines(surface, color, closed, pointlist, width) ——这个函数绘制了从一个点到下一个点的一系列的线条，这和 pygame.draw.polygon()函数非常相似。唯一的区别在于如果你给 closed 参数传递了 False，将不会有从 pointlist 中的最后一个点到第一个点的那条直线了。如果你传递了 True，将会绘制从最后一个点到第一个点的直线。
- pygame.draw.circle(surface, color, center_point, radius, width) ——该函数绘制一个圆。center_point 参数指定了圆的圆心。传递给 radius 参数的整数，确定了圆的大小。

圆的半径是从圆心到边的距离（圆的半径总是其直径的一半）。给 radius 参数传递 20，将会绘制半径为 20 个像素的一个圆。

- pygame.draw.ellipse(surface, color, bounding_rectangle, width) ——该函数绘制一个椭圆形（就像是一个压扁了或伸长了的圆）。该函数的参数都是常用的参数，但它们会告诉你绘制多大的椭圆，以及在何处绘制。必须指定椭圆的边界矩形。边界矩形（bounding rectangle）是围绕这一个形状所能绘制的最小的矩形。一个椭圆及其边界矩形的例子如图 2-7 所示。

图 2-7

bounding_rectangle 参数可以是一个 pygame.Rect 对象或者是 4 个整数的一个元组。注意，不必像在 pygame.draw.circle()函数中那样来指定椭圆形的圆心。

- pygame.draw.rect(surface, color, rectangle_tuple, width) ——该函数绘制一个矩形。rectangle_tuple 是 4 个整数的一个元组（分别表示左上角的 XY 坐标，以及宽度和高度），或者可以传递一个 pygame.Rect 对象来替代。如果 rectangle_tuple 的宽度和高度的大小相同，那就会绘制一个正方形。

2.15　pygame.PixelArray 对象

遗憾的是没有单个的函数可以调用来设置一个单个像素的颜色（除非你使用相同的起点和终点来调用 pygame.draw.line()）[1]。在向一个 Surface 对象绘制之前和之后，Pygame 框架需要在幕后运行一些代码。如果它必须针对想要设置的每一个像素来做这些事情，程序将会运行得慢很多（根据我的快速测试，用这种方式绘制像素，所需时间会是原来的两到三倍）。

相反，应该创建一个 Surface 对象的 pygame.PixelArray 对象（我们后面将其简称为 PixelArray 对象）。创建一个 Surface 对象的 PixelArray 对象，将会"锁定"该 Surface 对象。而当一个 Surface 对象锁定的时候，仍然能够在其上调用绘制函数，但是，不能够使用 blit()方法在其上绘制诸如 PNG 或 JPG 这样的图像（本章稍后将会介绍 blit()方法）。

如果想要查看一个 Surface 对象是否锁定了，使用 get_locked()方法，如果它锁定了，Surface 的 get_locked()方法将会返回 True，否则的话，返回 False。

由 pygame.PixelArray()返回的 PixelArray 对象，可以通过两个索引来访问，从而设置单个的像素。例如，第 28 行的 pixObj[480][380] = BLACK 将会把 X 坐标为 480、Y 坐标为 380 的点设置为黑色（别忘了，BLACK 变量存储的颜色元组是(0, 0, 0)）。

要告诉 Pygame 已经完成了单个像素的绘制，用一条 del 语句删除掉 PixelArray 对象，这就是第 33 行所做的事情。删除 PixelArray 对象，将会"解锁" Surface 对象，以便你可以再次在其上绘制图像。如果忘记了删除 PixelArray 对象，下一次尝试将一幅图像复制，即绘制到 Surface 上的时候，程序会导致一条如下所示的错误"pygame.error: Surfaces must not be locked during blit"。

2.16　pygame.display.update()函数

在调用了绘制函数以便让显示 Surface 对象看上去是你想要的方式之后，必须调用 pygame.display.update()让显示 Surface 真正地出现在用户的显示器上。

必须记住的一件事情是 pygame.display.update()将使得显示 Surface，即通过调用 pygame.display.set_mode()而返回的 Surface 对象，出现在屏幕上。如果想要让其他 Surface 对象上的图像出现在屏幕上，必须使用 blit()方法（我们将在 2.19 节中介绍）将其复制到显示 Surface 对象上。

2.17　动画

既然知道了如何让 Pygame 框架绘制到屏幕上，让我们来学习一下如何制作动画。只有

---

[1] 作者注：实际上，Surface 对象有 set_at( )和 get_at( )方法。这里最初的表述不严谨。

2.17 动画

静止的、不能移动的图像的游戏将会相当无聊（我的游戏"Look At This Rock"的销售情况就很令人失望）。动画的图像是做如下的事情所产生的结果：在屏幕上绘制图像，然后隔几秒后在屏幕上绘制一幅略为不同的图像。想象一下程序的窗口有 6 个像素宽和 1 个像素高，所有的像素都是白色，而只有位于 4,0 的一个像素是黑色，看上去如图 2-8 所示。

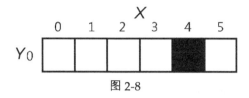

图 2-8

如果修改了窗口使得 3,0 的像素成为黑色，而 4,0 成为白色，看上去将会如图 2-9 所示。

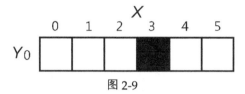

图 2-9

对于用户来说，看上去好像黑色的像素向左边"移动"了。如果重新绘制窗口，使得 2,0 的像素成为黑色，那么，看上去好像是黑色的像素继续向左移动了，如图 2-10 所示。

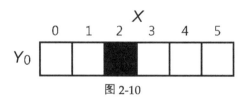

图 2-10

看上去好像是黑色的像素在移动，但这只是错觉。对于计算机来说，它只是显示了 3 幅不同的图像，而每一幅图像上恰好都有一个黑色的像素。考虑一下，如果如图 2-11 所示的 3 幅图像快速出现在屏幕上。

图 2-11

第 2 章 Pygame 基础知识

对于用户来说,看上去像是猫在朝着松鼠的方向移动。但是,对于计算机来说,它们只是一堆的像素。制作看上去逼真的动画的技巧在于让程序将一幅图片绘制到窗口上,等待数秒钟,然后绘制一幅略微不同的图片。

如下的示例程序展示了一个简单的动画。在 IDLE 的文件编辑器中输入这段代码,并且将其保存为 *catanimation.py*。还需要将图像文件 cat.png 放在与 *catanimation.py* 文件相同的目录下。可以从 http://invpy.com/cat.png 下载这个图像。从 http://invpy.com/catanimation.py 可以找到这段代码。

```
1. import pygame, sys
2. from pygame.locals import *
3.
4. pygame.init()
5.
6. FPS = 30 # frames per second setting
7. fpsClock = pygame.time.Clock()
8.
9. # set up the window
10. DISPLAYSURF = pygame.display.set_mode((400, 300), 0, 32)
11. pygame.display.set_caption('Animation')
12.
13. WHITE = (255, 255, 255)
14. catImg = pygame.image.load('cat.png')
15. catx = 10
16. caty = 10
17. direction = 'right'
18.
19. while True: # the main game loop
20.     DISPLAYSURF.fill(WHITE)
21.
22.     if direction == 'right':
23.         catx += 5
24.         if catx == 280:
25.             direction = 'down'
26.     elif direction == 'down':
27.         caty += 5
28.         if caty == 220:
29.             direction = 'left'
30.     elif direction == 'left':
31.         catx -= 5
32.         if catx == 10:
33.             direction = 'up'
34.     elif direction == 'up':
35.         caty -= 5
36.         if caty == 10:
37.             direction = 'right'
```

```
38.
39.     DISPLAYSURF.blit(catImg, (catx, caty))
40.
41.     for event in pygame.event.get():
42.         if event.type == QUIT:
43.             pygame.quit()
44.             sys.exit()
45.
46.     pygame.display.update()
47.     fpsClock.tick(FPS)
```

看一下，动画的猫实现了。这个程序比我的游戏"Look At This Rock 2: A Different Rock"在商业上要成功很多。

2.18 帧速率和 pygame.time.Clock 对象

帧速率（frame rate）或刷新速率（refresh rate）是程序每秒钟绘制的图像的数目，用 FPS 或帧/秒来度量（在计算机显示器上，FPS 常见的名称是赫兹。很多显示器的帧速率是 60Hz，或者说每秒 60 帧）。视频游戏中，较低的帧速率会使得游戏看上去抖动或卡顿。如果游戏包含的代码太多了，以至于无法运行来频繁地绘制到屏幕上，那么，FPS 会下降。但是，本书中的游戏都足够简单，甚至在较旧的计算机上也不会有问题。

pygame.time.Clock 对象可以帮助我们确保程序以某一个最大的 FPS 运行。Clock 对象将会在游戏循环的每一次迭代上都设置一个小小的暂停，从而确保游戏程序不会运行得太快。如果没有这些暂停，游戏程序可能会按照计算机所能够运行的速度去运行。这对玩家来说往往太快了，并且计算机越快，它们运行游戏也就越快。在游戏循环中调用一个 Clock 对象的 tick() 方法，可以确保不管计算机有多快，游戏都按照相同的速度运行。catanimation.py 程序的第 7 行创建了 Clock 对象。

```
7. fpsClock = pygame.time.Clock()
```

每次游戏循环的最后，在调用了 pygame.display.update() 之后，应该调用 Clock 对象的 tick() 方法。根据前一次调用 tick()（这在游戏循环的前一次迭代的末尾进行）之后经过了多长时间，来计算需要暂停多长时间（第一次调用 tick() 方法的时候，根本没有暂停）。在动画程序中，调用 tick() 是在第 47 行进行的，作为游戏循环中的最后一条指令。

你需要知道的是在游戏循环的每一次迭代中，应该在循环的末尾调用 tick() 方法一次。通常，这刚好在调用了 pygame.display.update() 之后进行。

```
47.     fpsClock.tick(FPS)
```

第 2 章　Pygame 基础知识

尝试修改 FPS 常量变量，以不同的帧速率来运行相同的游戏。将其设置为一个较低的值，就会使得程序运行得较慢。将其设置为一个较高的值，就会让程序运行得较快。

2.19 用 pygame.image.load()和 blit()绘制图像

如果你想要在屏幕上绘制简单的形状，绘制函数已经很好用了，但是，很多游戏都有图像（也叫作精灵，sprite）。Pygame 能够从 PNG、JPG、GIF 和 BMP 图像文件中，将图像加载到 Surface 对象上。这些图像文件格式的区别参见 http://invpy.com/formats。

猫的图像存储在一个名为 *cat.png* 的文件中。要加载这个文件的图像，将字符串'cat.png'传递给 pygame.image.load()函数。pygame.image.load()函数调用将会返回一个 Surface 对象，图像已经绘制于其上。这个 Surface 对象将会是和显示 Surface 对象不同的另一个 Surface 对象，因此，我们必须将图像的 Surface 对象复制到显示 Surface 对象上。位图复制（Blitting）就是将一个 Surface 的内容绘制到另一个 Surface 之上。这通过 blit() Surface 对象方法来完成。

如果在调用 pygame.image.load()的时候得到了一条错误消息，如"pygame. error: Couldn't open cat.png"，那么，在运行程序之前，请确保 cat.png 文件和 catanimation.py 位于同一文件夹之中。

```
39.        DISPLAYSURF.blit(catImg, (catx, caty))
```

动画程序的第 39 行使用 blit()方法把 catImg 复制到了 DISPLAYSURF。blit()方法有两个参数。第一个参数是源 Surface 对象，这是将要复制到 DISPLAYSURF Surface 对象上的内容。第 2 个参数是两个整数的一个元组，这两个整数表示图像应该复制到的位置的左上角的 X 和 Y 坐标。

如果 catx 和 caty 设置为 100 和 200，catImg 的宽度为 125，高度为 79，这个 blit()调用将会把该图像复制到 DISPLAYSURF 上，以使得 catImg 的左上角的 XY 坐标为(100, 200)，而其右下角的 XY 坐标为(225, 279)。

注意，不能复制当前"锁定"的一个 Surface（例如，通过其生成了一个 PixelArray 对象并且还没有删除该对象）。

游戏循环剩下的部分只是修改 catx、caty 和 direction 变量，以使得猫在窗口中移动。还有一个 pygame.event.get()调用负责处理 QUIT 事件。

2.20 字体

如果想要将文本绘制到屏幕上，也可以编写几个 pygame.draw.line()调用，来绘制出每个字母的线条。然而，录入所有那些 pygame.draw.line()调用并计算出所有的 XY 坐标，这

2.20 字体

将会是一件令人头疼的事情，并且看上去效果也不会很好，如图 2-12 所示。

上面的这条消息，可能需要调用 pygame.draw.line()函数 41 次才能产生。相反，Pygame 提供了一些非常简单的函数用于字体和文本创建。如下是使用 Pygame 的字体函数的一个较小的 Hello World 程序。在 IDLE 文件编辑器中输入代码，并且将其保存为 *fonttext.py*。

图 2-12

```
 1. import pygame, sys
 2. from pygame.locals import *
 3.
 4. pygame.init()
 5. DISPLAYSURF = pygame.display.set_mode((400, 300))
 6. pygame.display.set_caption('Hello World!')
 7.
 8. WHITE = (255, 255, 255)
 9. GREEN = (0, 255, 0)
10. BLUE = (0, 0, 128)
11.
12. fontObj = pygame.font.Font('freesansbold.ttf', 32)
13. textSurfaceObj = fontObj.render('Hello world!', True, GREEN, BLUE)
14. textRectObj = textSurfaceObj.get_rect()
15. textRectObj.center = (200, 150)
16.
17. while True: # main game loop
18.     DISPLAYSURF.fill(WHITE)
19.     DISPLAYSURF.blit(textSurfaceObj, textRectObj)
20.     for event in pygame.event.get():
21.         if event.type == QUIT:
22.             pygame.quit()
23.             sys.exit()
24.     pygame.display.update()
```

让文本显示到屏幕上，一共有 6 个步骤。

1. 创建一个 pygame.font.Font 对象（如第 12 行所示）。
2. 创建一个 Surface 对象，通过调用 Font 对象的 render()方法，将文本绘制于其上（如第 13 行所示）。
3. 通过调用 Surface 对象的 get_rect()方法，从 Surface 对象创建一个 Rect 对象（如第 14 行所示）。这个 Rect 对象将具有为文本而设置的正确的宽度和高度，但是，其 top 和 left 属性将为 0。
4. 通过修改 Rect 对象的属性之一，来设置其位置。在第 15 行，我们将 Rect 对象的中心设置为 200, 150。
5. 将带有文本的 Surface 对象复制到 pygame.display.set_mode()所返回的 Surface 对

象上（如第 19 行所示）。

6．调用 pygame.display.update()，使显示 Surface 出现在屏幕上（如第 24 行所示）。

pygame.font.Font()构造函数的参数是表示所要使用的字体文件的一个字符串，以及表示字体大小的一个整数（以点为单位，这和字处理程序度量字体大小的单位相同）。在第 12 行，我们传入了'freesansbold.ttf'（这是 Pygame 自带的一种字体）和整数 32（字体大小为 32 点）。

参见 http://invpy.com/usingotherfonts 了解使用其他字体的相关信息。

render()方法调用的参数是所要显示的文本的一个字符串，指定是否想要抗锯齿（本章稍后介绍）的一个 Boolean 值、文本的颜色，以及背景的颜色。如果想要一个透明的背景，那么，直接在方法调用中漏掉背景颜色参数即可。

2.21 抗锯齿

抗锯齿（Anti-aliasing）是一种图形技术，通过给文本和图形的边缘添加一些模糊效果，使其看上去不那么块状化。带有抗锯齿效果的绘制需要多花一些计算时间，因此，尽管图形看上去更好，但程序可能会运行得较慢（但只是略微慢一点）。

如果放大一条带有锯齿的线条和一条抗锯齿的线条，它们的样子如图 2-13 所示。

要对 Pygame 的文本使用抗锯齿效果，只需要给 render()方法的第二个参数传入 True。pygame. draw.aaline()和 pygame.draw.aalines()函数，分别和 pygame.draw.line()和 pygame.draw. lines()函数具有相同的参数，只不过，它们绘制抗锯齿（平滑的）线条，而不是带锯齿（块状化的）线条。

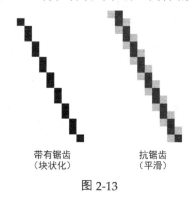

图 2-13

2.22 播放声音

播放存储在声音文件中的声音，甚至比显示图像文件中的图像还要简单。首先，必须通过调用 pygame.mixer.Sound()构造函数，来创建一个 pygame.mixer.Sound 对象（后面我们将其简称为 Sound 对象）。它接受一个字符串参数，这是声音文件的文件名。Pygame 可以加载 WAV、MP3 或 OGG 文件。http://invpy.com/formats 介绍了这些声音文件格式的区别。

要播放声音，调用 Sound 对象的 play()方法。如果想要立即停止 Sound 对象播放，调用 stop()方法。stop()方法没有参数。如下是一些示例代码。

```
soundObj = pygame.mixer.Sound('beeps.wav')
soundObj.play()
import time
time.sleep(1) # wait and let the sound play for 1 second
soundObj.stop()
```

可以从 http://invpy.com/beeps.wav 下载 *beeps.wav* 文件。

在调用 play()之后，程序会立即继续执行；在移动到下一行代码之前，它不会等待声音播放完成。

当玩家受到伤害、挥动一次剑，或收到一个硬币的时候，用 Sound 对象播放声音效果，这对游戏来说是很不错的。但是，如果不管游戏中发生了什么，都有一个背景音乐在播放，你的游戏可能会更好。Pygame 一次只能加载一个作为背景音乐播放的声音文件。要加载一个背景声音文件，调用 pygame.mixer.music.load()函数并且将要加载的声音文件作为一个字符串参数传递。这个文件可以是 WAV、MP3 或 MIDI 格式。

要开始把加载的声音文件作为背景音乐播放，调用 pygame.mixer.music.play(-1, 0.0)函数。当到达了声音文件的末尾的时候，-1 参数会使得背景音乐永远循环。如果将其设置为一个整数 0 或者更大，那么，音乐只能循环指定的那么多次数，而不是永远循环。0.0 意味着从头开始播放声音文件。如果这是一个较大的整数值或浮点值，音乐会开始播放直到声音文件中指定的那么多秒。例如，如果传入 13.5 作为第二个参数，声音文件会从开始处播放到第 13.5 秒的地方。

要立即停止背景音乐，调用 pygame.mixer.music.stop()函数。该函数没有参数。

如下是声音方法和函数的一些示例代码。

```
# Loading and playing a sound effect:
soundObj = pygame.mixer.Sound('beepingsound.wav')
soundObj.play()

# Loading and playing background music:
pygame.mixer.music.load(backgroundmusic.mp3')
pygame.mixer.music.play(-1, 0.0)
# ...some more of your code goes here...
pygame.mixer.music.stop()
```

2.23　本章小结

本章介绍了使用 Pygame 框架编写图形化游戏的基础知识。当然，只是了解这些函数，可能并不足以帮助你学习如何使用这些函数来编写游戏。本书中其他的各章，每一章都关注一个较小的、完整的游戏的源代码。这会帮助你认识到完整的游戏程序看上去是什么样的，以便随后你可以对如何编写自己的游戏程序有更多的思路。

和《Invent Your Own Computer Games with Python》一书不同，本书假设你了解基本的 Python 编程。如果你想不起来变量、函数、循环、if-else 语句和条件是如何工作的，你可能能够通过查看代码中的内容以及程序如何工作而搞清楚这些。但是，如果你仍然有困难，可以阅读《Invent Your Own Computer Games with Python》这本书（这本书针对完全的编程新手）。

第 3 章 Memory Puzzle 游戏

3.1 如何玩 Memory Puzzle 游戏

在 Memory Puzzle 游戏中，白色的方块盖住了几个图标，如图 3-1 所示。每种图标都有两个。玩家可以在两个方块上点击，看看方块背后的图标是什么。如果图标是一致的，那么，这些方块将保持打开的状态。当游戏板上的所有的方块都打开的时候，玩家就赢了。为了给玩家一个提示，在游戏每次开始的时候，所有方块都快速地打开一次。

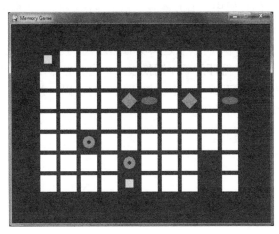

图 3-1

3.2 嵌套的 for 循环

在 Memory Puzzle（以及本书中的大多数游戏）中，你将会看到的一个概念是：在一个 for 循环之中使用一个 for 循环。这叫作嵌套的 for 循环（nested for loop）。嵌套的 for 循环对于遍历两个列表的每一种可能的组合很方便。在交互式 shell 中输入如下的代码。

```
>>> for x in [0, 1, 2, 3, 4]:
...     for y in ['a', 'b', 'c']:
...         print(x, y)
...
0 a
0 b
```

第3章　Memory Puzzle 游戏

```
0 c
1 a
1 b
1 c
2 a
2 b
2 c
3 a
3 b
3 c
4 a
4 b
4 c
>>>
```

　　在 Memory Puzzle 的代码中，有几次我们需要遍历游戏板上的 X 和 Y 坐标组合的每一种可能。我们将使用嵌套的 for 循环来确保得到每一个组合。注意，内部的 for 循环（位于另一个 for 循环之中的 for 循环）将会经过其所有的迭代，然后才会进入到外部的 for 循环的下一次迭代。如果我们将 for 循环的顺序调换过来，也会打印出相同的值，但是，会以不同的顺序将它们打印出来。在交互式 shell 中输入如下的代码，并将它打印值的顺序和前面的嵌套 for 循环示例中的顺序进行比较。

```
>>> for y in ['a', 'b', 'c']:
...     for x in [0, 1, 2, 3, 4]:
...         print(x, y)
...
0 a
1 a
2 a
3 a
4 a
0 b
1 b
2 b
3 b
4 b
0 c
1 c
2 c
3 c
4 c
>>>
```

3.3 Memory Puzzle 的源代码

可以从 http://invpy.com/memorypuzzle.py 下载 Memory Puzzle 的源代码。

继续学习并且先在 IDLE 的文件编辑器中录入整个程序，将其保存为 *memorypuzzle.py* 并运行它。如果得到任何的错误消息，看一下错误消息中所提到的行号，并且检查代码是否有任何的录入错误。也可以将代码复制并粘贴到位于 http://invpy.com/diff/memorypuzzle 的 Web 表单中，看看你的代码和书中的代码之间的差异。

录入一次代码，可能有助于了解程序是如何工作的。当你完成了录入，可以自己玩一玩游戏。

```
 1. # Memory Puzzle
 2. # By Al Sweigart al@inventwithpython.com
 3. # http://inventwithpython.com/pygame
 4. # Released under a "Simplified BSD" license
 5.
 6. import random, pygame, sys
 7. from pygame.locals import *
 8.
 9. FPS = 30 # frames per second, the general speed of the program
10. WINDOWWIDTH = 640 # size of window's width in pixels
11. WINDOWHEIGHT = 480 # size of windows' height in pixels
12. REVEALSPEED = 8 # speed boxes' sliding reveals and covers
13. BOXSIZE = 40 # size of box height & width in pixels
14. GAPSIZE = 10 # size of gap between boxes in pixels
15. BOARDWIDTH = 10 # number of columns of icons
16. BOARDHEIGHT = 7 # number of rows of icons
17. assert (BOARDWIDTH * BOARDHEIGHT) % 2 == 0, 'Board needs to have an even number of boxes for pairs of matches.'
18. XMARGIN = int((WINDOWWIDTH - (BOARDWIDTH * (BOXSIZE + GAPSIZE))) / 2)
19. YMARGIN = int((WINDOWHEIGHT - (BOARDHEIGHT * (BOXSIZE + GAPSIZE))) / 2)
20.
21. #            R    G    B
22. GRAY     = (100, 100, 100)
23. NAVYBLUE = ( 60,  60, 100)
24. WHITE    = (255, 255, 255)
25. RED      = (255,   0,   0)
26. GREEN    = (  0, 255,   0)
27. BLUE     = (  0,   0, 255)
28. YELLOW   = (255, 255,   0)
29. ORANGE   = (255, 128,   0)
30. PURPLE   = (255,   0, 255)
31. CYAN     = (  0, 255, 255)
32.
```

```
33. BGCOLOR = NAVYBLUE
34. LIGHTBGCOLOR = GRAY
35. BOXCOLOR = WHITE
36. HIGHLIGHTCOLOR = BLUE
37.
38. DONUT = 'donut'
39. SQUARE = 'square'
40. DIAMOND = 'diamond'
41. LINES = 'lines'
42. OVAL = 'oval'
43.
44. ALLCOLORS = (RED, GREEN, BLUE, YELLOW, ORANGE, PURPLE, CYAN)
45. ALLSHAPES = (DONUT, SQUARE, DIAMOND, LINES, OVAL)
46. assert len(ALLCOLORS) * len(ALLSHAPES) * 2 >= BOARDWIDTH * BOARDHEIGHT,
"Board is too big for the number of shapes/colors defined."
47.
48. def main():
49.     global FPSCLOCK, DISPLAYSURF
50.     pygame.init()
51.     FPSCLOCK = pygame.time.Clock()
52.     DISPLAYSURF = pygame.display.set_mode((WINDOWWIDTH, WINDOWHEIGHT))
53.
54.     mousex = 0 # used to store x coordinate of mouse event
55.     mousey = 0 # used to store y coordinate of mouse event
56.     pygame.display.set_caption('Memory Game')
57.
58.     mainBoard = getRandomizedBoard()
59.     revealedBoxes = generateRevealedBoxesData(False)
60.
61.     firstSelection = None # stores the (x, y) of the first box clicked.
62.
63.     DISPLAYSURF.fill(BGCOLOR)
64.     startGameAnimation(mainBoard)
65.
66.     while True: # main game loop
67.         mouseClicked = False
68.
69.         DISPLAYSURF.fill(BGCOLOR) # drawing the window
70.         drawBoard(mainBoard, revealedBoxes)
71.
72.         for event in pygame.event.get(): # event handling loop
73.             if event.type == QUIT or (event.type == KEYUP and event.key == K_ESCAPE):
74.                 pygame.quit()
75.                 sys.exit()
76.             elif event.type == MOUSEMOTION:
77.                 mousex, mousey = event.pos
```

```
78.            elif event.type == MOUSEBUTTONUP:
79.                mousex, mousey = event.pos
80.                mouseClicked = True
81.
82.          boxx, boxy = getBoxAtPixel(mousex, mousey)
83.          if boxx != None and boxy != None:
84.              # The mouse is currently over a box.
85.              if not revealedBoxes[boxx][boxy]:
86.                  drawHighlightBox(boxx, boxy)
87.              if not revealedBoxes[boxx][boxy] and mouseClicked:
88.                  revealBoxesAnimation(mainBoard, [(boxx, boxy)])
89.                  revealedBoxes[boxx][boxy] = True # set the box as
"revealed"
90.                  if firstSelection == None: # the current box was the first
box clicked
91.                      firstSelection = (boxx, boxy)
92.                  else: # the current box was the second box clicked
93.                      # Check if there is a match between the two icons.
94.                      icon1shape, icon1color = getShapeAndColor(mainBoard,
firstSelection[0], firstSelection[1])
95.                      icon2shape, icon2color = getShapeAndColor(mainBoard,
boxx, boxy)
96.
97.                      if icon1shape != icon2shape or icon1color !=
icon2color:
98.                          # Icons don't match. Re-cover up both selections.
99.                          pygame.time.wait(1000) # 1000 milliseconds = 1 sec
100.                         coverBoxesAnimation(mainBoard,
[(firstSelection[0], firstSelection[1]), (boxx, boxy)])
101.                         revealedBoxes[firstSelection[0]][firstSelection
[1]] = False
102.                         revealedBoxes[boxx][boxy] = False
103.                     elif hasWon(revealedBoxes): # check if all pairs found
104.                         gameWonAnimation(mainBoard)
105.                         pygame.time.wait(2000)
106.
107.                         # Reset the board
108.                         mainBoard = getRandomizedBoard()
109.                         revealedBoxes = generateRevealedBoxesData(False)
110.
111.                         # Show the fully unrevealed board for a second.
112.                         drawBoard(mainBoard, revealedBoxes)
113.                         pygame.display.update()
114.                         pygame.time.wait(1000)
115.
```

```
116.                        # Replay the start game animation.
117.                        startGameAnimation(mainBoard)
118.                        firstSelection = None # reset firstSelection variable
119.
120.            # Redraw the screen and wait a clock tick.
121.            pygame.display.update()
122.            FPSCLOCK.tick(FPS)
123.
124.
125.    def generateRevealedBoxesData(val):
126.        revealedBoxes = []
127.        for i in range(BOARDWIDTH):
128.            revealedBoxes.append([val] * BOARDHEIGHT)
129.        return revealedBoxes
130.
131.
132.    def getRandomizedBoard():
133.        # Get a list of every possible shape in every possible color.
134.        icons = []
135.        for color in ALLCOLORS:
136.            for shape in ALLSHAPES:
137.                icons.append( (shape, color) )
138.
139.        random.shuffle(icons) # randomize the order of the icons list
140.        numIconsUsed = int(BOARDWIDTH * BOARDHEIGHT / 2) # calculate how many icons are needed
141.        icons = icons[:numIconsUsed] * 2 # make two of each
142.        random.shuffle(icons)
143.
144.        # Create the board data structure, with randomly placed icons.
145.        board = []
146.        for x in range(BOARDWIDTH):
147.            column = []
148.            for y in range(BOARDHEIGHT):
149.                column.append(icons[0])
150.                del icons[0] # remove the icons as we assign them
151.            board.append(column)
152.        return board
153.
154.
155.    def splitIntoGroupsOf(groupSize, theList):
156.        # splits a list into a list of lists, where the inner lists have at
157.        # most groupSize number of items.
158.        result = []
159.        for i in range(0, len(theList), groupSize):
160.            result.append(theList[i:i + groupSize])
```

```
161.    return result
162.
163.
164. def leftTopCoordsOfBox(boxx, boxy):
165.    # Convert board coordinates to pixel coordinates
166.    left = boxx * (BOXSIZE + GAPSIZE) + XMARGIN
167.    top = boxy * (BOXSIZE + GAPSIZE) + YMARGIN
168.    return (left, top)
169.
170.
171. def getBoxAtPixel(x, y):
172.    for boxx in range(BOARDWIDTH):
173.        for boxy in range(BOARDHEIGHT):
174.            left, top = leftTopCoordsOfBox(boxx, boxy)
175.            boxRect = pygame.Rect(left, top, BOXSIZE, BOXSIZE)
176.            if boxRect.collidepoint(x, y):
177.                return (boxx, boxy)
178.    return (None, None)
179.
180.
181. def drawIcon(shape, color, boxx, boxy):
182.    quarter = int(BOXSIZE * 0.25) # syntactic sugar
183.    half =    int(BOXSIZE * 0.5)  # syntactic sugar
184.
185.    left, top = leftTopCoordsOfBox(boxx, boxy) # get pixel coords from board coords
186.    # Draw the shapes
187.    if shape == DONUT:
188.        pygame.draw.circle(DISPLAYSURF, color, (left + half, top + half), half - 5)
189.        pygame.draw.circle(DISPLAYSURF, BGCOLOR, (left + half, top + half), quarter - 5)
190.    elif shape == SQUARE:
191.        pygame.draw.rect(DISPLAYSURF, color, (left + quarter, top + quarter, BOXSIZE - half, BOXSIZE - half))
192.    elif shape == DIAMOND:
193.        pygame.draw.polygon(DISPLAYSURF, color, ((left + half, top), (left + BOXSIZE - 1, top + half), (left + half, top + BOXSIZE - 1), (left, top + half)))
194.    elif shape == LINES:
195.        for i in range(0, BOXSIZE, 4):
196.            pygame.draw.line(DISPLAYSURF, color, (left, top + i), (left + i, top))
197.            pygame.draw.line(DISPLAYSURF, color, (left + i, top + BOXSIZE - 1), (left + BOXSIZE - 1, top + i))
198.    elif shape == OVAL:
```

```
199.            pygame.draw.ellipse(DISPLAYSURF, color, (left, top + quarter,
BOXSIZE, half))
200.
201.
202. def getShapeAndColor(board, boxx, boxy):
203.     # shape value for x, y spot is stored in board[x][y][0]
204.     # color value for x, y spot is stored in board[x][y][1]
205.     return board[boxx][boxy][0], board[boxx][boxy][1]
206.
207.
208. def drawBoxCovers(board, boxes, coverage):
209.     # Draws boxes being covered/revealed. "boxes" is a list
210.     # of two-item lists, which have the x & y spot of the box.
211.     for box in boxes:
212.         left, top = leftTopCoordsOfBox(box[0], box[1])
213.         pygame.draw.rect(DISPLAYSURF, BGCOLOR, (left, top, BOXSIZE,
BOXSIZE))
214.         shape, color = getShapeAndColor(board, box[0], box[1])
215.         drawIcon(shape, color, box[0], box[1])
216.         if coverage > 0: # only draw the cover if there is an coverage
217.             pygame.draw.rect(DISPLAYSURF, BOXCOLOR, (left, top, coverage,
BOXSIZE))
218.     pygame.display.update()
219.     FPSCLOCK.tick(FPS)
220.
221.
222. def revealBoxesAnimation(board, boxesToReveal):
223.     # Do the "box reveal" animation.
224.     for coverage in range(BOXSIZE, - 1, - REVEALSPEED):
225.         drawBoxCovers(board, boxesToReveal, coverage)
226.
227.
228. def coverBoxesAnimation(board, boxesToCover):
229.     # Do the "box cover" animation.
230.     for coverage in range(0, BOXSIZE + REVEALSPEED, REVEALSPEED):
231.         drawBoxCovers(board, boxesToCover, coverage)
232.
233.
234. def drawBoard(board, revealed):
235.     # Draws all of the boxes in their covered or revealed state.
236.     for boxx in range(BOARDWIDTH):
237.         for boxy in range(BOARDHEIGHT):
238.             left, top = leftTopCoordsOfBox(boxx, boxy)
239.             if not revealed[boxx][boxy]:
240.                 # Draw a covered box.
241.                 pygame.draw.rect(DISPLAYSURF, BOXCOLOR, (left, top,
```

```
BOXSIZE, BOXSIZE))
242.             else:
243.                 # Draw the (revealed) icon.
244.                 shape, color = getShapeAndColor(board, boxx, boxy)
245.                 drawIcon(shape, color, boxx, boxy)
246.
247.
248. def drawHighlightBox(boxx, boxy):
249.     left, top = leftTopCoordsOfBox(boxx, boxy)
250.     pygame.draw.rect(DISPLAYSURF, HIGHLIGHTCOLOR, (left - 5, top - 5,
BOXSIZE + 10, BOXSIZE + 10), 4)
251.
252.
253. def startGameAnimation(board):
254.     # Randomly reveal the boxes 8 at a time.
255.     coveredBoxes = generateRevealedBoxesData(False)
256.     boxes = []
257.     for x in range(BOARDWIDTH):
258.         for y in range(BOARDHEIGHT):
259.             boxes.append( (x, y) )
260.     random.shuffle(boxes)
261.     boxGroups = splitIntoGroupsOf(8, boxes)
262.
263.     drawBoard(board, coveredBoxes)
264.     for boxGroup in boxGroups:
265.         revealBoxesAnimation(board, boxGroup)
266.         coverBoxesAnimation(board, boxGroup)
267.
268.
269. def gameWonAnimation(board):
270.     # flash the background color when the player has won
271.     coveredBoxes = generateRevealedBoxesData(True)
272.     color1 = LIGHTBGCOLOR
273.     color2 = BGCOLOR
274.
275.     for i in range(13):
276.         color1, color2 = color2, color1 # swap colors
277.         DISPLAYSURF.fill(color1)
278.         drawBoard(board, coveredBoxes)
279.         pygame.display.update()
280.         pygame.time.wait(300)
281.
282.
283. def hasWon(revealedBoxes):
284.     # Returns True if all the boxes have been revealed, otherwise False
285.     for i in revealedBoxes:
```

```
286.        if False in i:
287.            return False # return False if any boxes are covered.
288.    return True
289.
290.
291. if __name__ == '__main__':
292.     main()
```

3.4 声明和导入

```
1. # Memory Puzzle
2. # By Al Sweigart al@inventwithpython.com
3. # http://inventwithpython.com/pygame
4. # Released under a "Simplified BSD" license
5.
6. import random, pygame, sys
7. from pygame.locals import *
```

在程序的顶部，有一些注释，说明了该游戏是什么，谁开发了它，以及用户可以在哪里找到更多的信息。还有一条提示，说明源代码遵从"Simplified BSD"许可，可以自由地复制。Simplified BSD 许可是 Creative Common 许可的进一步引申，但是它们基本上意味着同样的事情：人们可以自由地复制和分享该游戏。关于许可的更多信息位于 http://invpy.com/licenses。

该程序用到了其他模块中的很多函数，因此，它在第 6 行导入了这些模块。第 7 行也是一条 from (module name) import *格式的 import 语句，这意味着，你不必在其前面输入模块名。pygame.locals 模块中没有函数，但是，其中有几个常量变量我们需要使用，如 MOUSEMOTION、KEYUP 和 QUIT。使用这种格式的 import 语句，我们只需要输入 MOUSEMOTION，而不是 pygame.locals.MOUSEMOTION。

3.5 幻数很糟糕

```
 9. FPS = 30 # frames per second, the general speed of the program
10. WINDOWWIDTH = 640 # size of window's width in pixels
11. WINDOWHEIGHT = 480 # size of windows' height in pixels
12. REVEALSPEED = 8 # speed boxes' sliding reveals and covers
13. BOXSIZE = 40 # size of box height & width in pixels
14. GAPSIZE = 10 # size of gap between boxes in pixels
```

本书中的游戏程序使用了很多的常量变量。你可能还没有意识到它们有多么的方便。例如，我们在代码中使用 BOXSIZE 变量，而不是直接录入整数 40。使用常量变量的原因有两点。首先，如果你随后想要修改每个方块的大小，不必查找整个程序以找出每个录入了 40 的地方进行替换。通过使用 BOXSIZE 常量，我们只需要在第 13 行做出修改，程序剩下的部分也就相应地更新了。特别是我们有可能会针对白色方块以外的其他某个地方使用整数值 40，如果直接修改 40 这个值，可能会意外地导致程序中出现 Bug，这时候使用常量的方法就要好很多了。其次，它使得代码更具有可读性。阅读本章的下一节，并注意第 18 行代码。这里为 XMARGIN 常量设置了一个计算式，即整个游戏板的边缘有多少个像素。这个表达式看上去有些复杂，但是，你可以仔细地拆解其含义。第 18 行如下所示。

```
XMARGIN = int((WINDOWWIDTH - (BOARDWIDTH * (BOXSIZE + GAPSIZE))) / 2)
```

但是，如果不使用常量变量的话，第 18 行将会如下所示。

```
XMARGIN = int((640 - (10 * (40 + 10))) / 2)
```

现在，你不可能记住程序员到底是什么意图。源代码中的这些未加说明的数字，通常叫作幻数（magic number）。当你发现自己输入了幻数的时候，应该考虑用一个常量变量来替代它。对于 Python 解释器来说，前面的两行代码都是相同的。但是，对于阅读源代码并试图理解其如何工作的人类程序员来说，第 18 行代码的第 2 个版本根本没有任何意义。常量确实有助于增加源代码的可读性。

当然，在使用常量变量替代数字上，过犹不及。看看下面的代码。

```
ZERO = 0
ONE = 1
TWO = 99999999
TWOANDTHREEQUARTERS = 2.75
```

不要编写出这样的代码。这看上去太傻了。

3.6　使用 assert 语句全面检查

```
15. BOARDWIDTH = 10 # number of columns of icons
16. BOARDHEIGHT = 7 # number of rows of icons
17. assert (BOARDWIDTH * BOARDHEIGHT) % 2 == 0, 'Board needs to have an even
number of boxes for pairs of matches.'
18. XMARGIN = int((WINDOWWIDTH - (BOARDWIDTH * (BOXSIZE + GAPSIZE))) / 2)
19. YMARGIN = int((WINDOWHEIGHT - (BOARDHEIGHT * (BOXSIZE + GAPSIZE))) / 2)
```

第 15 行的 assert 语句确保了我们所选择的游戏板的宽度和高度会产生偶数个方块（因

第 3 章　Memory Puzzle 游戏

为游戏中将要有成对的图标）。一条 assert 语句由 3 个部分组成：assert 关键字，一个表达式，如果该表达式为 False 的话，会导致程序崩溃。第三个部分（在表达式的逗号之后）是一个字符串，如果程序由于断言而崩溃的话，会显示该字符串。

带有一个表达式的 assert 语句基本上是在说"程序员断言该表达式必须为 True，否则让程序崩溃"。这是给程序添加一个全面检查的一种方式，可以确保如果执行通过了一个断言的话，我们至少知道代码可以像预期的那样工作。

3.7　判断一个数字是偶数还是奇数

如果游戏板的宽度和高度的乘积除以 2 并且余数为 0（模除操作符%计算余数是多少），那么，这个数字是偶数。偶数除以 2 的余数总是为 0。奇数除以 2 总是会余 1。如果你的代码需要判断一个数字是偶数还是奇数的话，记住这是一个好办法。

```
>>> isEven = someNumber % 2 == 0
>>> isOdd = someNumber % 2 != 0
```

在上面的例子中，如果 someNumber 中的整数是偶数，那么 isEven 将为 True。如果它是奇数，那么 isOdd 将为 True。

3.8　较早崩溃和经常崩溃

让程序崩溃是一件糟糕的事情。当你的程序代码中有某些错误并且无法继续的时候，会发生这种情况。但是，在有些情况下，程序较早的崩溃可以避免后面更糟糕的 Bug。

如果我们在第 15 行和第 16 行为 BOARDWIDTH 和 BOARDHEIGHT 所选择的值导致了方块的数目为奇数（例如，宽度为 3，而高度 5），那么，总是会剩下一个图标无法配对。这将会导致程序后面产生 Bug，并且可能要花费很多的调试工作才能搞清楚 Bug 的源头在程序刚开始的位置。实际上，可以尝试一下注释掉断言以使其无法运行，然后，将 BOARDWIDTH 和 BOARDHEIGHT 常量都设置为奇数。当运行该程序的时候，会立即显示在 *memorypuzzle.py* 的第 149 行出现了一个错误，也就是在 getRandomizedBoard()函数中。

```
Traceback (most recent call last):
  File "C:\book2svn\src\memorypuzzle.py", line 292, in <module>
    main()
  File "C:\book2svn\src\memorypuzzle.py", line 58, in main
    mainBoard = getRandomizedBoard()
  File "C:\book2svn\src\memorypuzzle.py", line 149, in getRandomizedBoard
    columns.append(icons[0])
IndexError: list index out of range
```

可能要花很多的时间查看 getRandomizedBoard() 以尝试搞清楚哪里出错了，然后才意识到 getRandomizedBoard() 并没问题，Bug 真正的根源在第 15 行和第 16 行我们设置 BOARDWIDTH 和 BOARDHEIGHT 常量的地方。断言确保了不会发生这种情况。如果代码崩溃了，我们想要让代码一发现某些可怕的错误就尽快崩溃，否则的话，Bug 可能等到在程序中很晚的时候才出现。还是尽早崩溃吧！

当程序中有某些条件总是必须为 True 的时候，你想要添加 assert 语句。崩溃是常事！

你不必矫枉过正，并且随处放置 assert 语句，但是断言引发的崩溃往往需要花很长的时间才能检测到 Bug 的真正来源。因此，还是尽早崩溃，并经常崩溃吧！

3.9 让源代码更好看一些

```
21. #              R    G    B
22. GRAY        = (100, 100, 100)
23. NAVYBLUE   = ( 60,  60, 100)
24. WHITE      = (255, 255, 255)
25. RED        = (255,   0,   0)
26. GREEN      = (  0, 255,   0)
27. BLUE       = (  0,   0, 255)
28. YELLOW     = (255, 255,   0)
29. ORANGE     = (255, 128,   0)
30. PURPLE     = (255,   0, 255)
31. CYAN       = (  0, 255, 255)
32.
33. BGCOLOR = NAVYBLUE
34. LIGHTBGCOLOR = GRAY
35. BOXCOLOR = WHITE
36. HIGHLIGHTCOLOR = BLUE
```

还记得吧，在 Pygame 中，颜色是使用从 0~255 的 3 个整数的一个元组来表示的。这 3 个整数分别表示颜色之中的红色、绿色和蓝色的量，因此，这些元组也称为 RGB 值。注意，在第 22 行到第 31 行代码之中，元组之中的空格使得 R、G 和 B 整数是对齐的。在 Python 中，缩进（也就是代码行开始的空格）需要是准确的，但是，剩下的代码行中的空格是不需要那么严格的。通过将元组中的整数之间的空格对齐，我们可以清楚地看到 RGB 值彼此之间的对比（http://invpy.com/whitespace 给出了关于空格和缩进的更多信息）。

通过这种方式让代码变得更加可读，是一件很好的事情，但不要那么麻烦地花太多时间来做这件事情。代码并不一定是要漂亮才能够工作。在某些情况下，你用来输入空格的时间，比通过让元组变得可读而节省下来的时间还要多。

3.10 使用常量变量而不是字符串

```
38. DONUT = 'donut'
39. SQUARE = 'square'
40. DIAMOND = 'diamond'
41. LINES = 'lines'
42. OVAL = 'oval'
```

该程序还为一些字符串设置了常量变量。这些常量将用于游戏板的数据结构中，以记录游戏板上的哪一个空间拥有哪个图标。使用常量变量，而不是字符串值，这是一个好主意。看看下面的代码，这是第 187 行的代码。

```
if shape == DONUT:
```

shape 变量将设置为如下的字符串之一：'donut'、'square'、'diamond'、'lines'或'oval'，并且随后和 DONUT 常量进行比较。如果在编写第 187 行的时候，有一个录入错误，如下所示：

```
if shape == DUNOT:
```

那么，Python 就会崩溃，并且给出一条错误消息，表示不存在一个名为 DUNOT 的变量。这很好。因为程序在第 187 行崩溃了，我们检查该行的时候，能够很容易看到这个 Bug 是由于输入错误而导致的。然而，如果我们使用了字符串，而不是常量变量，并且犯了这个录入错误，第 187 行会如下所示：

```
if shape == 'dunot':
```

这完全是可以接受的 Python 代码，因此，当第一次运行它的时候，它不会崩溃。然而，这将会导致程序随后出现奇怪的 Bug。由于代码并不会在引发问题的地方立即崩溃，这个 Bug 可能较难找到。

3.11 确保有足够的图标

```
44. ALLCOLORS = (RED, GREEN, BLUE, YELLOW, ORANGE, PURPLE, CYAN)
45. ALLSHAPES = (DONUT, SQUARE, DIAMOND, LINES, OVAL)
46. assert len(ALLCOLORS) * len(ALLSHAPES) * 2 >= BOARDWIDTH * BOARDHEIGHT,
"Board is too big for the number of shapes/colors defined."
```

为了让游戏程序能够创建每一种可能的颜色和形状的组合而成的图标，我们需要有一个元组来保存所有这些值。第 46 行还有另一个断言，以确保针对我们的游戏板的大小，有足够的颜色、形状的组合。如果没有那么多组合，程序会在第 46 行崩溃，并且我们将会知道，要么是添加了过多的颜色和形状，要么是游戏的宽度和高度太小了。通过 7 种颜色和 5 种形状，我们可以生成 35 个（7×5）不同的图标。并且，由于我们的每一个图标都是成对的，这意味着必须拥有一个总共有 70 个方块空间的游戏板（35×2 或 7×5×2）。

3.12 元组 vs.列表，不可变 vs.可变

你可能已经注意到了，ALLCOLORS 和 ALLSHAPES 变量是元组，而不是列表。那么，何时想要使用元组，而何时又想要使用列表呢？二者之间有什么区别呢？

元组和列表在很多方面是相同的，只有两个方面例外：元组使用圆括号，而不是方括号，并且元组中的项是不能修改的（但是列表中的项可以修改）。我们通常说列表是可变的（mutable，这意味着可以修改），而元组是不可变的（immutable，意味着不可修改）。

举一个试图修改列表和元组中的值的例子，如下面的代码所示。

```
>>> listVal = [1, 1, 2, 3, 5, 8]
>>> tupleVal = (1, 1, 2, 3, 5, 8)
>>> listVal[4] = 'hello!'
>>> listVal
[1, 1, 2, 3, 'hello!', 8]
>>> tupleVal[4] = 'hello!'
Traceback (most recent call last):
  File "<stdin>", line 1, in <module>
TypeError: 'tuple' object does not support item assignment
>>> tupleVal
(1, 1, 2, 3, 5, 8)
>>> tupleVal[4]
5
```

注意，当我们试图修改元组中索引为 4 的项，Python 会给出一条错误消息，表明元组对象不支持给项赋值。

元组的不可变性有一个直接的好处和一个重要的好处。直接的好处是使用元组的代码比使用列表的代码要稍微快一些（Python 能够进行一些优化，它知道一个元组中的值是不能修改的）。但是，让代码运行快上几个纳秒无济于事。

使用元组的重要的优点和使用常量的有点类似，即这是一个标记，表明了元组中的值是不能够修改的，因此，随后阅读代码的任何人都知道"我可以认为这个元组永远是相同的。否则的话，程序员就会使用列表了"。这也会让将来读到你的代码的程序员这么认为，即"如果我看到一个列表值，我知道它将会在这个程序中的某个地方修改，否则的话，编写这段代码的程序员就会使用一个元组了"。

我们仍然可以将一个新的元组值赋值给一个变量。

```
>>> tupleVal = (1, 2, 3)
>>> tupleVal = (1, 2, 3, 4)
```

这段代码之所以能够工作，是因为第 2 行的代码并没有修改(1, 2, 3)元组。它将一个全新的元组(1, 2, 3, 4)赋值给了 tupleVal，并且覆盖了旧的元组值。然而，你不能使用方括号来修改一个元组中的项。字符串也是不可变的数据类型。你可以使用方括号来读取字符串中单个的字符，但是，不能修改一个字符串中的单个字符。

```
>>> strVal = 'Hello'
>>> strVal[1]
'e'
>>> strVal[1] = 'X'
Traceback (most recent call last):
  File "<stdin>", line 1, in <module>
TypeError: 'str' object does not support item assignment
```

3.13 单项元组需要一个结尾的逗号

此外，对于元组还有一个小细节需要注意：如果你确实需要编写关于只有一个值的元组的代码，那么，这个元组中需要有一个结尾的逗号，如下所示。

```
oneValueTuple = (42, )
```

如果忘记了这个逗号（并且很容易会忘记它），那么，Python 将会混淆，认为其只是改变运算顺序的一对圆括号区。例如，看一下如下的两行代码。

```
variableA = (5 * 6)
variableB = (5 * 6, )
```

variableA 中存储的值只是整数 30。然而，variableB 赋值语句中的表达式是一个单项的元组值(30,)。空白元组值中不需要逗号，它们本身只是一对圆括号()。

3.14 在列表和元组之间转换

可以在列表和元组值之间转换，就像在字符串和整数值之间转换一样。只需要将一个元组值传递给 list()函数，就可以转换为元组的值所构成的一个列表。或者将值的一个列表传递给 tuple()，它将会转换为该列表值所构成的一个元组。尝试在交互式 shell 中输入如下的代码。

```
>>> spam = (1, 2, 3, 4)
>>> spam = list(spam)
>>> spam
[1, 2, 3, 4]
>>> spam = tuple(spam)
>>> spam
(1, 2, 3, 4)
>>>
```

3.15　global 语句以及为什么全局变量是罪恶的

```
48. def main():
49.     global FPSCLOCK, DISPLAYSURF
50.     pygame.init()
51.     FPSCLOCK = pygame.time.Clock()
52.     DISPLAYSURF = pygame.display.set_mode((WINDOWWIDTH, WINDOWHEIGHT))
53.
54.     mousex = 0 # used to store x coordinate of mouse event
55.     mousey = 0 # used to store y coordinate of mouse event
56.     pygame.display.set_caption('Memory Game')
```

这是 main() 函数的开始几行，也是游戏代码的主要部分所在（这一点很奇怪）。在 main() 函数中调用的函数，我们将会在本章后面介绍。

第 49 行是一条 global 语句。global 语句是 global 关键字后面跟着用逗号隔开的变量名的一个列表。这些变量名随后标记为全局变量。在 main() 函数中，这些名称并不是用于那些恰好和全局变量具有相同的名称的局部变量。它们就是全局变量。在 main() 函数中，赋给它们的任何值，都会在 main() 函数之外持久化。我们将 FPSCLOCK 和 DISPLAYSURF 变量标记为全局的，因为在程序的其他几个函数中也会用到它们（更多信息参见 http://invpy.com/scope）。

判断一个变量是局部的，还是全局的，有以下 4 条简单的规则。
1. 如果在函数的开始处，有一条针对一个变量的全局语句，那么，该变量是全局的。
2. 如果函数中一个变量的名称与一个全局变量的名称相同，并且该函数没有给该变量赋一个值，那么，该变量是全局变量。
3. 如果函数中一个变量的名称与一个全局变量的名称相同，并且该函数确实给该变量赋了一个值，那么，该变量是局部变量。
4. 如果没有一个全局变量与函数中的该变量具有相同的名称，那么该变量显然是一个局部变量。

通常，我们想要避免在函数中使用全局变量。函数被认为是你的程序中的一个小程序，它具有特定的输入（参数）和输出（返回值）。但是，读写全局变量的函数，拥有额外的输入和输出。在调用函数之前，全局变量可能已经在很多地方修改过，因此，如果一个 Bug 和在全局变量中设置了糟糕的值有关，我们将很难找到它。把一个函数作为没有使用全局变量的一个单独的小程序，这使得我们很容易找出代码中的 Bug，因为，函数的参数是清晰可见的。这也使得修改函数中的代码更为容易，因为如果新的函数能够使用相同的参数并给出相同的返回值，那么，它将能够像旧的函数一样自动地和程序其他部分一起工作。

基本上，使用全局变量使得编写程序更容易，但是通常会导致难以调试。

在本书中的游戏中，全局变量主要用于那些不会修改的全局性的常量，但是需要先调用 pygame.init()函数。由于这个调用在 main()函数中进行，我们会在 main()函数中设置全局变量，并且这些变量必须对查看它们的其他函数来说是全局的。但是，全局变量用作了常量，并且不可修改，因此它们引起令人混淆的 Bug 的可能性比较小。

如果不理解这些，也不要担心。作为一条通用的规则，在编写自己的代码的时候，给函数传入值，而不是让函数读取全局变量。

3.16 数据结构和 2D 列表

```
58.        mainBoard = getRandomizedBoard()
59.        revealedBoxes = generateRevealedBoxesData(False)
```

getRandomizedBoard() 函数返回了一个数据结构，表示游戏板的状态。generateRevealedBoxesData()函数返回了一个数据结构，分别表示哪个方块被盖住了。这些函数的返回值是 2D 列表，或者说是列表的列表。值的列表的列表的列表，就是一个 3D 列表。表示两维或更多维的列表的另一个术语是多维列表（multidimensional list）。

如果将一个列表存储到了名为 spam 的变量中，可以使用方括号，如 spam[2]来获取列表中的第 3 个值。如果 spam[2]的值本身就是一个列表，那么，我们可以使用另一对方括号来获取该列表之中的值。这看上去像是这样的 spam[2][4]，它将会访问列表的第 5 个值，而这个列表是 spam 的第 3 个值。由于 mainBoard 变量将会在其中存储图标，如果想要获取游戏板上位于(4, 5) 的图标，那么，我们只需要使用表达式 mainBoard[4][5]。由于图标本身是存储在带有形状和颜色的一个二项元组之中的，完整的数据结构是二项元组的列表的一个列表。

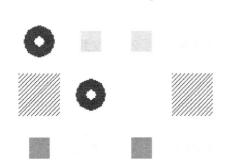

图 3-2

如下是一个小例子。假设游戏板如图 3-2 所示。
相应的数据结构是：

```
mainBoard = [[(DONUT, BLUE), (LINES, BLUE), (SQUARE, ORANGE)], [(SQUARE,
GREEN), (DONUT, BLUE), (DIAMOND, YELLOW)], [(SQUARE, GREEN), (OVAL, YELLOW),
(SQUARE, ORANGE)], [(DIAMOND, YELLOW), (LINES, BLUE), (OVAL, YELLOW)]]
```

如果你的图书是黑白印刷的,你可以通过 http://invpy.com/memoryboard 查看上面的图片的一个彩色版本。你将会注意到,mainBoard[x][y]和游戏板上位于(X, Y)坐标处的图标相对应。

同时,"揭开的方块"这一数据结构也是一个 2D 列表,只不过它拥有 Boolean 值,而不是像游戏板数据结构那样的二项元组,如果位于 X,Y 坐标的方块是揭开的,这个 Boolean 值为 True,如果方块是盖住的,值为 False。给 generateRevealedBoxesData()函数传递 False,从而将所有的 Boolean 值都设置为 False(稍后将详细介绍该函数)。

这两个数据结构用来记录游戏板的状态。

3.17 "开始游戏"动画

```
61.     firstSelection = None # stores the (x, y) of the first box clicked.
62.
63.     DISPLAYSURF.fill(BGCOLOR)
64.     startGameAnimation(mainBoard)
```

第 61 行使用值 None 设置了一个名为 firstSelection 的变量(None 是表示缺少值的一个值。它也是该数据类型 NoneType 的唯一一个值。更多信息参见 http://invpy.com/None)。当玩家点击游戏板上的一个图标的时候,程序需要记录这是在游戏板上点击的成对的图标中的第一个还是第二个。如果 firstSelection 为 None,那么这次点击的是第一个图标,我们将 XY 坐标作为两个整数的一个元组(一个整数表示 X 值,一个整数表示 Y 值),存储到 firstSelection 变量中。

在第二次点击的时候,这个值将会是这个元组,而不再是 None,程序由此可以记录这是第二次图标点击。第 63 行用背景颜色填充了整个 Surface。这将会通过绘制覆盖曾经在该 Surface 上的所有内容,给我们一个干净的游戏板以便开始在上面绘制图形。

如果你玩过 Memory Puzzle 游戏,将会注意到,在游戏开始的时候,所有的方块都快速地、随机地盖住和翻开,以便让玩家瞄一眼哪个方块下面是哪一个图标。这在 startGameAnimation()函数中发生,我们将在本章后面介绍该函数。

让玩家瞄一眼,这一点很重要(但是不要看的时间太长了,否则玩家很容易记住图标的位置),因为不这样做的话,玩家就对任何图标的位置都没有了线索。和给出一点提示相比,盲目地点击图标并不会更有趣。

3.18 游戏循环

```
66.    while True: # main game loop
67.        mouseClicked = False
68.
69.        DISPLAYSURF.fill(BGCOLOR) # drawing the window
70.        drawBoard(mainBoard, revealedBoxes)
```

游戏循环是从第 66 行开始的一个无限循环，只要游戏还在运行中，该循环就持续迭代。还记得吧，游戏循环负责处理事件、更新游戏状态，并且将游戏状态绘制到屏幕上。

Memory Puzzle 程序的游戏状态存储在如下的变量中。

- mainBoard
- revealedBoxes
- firstSelection
- mouseClicked
- mousex
- mousey

在 Memory Puzzle 游戏中的游戏循环的每一次迭代中，mouseClicked 变量都存储了一个 Boolean 值，如果玩家在游戏循环的该次迭代中点击了鼠标的话，这个值为 True。

在第 69 行，使用背景颜色绘制到 Surface 上，以擦除掉之前在其上绘制的任何内容。然后，程序调用 drawBoard()，根据我们传递给它的游戏板（mainBoard）和"揭开的方块"（revealedBoxes）数据结构来绘制游戏板（这些代码行是负责绘制和更新屏幕的部分）。

记住，我们的绘制函数只是在内存中的显示 Surface 对象上绘制。在调用 pygame.display.update() 函数之前，该 Surface 对象并不会真地出现在屏幕上，而我们将在游戏循环末尾的第 121 行调用该函数。

3.19 事件处理循环

```
72.    for event in pygame.event.get(): # event handling loop
73.        if event.type == QUIT or (event.type == KEYUP and event.key == K_ESCAPE):
74.            pygame.quit()
75.            sys.exit()
76.        elif event.type == MOUSEMOTION:
77.            mousex, mousey = event.pos
78.        elif event.type == MOUSEBUTTONUP:
```

3.20 检查鼠标光标在哪一个方块之上

```
79.            mousex, mousey = event.pos
80.            mouseClicked = True
```

第 72 行的 for 循环针对游戏循环的最后一次迭代之后所发生的每一次事件来执行代码。这个循环叫作事件处理循环（event handling loop，它和游戏循环不同，尽管它位于游戏循环之中），并且它会遍历调用 pygame.event.get() 所返回的 pygame.Event 的列表。

如果事件对象是一个 QUIT 事件或针对 ESC 键的一个 KEYUP 事件，那么，程序应该终止。否则，在一次 MOUSEMOTION 事件（鼠标光标移动）或 MOUSEBUTTONUP（鼠标按键先按下并且现在释放按键）中，鼠标光标的位置应该存储到 mousex 和 mousey 变量中。如果这是一个 MOUSEBUTTONUP 事件，mouseClicked 还应该设置为 True。一旦处理了所有这些事件，mousex、mousey 和 mouseClicked 中所存储的值将会告诉我们玩家所给出的任何输入。现在，应该更新游戏状态并且将结果绘制到屏幕上。

3.20 检查鼠标光标在哪一个方块之上

```
82.        boxx, boxy = getBoxAtPixel(mousex, mousey)
83.        if boxx != None and boxy != None:
84.            # The mouse is currently over a box.
85.            if not revealedBoxes[boxx][boxy]:
86.                drawHighlightBox(boxx, boxy)
```

getBoxAtPixel() 函数将会返回两个整数的一个元组。这两个整数表示鼠标位于其上的方块在游戏板上的 XY 坐标。稍后将介绍 getBoxAtPixel() 如何做到这一点。现在，我们需要知道的是如果 mousex 和 mousey 坐标在一个方块之上，该函数返回 XY 游戏板坐标的一个元组，并且存储到 boxx 和 boxy 中。如果鼠标光标不在任何方块之上（例如，如果它超出了游戏板的边界，或者它位于两个方块之间的间隙上），那么该函数返回元组 (None, None)，并且 boxx 和 boxy 中都存储为 None。

我们只是对 boxx 和 boxy 中都不是 None 的情况感兴趣，因此，第 83 行的 if 语句之后的语句块中的几行代码负责检查这种情况。如果执行到这个语句块中，我们知道用户把鼠标光标放到了一个方块之上（并且根据 mouseClicked 中存储的值，用户可能还点击了鼠标）。

第 85 行的 if 语句，通过读取 revealedBoxes[boxx][boxy] 中存储的值，检查这个方块是否被盖住了。如果该值为 false，我们知道方块被盖住了。无论何时，当鼠标位于一个盖住的方块之上，我们想要围绕方块绘制一个蓝色的高亮的边框，以通知玩家他们可以在其上点击。对于已经打开的方块，不会绘制这个高亮边框。高亮边框的绘制由 drawHighlightBox() 函数来完成，我们将在后面介绍该函数。

第 3 章　Memory Puzzle 游戏

```
87.             if not revealedBoxes[boxx][boxy] and mouseClicked:
88.                 revealBoxesAnimation(mainBoard, [(boxx, boxy)])
89.                 revealedBoxes[boxx][boxy] = True # set the box as
"revealed"
```

在第 87 行，我们检查鼠标光标是否不仅在一个盖住的方框上，而且已经点击了鼠标。在这种情况下，我们将要调用 revealBoxesAnimation() 函数（和 main() 中调用的所有其他函数一样，将在本章后面介绍），以播放该方块"揭开"的动画。你应该注意，调用该函数只是绘制打开方块的动画。直到第 89 行，我们才设置 revealedBoxes[boxx][boxy] = True，这是负责记录游戏状态已经更新的数据结构。

如果注释掉第 89 行，然后再运行程序，将会注意到，在点击了一个方框之后，会播放揭开方框的动画，然后，方框立即再次显示为盖住的状态。这是因为 revealedBoxes[boxx][boxy] 仍然设置为 False，在游戏循环的下一次迭代中，将该盖住的方块绘制到了游戏板之上。如果没有第 89 行的代码，将会导致程序中一个安静而奇怪的 Bug。

3.21　处理第一次点击的方块

```
90.             if firstSelection == None: # the current box was the first
box clicked
91.                 firstSelection = (boxx, boxy)
92.             else: # the current box was the second box clicked
93.                 # Check if there is a match between the two icons.
94.                 icon1shape, icon1color = getShapeAndColor(mainBoard,
firstSelection[0], firstSelection[1])
95.                 icon2shape, icon2color = getShapeAndColor(mainBoard,
boxx, boxy)
```

在执行进入到游戏循环之前，firstSelection 变量设置为 None。我们的程序会将此解释为没有方块被点击，因此，如果第 90 行的条件为 True，这意味着这是点击的两个可能匹配的方块中的第一个。我们想要播放该方块揭开的动画，并且随后保持该方块为盖住的。我们还将 firstSelection 变量设置为所点击的方块的方块坐标。如果这是玩家所点击的第二个方块，我们想要为该方块播放揭开动画，随后检查两个方块下面的两个图标是否一致。getShapeAndColor() 函数（稍后解释）将会获取该图标的形状和颜色值（这些值将会是 ALLCOLORS 和 ALLSHAPES 元组中值之一）。

3.22 处理不一致的一对图标

```
 97.                if icon1shape != icon2shape or icon1color != icon2color:
 98.                    # Icons don't match. Re-cover up both selections.
 99.                    pygame.time.wait(1000) # 1000 milliseconds = 1 sec
100.                    coverBoxesAnimation(mainBoard, [(firstSelection[0], firstSelection[1]), (boxx, boxy)])
101.                    revealedBoxes[firstSelection[0]][firstSelection[1]] = False
102.                    revealedBoxes[boxx][boxy] = False
```

第 97 行的 if 语句检查两个图标的形状或颜色是否一致。如果是不一致的，那么，我们将会调用 pygame.time.wait(1000)，暂停游戏 1000 毫秒（这等于 1 秒钟），从而使得玩家能够注意到两个图标不一致。然后，为两个方块播放"盖住"的动画。我们还想要更新游戏状态，以将两个方块标记为非揭开的（盖住的）状态。

3.23 处理玩家获胜

```
103.                elif hasWon(revealedBoxes): # check if all pairs found
104.                    gameWonAnimation(mainBoard)
105.                    pygame.time.wait(2000)
106.
107.                    # Reset the board
108.                    mainBoard = getRandomizedBoard()
109.                    revealedBoxes = generateRevealedBoxesData(False)
110.
111.                    # Show the fully unrevealed board for a second.
112.                    drawBoard(mainBoard, revealedBoxes)
113.                    pygame.display.update()
114.                    pygame.time.wait(1000)
115.
116.                    # Replay the start game animation.
117.                    startGameAnimation(mainBoard)
118.                firstSelection = None # reset firstSelection variable
```

否则，如果第 97 行的条件为 False，那么，两个图标必须是一致的。此时，程序不必对方块做任何其他的事情，它只需要让这两个方块都处于打开的状态。然而，程序应该检查这是否是游戏板上最后一对一致的图标。这在 hasWon() 函数中实现，如果游戏板处于一个获胜的状态（所有的方块都是打开的），该函数返回 True。

第 3 章 Memory Puzzle 游戏

如果是这种情况，我们想要调用 gameWonAnimation() 来播放"游戏获胜"动画，稍作暂停让玩家庆祝获胜，然后，重新设置 mainBoard 和 revealedBoxes 中的数据结构以启动一次新的游戏。

第 117 行再次播放了"开始游戏"的动画。此后，程序执行将会像通常那样进入游戏循环，玩家可以继续玩游戏直到他们退出程序。

不管两个方块是否一致，在第 118 行点击了第二个方块之后，都会将 firstSelection 变量设置回到 None，以便玩家点击下一个方块的时候，会被当作可能匹配的图标对中第一次点击的方块。

3.24 将游戏状态绘制到屏幕

```
120.        # Redraw the screen and wait a clock tick.
121.        pygame.display.update()
122.        FPSCLOCK.tick(FPS)
```

此时，游戏状态已经根据玩家的输入而更新了，并且，最新的游戏状态已经绘制到 DISPLAYSURF 显示 Surface 对象上了。我们已经到达了游戏循环的末尾，因此，调用 pygame.display.update()将 DISPLAYSURF Surface 对象绘制到计算机屏幕上。

第 9 行将 FPS 常量设置为整数值 30，意味着我们想要让游戏以（最大的）每秒 30 帧的速度运行。如果想要让程序运行得更快，可以增加这个数字。如果想要让程序运行得慢一些，减少这个值。甚至可以将其设置为一个 0.5 这样的浮点数，这将会以每秒半帧的速度运行程序，也就是每 2 秒钟一帧。

为了达到每秒运行 30 帧，每帧必须在 1/30 秒钟绘制出来。这意味着，pygame.display.update()以及游戏循环中的所有代码，必须在 33.3 毫秒之内执行。任何现代的计算机都能够很容易地做到这一点，并且余下足够的时间。为了防止程序运行得过快，我们调用 FPSCLOCK 中的 pygame.Clock 对象的 tick()方法，从而让程序在 33.3 毫秒之外的剩余时间中暂停。

由于这是在游戏循环的最后进行的，这确保了游戏循环的每一次迭代都会花（至少）33.3 毫秒。如果由于某些原因，pygame.display.update()调用和游戏循环中的代码花费了超过 33.3 毫秒的时间，那么 tick()方法根本不会等待，而是立即返回。

我一直在说，其他的函数将会在本章的后面介绍。既然已经介绍完了 main()函数，并且对于一般的程序如何工作已经有了一些了解，让我们深入地看看通过 main()调用的所有其他函数的细节。

3.25 创建"揭开的方块"数据结构

```
125. def generateRevealedBoxesData(val):
126.     revealedBoxes = []
127.     for i in range(BOARDWIDTH):
128.         revealedBoxes.append([val] * BOARDHEIGHT)
129.     return revealedBoxes
```

generateRevealedBoxesData()函数需要创建 Boolean 值的列表的一个列表。该 Boolean 值将作为一个 val 参数传递给函数。一开始的时候，这个数据结构是 revealed Boxes 变量中的一个空的列表。

为了让该数据结构拥有 revealedBoxes[x][y]的结构，我们需要确保内部的列表表示的是游戏板的垂直列，而不是水平行。否则的话，数据结构将会是一个 revealedBoxes[y] [x]结构。这个 for 循环将会创建列，然后将它们添加到 revealedBoxes 中。使用列表复制功能来创建列，以便列列表拥有 BOARDHEIGHT 所指定的那么多个 val 值。

3.26 创建游戏板数据结构：第 1 步——获取所有可能的图标

```
132. def getRandomizedBoard():
133.     # Get a list of every possible shape in every possible color.
134.     icons = []
135.     for color in ALLCOLORS:
136.         for shape in ALLSHAPES:
137.             icons.append( (shape, color) )
```

游戏板数据结构只是元组的列表的一个列表，其中，每个元组都有两个值：一个用于图标的形状，另一个用于图标的颜色。但是，创建这一数据结构有点复杂。我们需要确保确实有游戏板上的方块数目那么多个图标，并且还要确保每种类型的图标有且只有两个。

做到这一点的第一步，是创建带有形状和颜色的所有可能组合的一个列表。还记得吧，我们分别在 ALLCOLORS 和 ALLSHAPES 中拥有每种颜色和形状的一个列表，因此，第 135 行和第 136 行的嵌套 for 循环将会针对每种可能的颜色来遍历每一种可能的形状。这些就是在第 137 行添加到 icons 变量中的列表的每一项内容。

3.27　第 2 步——打乱并截取所有图标的列表

```
139.        random.shuffle(icons) # randomize the order of the icons list
140.        numIconsUsed = int(BOARDWIDTH * BOARDHEIGHT / 2) # calculate how many
icons are needed
141.        icons = icons[:numIconsUsed] * 2 # make two of each
142.        random.shuffle(icons)
```

但是记住，可能的组合比游戏板上的空格要多得多。我们需要通过将 BOARDWIDTH 和 BOARDHEIGHT 相乘，从而计算游戏板上的空格的数目。然后，我们将这个数目除以 2，因为我们需要有成对的图标。在一块具有 70 个空格的游戏板上，我们只需要 35 种不同的图标，因为每种图标都有两个。这个数字将会存储在 numIconsUsed 中。

第 141 行使用列表切片来抓取列表中的前 numIconsUsed 个图标（如果你忘记了列表切片是如何工作的，请查看 http://invpy.com/slicing）。这个列表已经在第 139 行打乱了，因此，它在每次游戏的时候都会使用不同的图标。然后，使用*操作符复制该列表，以便每种图标都有两个。这个新的复制后翻倍的列表，将会覆盖 icons 变量中的旧的列表。由于这个新的列表的前一半和后一半是相同的，我们再次调用 shuffle()方法来随机地打乱图标的顺序。

3.28　第 3 步——将图标放置到游戏板上

```
144.    # Create the board data structure, with randomly placed icons.
145.    board = []
146.    for x in range(BOARDWIDTH):
147.        column = []
148.        for y in range(BOARDHEIGHT):
149.            column.append(icons[0])
150.            del icons[0] # remove the icons as we assign them
151.        board.append(column)
152.    return board
```

现在，我们需要创建游戏板的数据结构的一个列表。我们可以使用嵌套的 for 循环来做到这一点，就像 generateRevealedBoxesData()函数所做的一样。对于游戏板上的每一列，我们将创建随机选取的图标的一个列表。随着在第 149 行为该列添加图标，我们随后将在第 150 行从 icons 列表的前面删除它们。通过这种方法，随着 icons 列表变得越来越短，icons[0]的位置上将会有一个不同的图标添加到该列之上。为了更好地描述这一点，在交互式 shell 中输入如下代码。注意 del 语句是如何修改 myList 列表的。

3.29 将一个列表分割为列表的列表

```
>>> myList = ['cat', 'dog', 'mouse', 'lizard']
>>> del myList[0]
>>> myList
['dog', 'mouse', 'lizard']
>>> del myList[0]
>>> myList
['mouse', 'lizard']
>>> del myList[0]
>>> myList
['lizard']
>>> del myList[0]
>>> myList
[]
>>>
```

由于我们删除了列表前面的项，其他的项将会向前移动，以使得列表中的下一项变为新的"第一"项。这和第 150 行代码的工作方式相同。

3.29 将一个列表分割为列表的列表

```
155. def splitIntoGroupsOf(groupSize, theList):
156.     # splits a list into a list of lists, where the inner lists have at
157.     # most groupSize number of items.
158.     result = []
159.     for i in range(0, len(theList), groupSize):
160.         result.append(theList[i:i + groupSize])
161.     return result
```

splitIntoGroupsOf()函数(它由 startGameAnimation()函数调用)将一个列表分隔为列表的列表，其中，内部的列表中拥有 groupSize 个项（如果剩下的项少于 groupSize 个，那么，最后的那个列表将会比较小一点）。

在第 159 行，使用带有 3 个参数的形式来调用 range()（如果你不熟悉这种形式，参考一下 http://invpy.com/range）。我们来看一个例子。如果列表的长度是 20，而 groupSize 参数是 8，那么 range(0, len(theList), groupSize)将会计算为 range(0, 20, 8)。这将在 for 循环的 3 次迭代中给 i 变量分别赋值为 0、8 和 16。

第 160 行的列表分片使用了 theList[i:i + groupSize]，创建了添加到 result 列表中的列表。在每次迭代中，i 分别是 0、8 和 16(groupSize 是 8)，列表分片表达式将会是 theList[0:8]，然后在第二次迭代中是 theList[8:16]，在第三次迭代中是 theList[16:24]。

注意，即便在我们的例子中 theList 最大的索引是 19，theList[16:24]也不会导致一个 IndexError 错误，即便 24 比 19 要大。它只是使用列表中剩下的项来创建一个列表分片。列

第 3 章 Memory Puzzle 游戏

表分片不会销毁或修改 theList 中所存储的最初的列表。它只是将列表的一部分复制，以求得一个新的列表值。这个新的列表值，就是在第 160 行附加到了 result 变量中的列表的后面的那个列表。因此，当我们在函数的末尾返回 result，我们将返回列表的一个列表。

3.30 不同的坐标系

```
164. def leftTopCoordsOfBox(boxx, boxy):
165.     # Convert board coordinates to pixel coordinates
166.     left = boxx * (BOXSIZE + GAPSIZE) + XMARGIN
167.     top = boxy * (BOXSIZE + GAPSIZE) + YMARGIN
168.     return (left, top)
```

你应该熟悉笛卡尔坐标系（如果你想要回顾这一主题，请阅读 http://invpy.com/coordinates）。在我们的大多数游戏中，将要使用多个笛卡尔坐标系。在 Memory Puzzle 中使用的一种坐标系统，就是像素坐标系或屏幕坐标系。但是，我们还将针对方块使用另一种坐标系统。这是因为，使用(3, 2)可以更容易地表示从左向右的第 3 个和从上到下的第 4 个方块（记住，这个数字是从 0 开始的，而不是从 1 开始的），而使用方块的左上角的像素坐标系(220, 165)则不然。然而，我们需要一种方式在这两种坐标系之间转换。

图 3-3 展示了游戏和两种不同的坐标系统。记住，窗口的宽度为 640 像素宽和 480 像素高，因此(639, 479)位于右下角（因为左上角的像素位置是(0, 0)，而不是(1, 1)）。

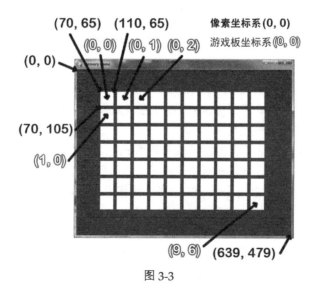

图 3-3

leftTopCoordsOfBox()函数将接受方块坐标并且返回像素坐标。由于一个方块占据了屏幕上的多个像素，我们将总是返回位于方块左上角的单个像素。这个值会作为两个整数的一个元组返回。当我们需要像素坐标以绘制这些方块的时候，经常会用到 leftTop CoordsOfBox() 函数。

3.31 从像素坐标转换为方块坐标

```
171. def getBoxAtPixel(x, y):
172.     for boxx in range(BOARDWIDTH):
173.         for boxy in range(BOARDHEIGHT):
174.             left, top = leftTopCoordsOfBox(boxx, boxy)
175.             boxRect = pygame.Rect(left, top, BOXSIZE, BOXSIZE)
176.             if boxRect.collidepoint(x, y):
177.                 return (boxx, boxy)
178.     return (None, None)
```

我们还需要一个函数将像素坐标（也就是鼠标点击和鼠标移动事件所使用的坐标）转换为方块坐标（以便能够搞清楚鼠标事件发生在哪一个方块上）。Rect 对象有一个 collidepoint() 方法，也可以传递 X 和 Y 坐标给它，如果该坐标位于 Rect 对象的区域之中（也就是说，发生了碰撞），它会返回 True。

为了搞清楚鼠标坐标在哪一个方块上，我们将遍历每一个方块的坐标，并且使用该坐标在一个 Rect 对象上调用 collidepoint() 方法。当 collidepoint() 方法返回 True 的时候，我们知道已经找到了在其上点击或者鼠标移动到其上的方块，并且将返回方块坐标。如果没有返回 True，那么 getBoxAtPixel() 函数将会返回 (None, None)。之所以返回这个元组，而不是直接返回 None，是因为 getBoxAtPixel() 的调用者期待返回两个值的一个元组。

3.32 绘制图标及语法糖

```
181. def drawIcon(shape, color, boxx, boxy):
182.     quarter = int(BOXSIZE * 0.25) # syntactic sugar
183.     half =    int(BOXSIZE * 0.5)  # syntactic sugar
184.
185.     left, top = leftTopCoordsOfBox(boxx, boxy) # get pixel coords from board coords
```

drawIcon() 函数将在 boxx 和 boxy 参数所指定的坐标的空格中绘制一个图标（使用指定的 shape 和 color）。每种可能的形状都会有一组不同的 Pygame 调用，因此，我们必须有很大的一组 if 和 elif 语句，以区分它们（这些语句在第 187 行到第 198 行）。

通过调用 leftTopCoordsOfBox() 来获取方块的左上角的 X 和 Y 坐标。方块的宽度和高度都是在 BOXSIZE 常量中设置的。然而，很多形状绘制函数调用也使用了方块的中点和四分之一点。我们可以计算这些点，并且将其存储到 quarter 变量和 half 变量中。我们本来也可以很容易地使用代码 int(BOXSIZE * 0.25)，而不是使用 quarter 变量，但是，在代码中使用变量的这种方式会更容易阅读，因为 quarter 的含义很明显，而 int (BOXSIZE * 0.25) 则不然。

这样的变量也是语法糖的一个例子。语法糖（Syntactic sugar）是这样的一种代码，我们本来可以用另外一种方式来编写它（可能会使用较少的代码和变量），但是，这么做确实会使得源代码容易阅读。常量变量就是语法糖的一种形式。预先计算一个值并且将其存储到一个变量中，这是另外一种形式的语法糖（例如，在 getRandomizedBoard()函数中，我们可以很容易地将第 140 行和第 141 行的代码合并成单个的一行代码。但是，写成两行分开的代码会更易于阅读）。我们并不需要额外的 quarter 和 half 变量，但是，有了它们，代码更容易阅读。易于阅读的代码，在将来也会易于调试和升级。

```
186.    # Draw the shapes
187.    if shape == DONUT:
188.        pygame.draw.circle(DISPLAYSURF, color, (left + half, top + half), half - 5)
189.        pygame.draw.circle(DISPLAYSURF, BGCOLOR, (left + half, top + half), quarter - 5)
190.    elif shape == SQUARE:
191.        pygame.draw.rect(DISPLAYSURF, color, (left + quarter, top + quarter, BOXSIZE - half, BOXSIZE - half))
192.    elif shape == DIAMOND:
193.        pygame.draw.polygon(DISPLAYSURF, color, ((left + half, top), (left + BOXSIZE - 1, top + half), (left + half, top + BOXSIZE - 1), (left, top + half)))
194.    elif shape == LINES:
195.        for i in range(0, BOXSIZE, 4):
196.            pygame.draw.line(DISPLAYSURF, color, (left, top + i), (left + i, top))
197.            pygame.draw.line(DISPLAYSURF, color, (left + i, top + BOXSIZE - 1), (left + BOXSIZE - 1, top + i))
198.    elif shape == OVAL:
199.        pygame.draw.ellipse(DISPLAYSURF, color, (left, top + quarter, BOXSIZE, half))
```

每一个多纳圈、方块、菱形、线条、椭圆形的绘制函数，都需要调用不同的绘制图元函数，才能够绘制出来。

3.33 获取游戏板控件的图标的形状和颜色的语法糖

```
202. def getShapeAndColor(board, boxx, boxy):
203.     # shape value for x, y spot is stored in board[x][y][0]
204.     # color value for x, y spot is stored in board[x][y][1]
205.     return board[boxx][boxy][0], board[boxx][boxy][1]
```

getShapeAndColor()函数只有一行代码。你可能会感到奇怪，为什么要使用一个函数而不是直接使用这一行代码呢？只要在需要的任何地方输入这行代码不就行了吗？这么做的原因和使用常量变量的原因相同，即提高代码的可读性。

我们很容易搞清楚诸如 shape, color = getShapeAndColor()这样的代码是要做什么。但是，如果看到诸如 shape, color = board[boxx][boxy][0], board[boxx] [boxy][1]这样的代码，你可能有点难以理解其含义。

3.34 绘制盖住的方块

```
208. def drawBoxCovers(board, boxes, coverage):
209.     # Draws boxes being covered/revealed. "boxes" is a list
210.     # of two-item lists, which have the x & y spot of the box.
211.     for box in boxes:
212.         left, top = leftTopCoordsOfBox(box[0], box[1])
213.         pygame.draw.rect(DISPLAYSURF, BGCOLOR, (left, top, BOXSIZE, BOXSIZE))
214.         shape, color = getShapeAndColor(board, box[0], box[1])
215.         drawIcon(shape, color, box[0], box[1])
216.         if coverage > 0: # only draw the cover if there is an coverage
217.             pygame.draw.rect(DISPLAYSURF, BOXCOLOR, (left, top, coverage, BOXSIZE))
218.     pygame.display.update()
219.     FPSCLOCK.tick(FPS)
```

drawBoxCovers()函数有 3 个参数：游戏板数据结构，(X, Y)元组的一个列表表示应该绘制盖住的方块的每一个位置，然后是针对方块绘制的覆盖量。

由于想要对 boxes 参数中的每一个方块使用相同的绘制代码，我们将在第 211 行使用一个 for 循环，以便对 boxes 列表中的每个方块都执行相同的代码。在这个 for 循环中，代码应该做 3 件事情：绘制背景颜色（这将会覆盖之前存在的任何内容），绘制图标，然后绘制覆盖在图标之上的、所需的任意大小的白色方块。leftTopCoordsOfBox()函数将会返回方块的左上角的像素坐标。第 216 行的 if 语句会确保如果 coverage 中的数值是否恰好小于 0，我们将不会调用 pygame.draw.rect()函数。

当 coverage 参数为 0 时,根本没有覆盖。当 coverage 设置为 20 时,将会有一个宽度为 20 像素的白色方块覆盖住图标。我们想要让 coverage 允许设置的最大的值就是 BOXSIZE 中的数字,也就是说刚好能将整个图标盖住。

drawBoxCovers()将要从一个单独的循环而不是从游戏循环中调用,因此,drawBoxCovers()需要自行调用 pygame.display.update()和 FPSCLOCK.tick (FPS)以显示动画(这确实意味着尽管这个函数会在循环内部调用,但该循环中没有负责处理所生成的任何事件的代码。这很好,因为覆盖和揭开动画只需要花 1 秒左右的时间播放)。

3.35 处理揭开和覆盖动画

```
222. def revealBoxesAnimation(board, boxesToReveal):
223.     # Do the "box reveal" animation.
224.     for coverage in range(BOXSIZE, - 1, - REVEALSPEED):
225.         drawBoxCovers(board, boxesToReveal, coverage)
226.
227.
228. def coverBoxesAnimation(board, boxesToCover):
229.     # Do the "box cover" animation.
230.     for coverage in range(0, BOXSIZE + REVEALSPEED, REVEALSPEED):
231.         drawBoxCovers(board, boxesToCover, coverage)
```

还记得吧,动画只是将不同的图像显示较短的时间,整合起来,好像是物体在屏幕上移动一样。revealBoxesAnimation()和 coverBoxesAnimation()只需要绘制被白色方块遮盖住不同的量的一个图标。我们可以编写一个名为 drawBoxCovers()的单个函数来做到这一点,然后,让动画函数针对动画的每一帧调用 drawBoxCovers()。正如我们在上一小节中看到的,drawBoxCovers()自己会调用 pygame.display.update()和 FPSCLOCK. tick(FPS)。

为了做到这一点,我们将创建一个 for 循环来递减(在 revealBoxesAnimation()的情况下)或递增(在 coverBoxesAnimation()的情况下)coverage 参数的值。coverage 变量的量将会根据 REVEALSPEED 常量中的数字,来递减或递增。在第 12 行,我们将这个常量设置为 8,意味着对于 drawBoxCovers()的每一次调用,白色的方块都在每一次迭代上减少、增加 8 个像素。如果我们增加这个数字,那么,每次调用将会绘制更多的像素,意味着白色的方块的大小将会递减或递增得更快。如果我们将其设置为 1,那么,每次迭代中白色的方块只会显示为递减或递增 1 个像素,这使得整个揭开或盖上的动画要花较长的时间。

这就像是爬楼梯。如果你每一步都是爬一阶楼梯,那么,爬上整个楼梯所需的时间是常规的时间。但是,如果你每一步爬两阶楼梯(并且每一步所用的时间和前面爬一阶楼梯是一样的），你就能够更快地爬完整个楼梯。如果你一次爬 8 阶楼梯，那么，你爬完整个楼梯的速度会快 8 倍。

3.36 绘制整个游戏板

```
234. def drawBoard(board, revealed):
235.     # Draws all of the boxes in their covered or revealed state.
236.     for boxx in range(BOARDWIDTH):
237.         for boxy in range(BOARDHEIGHT):
238.             left, top = leftTopCoordsOfBox(boxx, boxy)
239.             if not revealed[boxx][boxy]:
240.                 # Draw a covered box.
241.                 pygame.draw.rect(DISPLAYSURF, BOXCOLOR, (left, top, BOXSIZE, BOXSIZE))
242.             else:
243.                 # Draw the (revealed) icon.
244.                 shape, color = getShapeAndColor(board, boxx, boxy)
245.                 drawIcon(shape, color, boxx, boxy)
```

drawBoard() 函数针对游戏板上的每一个方块调用一次 drawIcon()。第 236 行和第 237 行的嵌套的 for 循环，将会针对方块遍历每一个可能的 X 和 Y 坐标，然后要么在该位置绘制图标，要么绘制一个白色的正方形（表示一个盖住的方块）。

3.37 绘制高亮边框

```
248. def drawHighlightBox(boxx, boxy):
249.     left, top = leftTopCoordsOfBox(boxx, boxy)
250.     pygame.draw.rect(DISPLAYSURF, HIGHLIGHTCOLOR, (left - 5, top - 5, BOXSIZE + 10, BOXSIZE + 10), 4)
```

为了帮助玩家识别他们能够在一个盖住的方块上点击以揭开它，我们将围绕一个方块绘制一个蓝色的边框以突出显示该方块。这个边框通过调用 pygame.draw.rect() 来绘制，以生成一个宽度为 4 像素的矩形。

3.38 "开始游戏"动画

```
253. def startGameAnimation(board):
254.     # Randomly reveal the boxes 8 at a time.
255.     coveredBoxes = generateRevealedBoxesData(False)
256.     boxes = []
257.     for x in range(BOARDWIDTH):
258.         for y in range(BOARDHEIGHT):
259.             boxes.append( (x, y) )
260.     random.shuffle(boxes)
261.     boxGroups = splitIntoGroupsOf(8, boxes)
```

这个动画在游戏开始的时候播放，针对所有的图标的位置，给玩家一个快速的提示。为了生成这个动画，我们必须一组一组地揭开和盖住方块。为了做到这一点，首先创建游戏板上每一个可能的空格的一个列表。第 257 行和第 258 行的嵌套 for 循环将会把(X, Y)元组添加到 boxes 变量中的一个列表中。

我们将要揭开或盖住该列表中的前 8 个方块，然后是接下来的 8 个方块，然后是其后面的 8 个方块，以此类推。然而，由于 boxes 中的(X, Y)元组的顺序可能每次都是相同的，那么方块显示的顺序也会是相同的（尝试注释掉第 260 行，然后运行该程序几次，以看看效果）。为了在每次游戏开始的时候都变化方块，我们调用 random.shuffle()函数以随机地打乱 boxes 列表中的元组的顺序。然后，当揭开或盖住列表中的前 8 个方块（以及每次组织随后的 8 个方块）的时候，它将会随机地组织 8 个方块。

为了得到 8 个方块的列表，我们调用 splitIntoGroupsOf()函数，传入 8 以及 boxes 中的列表。该函数所返回的列表的列表将会存储在一个名为 boxGroups 的变量中。

3.39 揭开和盖住成组的方块

```
263.     drawBoard(board, coveredBoxes)
264.     for boxGroup in boxGroups:
265.         revealBoxesAnimation(board, boxGroup)
266.         coverBoxesAnimation(board, boxGroup)
```

首先，我们绘制了游戏板。由于 coveredBoxes 中的每个值都设置为 False，这次对 drawBoard()调用最终只是绘制了盖住的白色方块。revealBoxesAnimation()和 coverBoxesAnimation()函数将会在这些白色方块的空格之上绘制。

for 循环将会遍历 boxGroups 列表中的每一个内部列表。我们将其传递给 revealBoxesAnimation()，该函数将执行将白色方块移走以显示其下面的图标的动画。调用

coverBoxesAnimation()实现扩展白色方块以盖住图标的动画。然后，该 for 循环进入下一次迭代，以便对接下来的一组 8 个方块实现动画。

3.40 "游戏获胜"动画

```
269. def gameWonAnimation(board):
270.     # flash the background color when the player has won
271.     coveredBoxes = generateRevealedBoxesData(True)
272.     color1 = LIGHTBGCOLOR
273.     color2 = BGCOLOR
274.
275.     for i in range(13):
276.         color1, color2 = color2, color1 # swap colors
277.         DISPLAYSURF.fill(color1)
278.         drawBoard(board, coveredBoxes)
279.         pygame.display.update()
280.         pygame.time.wait(300)
```

当玩家通过匹配游戏板上的每一对图标而揭开了所有的方块的时候，我们想要通过闪烁的背景颜色来祝贺他们。这个 for 循环将会用变量 color1 中的颜色来绘制背景颜色，然后在其上绘制游戏板。然而，在 for 循环的每一次迭代中，color1 和 color2 的值都会在第 276 行相互交换。通过这种方式，程序将交替地绘制两种不同的背景颜色。

记住，该函数需要调用 pygame.display.update()，以真正地让 DISPLAYSURF Surface 显示到屏幕上。

3.41 判断玩家是否已经获胜

```
283. def hasWon(revealedBoxes):
284.     # Returns True if all the boxes have been revealed, otherwise False
285.     for i in revealedBoxes:
286.         if False in i:
287.             return False # return False if any boxes are covered.
288.     return True
```

当所有成对的图标都已经匹配的时候，玩家获得了游戏的胜利。由于当图标匹配的时候，"揭开"数据结构将其中的值都设置为 True，我们可以直接遍历 revealedBoxes 中的每一个空格，以查找一个 False 值。即便 revealedBoxes 中有一个 False 值，那我们也可以知道，游戏板上仍然还有没有匹配的图标。

注意，由于 revealedBoxes 是列表的列表，第 285 行的 for 循环将会把内部的列表设置为 i 的值。但是，我们可以使用一个 in 操作符来在整个内部列表中查找一个 False 值。通过这种方式，我们不需要编写额外的代码行，就拥有了像下面这样的两个嵌套的 for 循环。

```
for x in revealedBoxes:
    for y in revealedBoxes[x]:
        if False == revealedBoxes[x][y]:
            return False
```

3.42 为何要那么麻烦地使用 main()函数

```
291. if __name__ == '__main__':
292.     main()
```

拥有一个 main()函数似乎毫无意义，因为你可以将这些代码放到程序的底部，使其具备全局作用域，并且这些代码也会完全一样地执行。然而，将其放在 main()函数中有两个很好的理由。

首先，这使得我们拥有局部变量，否则的话，main()函数中的局部变量将会变成全局变量。限制全局变量的数目，这是保持代码简单且易于调试的一种好办法（参见本章 3.15 节）。

其次，这还允许我们导入程序，以便可以调用和测试单个的函数。如果 *memorypuzzle.py* 文件位于 C:\Python32 文件夹下，那么我们可以通过交互式 shell 来导入它。输入如下内容可以测试 splitIntoGroupsOf()和 getBoxAtPixel()函数，以确保它们返回正确的返回值。

```
>>> import memorypuzzle
>>> memorypuzzle.splitIntoGroupsOf(3, [0,1,2,3,4,5,6,7,8,9])
[[0, 1, 2], [3, 4, 5], [6, 7, 8], [9]]
>>> memorypuzzle.getBoxAtPixel(0, 0)
(None, None)
>>> memorypuzzle.getBoxAtPixel(150, 150)
(1, 1)
```

当导入一个模块的时候，其中所有的代码都会运行。如果我们没有 main()函数，并且让其代码都处于全局作用域中，那么，只要我们导入了它，游戏就会自动启动，我们真地没有办法调用其中单个的函数了。

这就是为什么将这些代码放到一个名为 main()的单个函数中。随后，我们就可以检查内建的 Python 变量__name__，看看是否应该调用 main()函数。如果程序自身在运行之中的话，这个变量由 Python 解释器自动地设置为字符串'__main__'，如果程序在导入中的话，则设置为'memorypuzzle'。这就是为什么当我们在交互式 shell 中执行 import memorypuzzle 的时候，main()函数没有运行。

这是一种方便的技术，能够从交互式的 shell 导入你所操作的程序，通过一次性调用函数来确保单个函数返回正确的值。

3.43 为什么要为可读性操心

本章中的很多建议，与其说和如何编写计算机能够运行的程序相关，还不如说是和如何编写程序员能够阅读的程序更为相关。你可能不理解为什么后者很重要。毕竟，只要代码能够工作，谁会在乎人类程序员阅读起来是困难，还是容易呢？

然而，重要的事情是要意识到软件很少是放在那里置之不理的。当你创建了自己的游戏，你很少会"完成"了程序。你总是会产生新的思路，如想要添加游戏功能，或者从程序中找到新的 Bug。因此，保持程序的可读性很重要，这样一来，你可以阅读代码并理解代码。并且，理解代码是修改它以添加更多的代码或修复 Bug 的第一步。

例如，下面是 Memory Puzzle 程序的一个混淆后的版本，这使得它完全不可读。如果你录入它（或者从 http://invpy.com/memorypuzzle_obfuscated.py 下载它）并且运行它，你会发现，它和本章开始给出的代码完全相同地运行。但是，如果这段代码中有一个 Bug，我们不能够阅读代码并理解发生了什么，更不要说是修改 Bug 了。

```
import random, pygame, sys
from pygame.locals import *
def hhh():
    global a, b
    pygame.init()
    a = pygame.time.Clock()
    b = pygame.display.set_mode((640, 480))
    j = 0
    k = 0
    pygame.display.set_caption('Memory Game')
    i = c()
    hh = d(False)
    h = None
    b.fill((60, 60, 100))
    g(i)
    while True:
        e = False
        b.fill((60, 60, 100))
        f(i, hh)
        for eee in pygame.event.get():
            if eee.type == QUIT or (eee.type == KEYUP and eee.key == K_ESCAPE):
                pygame.quit()
```

第 3 章　Memory Puzzle 游戏

```
                sys.exit()
            elif eee.type == MOUSEMOTION:
                j, k = eee.pos
            elif eee.type == MOUSEBUTTONUP:
                j, k = eee.pos
                e = True
        bb, ee = m(j, k)
        if bb != None and ee != None:
            if not hh[bb][ee]:
                n(bb, ee)
            if not hh[bb][ee] and e:
                o(i, [(bb, ee)])
                hh[bb][ee] = True
                if h == None:
                    h = (bb, ee)
                else:
                    q, fff = s(i, h[0], h[1])
                    r, ggg = s(i, bb, ee)
                    if q != r or fff != ggg:
                        pygame.time.wait(1000)
                        p(i, [(h[0], h[1]), (bb, ee)])
                        hh[h[0]][h[1]] = False
                        hh[bb][ee] = False
                    elif ii(hh):
                        jj(i)
                        pygame.time.wait(2000)
                        i = c()
                        hh = d(False)
                        f(i, hh)
                        pygame.display.update()
                        pygame.time.wait(1000)
                        g(i)
                    h = None
        pygame.display.update()
        a.tick(30)
def d(ccc):
    hh = []
    for i in range(10):
        hh.append([ccc] * 7)
    return hh
def c():
    rr = []
    for tt in ((255, 0, 0), (0, 255, 0), (0, 0, 255), (255, 255, 0), (255, 128, 0), (255, 0, 255), (0, 255, 255)):
        for ss in ('a', 'b', 'c', 'd', 'e'):
            rr.append( (ss, tt) )
    random.shuffle(rr)
```

```
        rr = rr[:35] * 2
        random.shuffle(rr)
        bbb = []
        for x in range(10):
            v = []
            for y in range(7):
                v.append(rr[0])
                del rr[0]
            bbb.append(v)
        return bbb
    def t(vv, uu):
        ww = []
        for i in range(0, len(uu), vv):
            ww.append(uu[i:i + vv])
        return ww
    def aa(bb, ee):
        return (bb * 50 + 70, ee * 50 + 65)
    def m(x, y):
        for bb in range(10):
            for ee in range(7):
                oo, ddd = aa(bb, ee)
                aaa = pygame.Rect(oo, ddd, 40, 40)
                if aaa.collidepoint(x, y):
                    return (bb, ee)
        return (None, None)
    def w(ss, tt, bb, ee):
        oo, ddd = aa(bb, ee)
        if ss == 'a':
            pygame.draw.circle(b, tt, (oo + 20, ddd + 20), 15)
            pygame.draw.circle(b, (60, 60, 100), (oo + 20, ddd + 20), 5)
        elif ss == 'b':
            pygame.draw.rect(b, tt, (oo + 10, ddd + 10, 20, 20))
        elif ss == 'c':
            pygame.draw.polygon(b, tt, ((oo + 20, ddd), (oo + 40 - 1, ddd + 20),
(oo + 20, ddd + 40 - 1), (oo, ddd + 20)))
        elif ss == 'd':
            for i in range(0, 40, 4):
                pygame.draw.line(b, tt, (oo, ddd + i), (oo + i, ddd))
                pygame.draw.line(b, tt, (oo + i, ddd + 39), (oo + 39, ddd + i))
        elif ss == 'e':
            pygame.draw.ellipse(b, tt, (oo, ddd + 10, 40, 20))
    def s(bbb, bb, ee):
        return bbb[bb][ee][0], bbb[bb][ee][1]
    def dd(bbb, boxes, gg):
        for box in boxes:
            oo, ddd = aa(box[0], box[1])
            pygame.draw.rect(b, (60, 60, 100), (oo, ddd, 40, 40))
```

第 3 章　Memory Puzzle 游戏

```
            ss, tt = s(bbb, box[0], box[1])
            w(ss, tt, box[0], box[1])
            if gg > 0:
                pygame.draw.rect(b, (255, 255, 255), (oo, ddd, gg, 40))
    pygame.display.update()
    a.tick(30)
def o(bbb, cc):
    for gg in range(40, (-8) - 1, -8):
        dd(bbb, cc, gg)
def p(bbb, ff):
    for gg in range(0, 48, 8):
        dd(bbb, ff, gg)
def f(bbb, pp):
    for bb in range(10):
        for ee in range(7):
            oo, ddd = aa(bb, ee)
            if not pp[bb][ee]:
                pygame.draw.rect(b, (255, 255, 255), (oo, ddd, 40, 40))
            else:
                ss, tt = s(bbb, bb, ee)
                w(ss, tt, bb, ee)
def n(bb, ee):
    oo, ddd = aa(bb, ee)
    pygame.draw.rect(b, (0, 0, 255), (oo - 5, ddd - 5, 50, 50), 4)
def g(bbb):
    mm = d(False)
    boxes = []
    for x in range(10):
        for y in range(7):
            boxes.append( (x, y) )
    random.shuffle(boxes)
    kk = t(8, boxes)
    f(bbb, mm)
    for nn in kk:
        o(bbb, nn)
        p(bbb, nn)
def jj(bbb):
    mm = d(True)
    tt1 = (100, 100, 100)
    tt2 = (60, 60, 100)
    for i in range(13):
        tt1, tt2 = tt2, tt1
        b.fill(tt1)
        f(bbb, mm)
        pygame.display.update()
        pygame.time.wait(300)
def ii(hh):
```

```
    for i in hh:
        if False in i:
            return False
    return True
if __name__ == '__main__':
    hhh()
```

绝不要编写这样的代码。如果你的程序像这样，那么，当你关掉灯光照镜子的时候，Ada Lovelace 的灵魂会从镜子里走出来并把你扔到转动的织布机中。

3.44 本章小结

本章完整地讲解了 Memory Puzzle 程序是如何工作的。请再次阅读本章内容以及源代码，以更好地理解内容。本书中的很多其他的游戏程序也使用了相同的编程概念（例如，嵌套的 for 循环、语法糖以及在同一程序中使用不同的坐标系统），因此，为了节省篇幅，后面不会再重复介绍这些概念。

试图理解代码是如何工作的一种方法是随机地注释掉一些代码行，有意地破坏程序。注释掉一些代码行可能会导致语法错误，从而使得脚本根本无法运行。但是，注释掉另外一些代码行，可能会导致奇怪的 Bug 和其他的一些很酷的效果。尝试这么做，然后，搞清楚为什么程序会有这些 Bug。这也是能够给程序添加自己的秘密作弊代码或 hack 技巧的第一步。通过将正常运行的程序破坏掉，你可以学习到如何修改程序以执行一些漂亮的效果（例如，悄悄地给你一些关于如何解决谜题的提示）。请随心所欲地进行体验。如果你想要再次正常地玩某个游戏，你总是可以在一个不同的文件中保存一份未经修改的源代码。

实际上，如果你想要做一些练习来修正 Bug，这款游戏的源代码还有几个版本，其中带有一些小的 Bug。你可以从 http://invpy.com/buggy/memorypuzzle 下载这些带 Bug 的版本。尝试运行这些程序，以搞清楚 Bug 在哪里，以及为什么程序会那样工作。

第 4 章　Slide Puzzle

4.1　如何玩 Slide Puzzle

Slide Puzzle 的游戏板是一个 4×4 的栅格，带有 15 个贴片（从左到右，编号依次是 1 到 15）和一个空白空格，如图 4-1 所示。贴片最初处于随机的位置，玩家必须来回滑动贴片，直到贴片都按照其原本的顺序排列好。

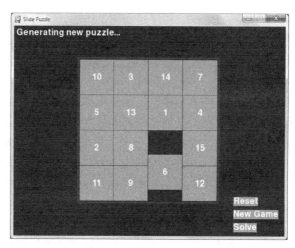

图 4-1

4.2　Slide Puzzle 的源代码

源代码可以从 http://invpy.com/slidepuzzle.py 下载。如果你得到了错误消息，查看一下错误消息中提到的行号，并且检查你的源代码是否有录入错误。你也可以将你的代码复制并粘贴到位于 http://invpy.com/diff/slidepuzzle 的 Web 表单中，以检查你的代码和本书中的代码之间的差异。

```
1. # Slide Puzzle
2. # By Al Sweigart al@inventwithpython.com
3. # http://inventwithpython.com/pygame
4. # Creative Commons BY-NC-SA 3.0 US
```

```
5.
6. import pygame, sys, random
7. from pygame.locals import *
8.
9. # Create the constants (go ahead and experiment with different values)
10. BOARDWIDTH = 4  # number of columns in the board
11. BOARDHEIGHT = 4 # number of rows in the board
12. TILESIZE = 80
13. WINDOWWIDTH = 640
14. WINDOWHEIGHT = 480
15. FPS = 30
16. BLANK = None
17.
18. #                      R    G    B
19. BLACK =             (  0,   0,   0)
20. WHITE =             (255, 255, 255)
21. BRIGHTBLUE =        (  0,  50, 255)
22. DARKTURQUOISE = (  3,  54,  73)
23. GREEN =             (  0, 204,   0)
24.
25. BGCOLOR = DARKTURQUOISE
26. TILECOLOR = GREEN
27. TEXTCOLOR = WHITE
28. BORDERCOLOR = BRIGHTBLUE
29. BASICFONTSIZE = 20
30.
31. BUTTONCOLOR = WHITE
32. BUTTONTEXTCOLOR = BLACK
33. MESSAGECOLOR = WHITE
34.
35. XMARGIN = int((WINDOWWIDTH - (TILESIZE * BOARDWIDTH + (BOARDWIDTH - 1))) / 2)
36. YMARGIN = int((WINDOWHEIGHT - (TILESIZE * BOARDHEIGHT + (BOARDHEIGHT - 1))) / 2)
37.
38. UP = 'up'
39. DOWN = 'down'
40. LEFT = 'left'
41. RIGHT = 'right'
42.
43. def main():
44.     global FPSCLOCK, DISPLAYSURF, BASICFONT, RESET_SURF, RESET_RECT, NEW_SURF, NEW_RECT, SOLVE_SURF, SOLVE_RECT
45.
46.     pygame.init()
```

第4章 Slide Puzzle

```
47.     FPSCLOCK = pygame.time.Clock()
48.     DISPLAYSURF = pygame.display.set_mode((WINDOWWIDTH, WINDOWHEIGHT))
49.     pygame.display.set_caption('Slide Puzzle')
50.     BASICFONT = pygame.font.Font('freesansbold.ttf', BASICFONTSIZE)
51.
52.     # Store the option buttons and their rectangles in OPTIONS.
53.     RESET_SURF, RESET_RECT = makeText('Reset',    TEXTCOLOR, TILECOLOR, WINDOWWIDTH - 120, WINDOWHEIGHT - 90)
54.     NEW_SURF,   NEW_RECT   = makeText('New Game', TEXTCOLOR, TILECOLOR, WINDOWWIDTH - 120, WINDOWHEIGHT - 60)
55.     SOLVE_SURF, SOLVE_RECT = makeText('Solve',    TEXTCOLOR, TILECOLOR, WINDOWWIDTH - 120, WINDOWHEIGHT - 30)
56.
57.     mainBoard, solutionSeq = generateNewPuzzle(80)
58.     SOLVEDBOARD = getStartingBoard() # a solved board is the same as the board in a start state.
59.     allMoves = [] # list of moves made from the solved configuration
60.
61.     while True: # main game loop
62.         slideTo = None # the direction, if any, a tile should slide
63.         msg = '' # contains the message to show in the upper left corner.
64.         if mainBoard == SOLVEDBOARD:
65.             msg = 'Solved!'
66.
67.         drawBoard(mainBoard, msg)
68.
69.         checkForQuit()
70.         for event in pygame.event.get(): # event handling loop
71.             if event.type == MOUSEBUTTONUP:
72.                 spotx, spoty = getSpotClicked(mainBoard, event.pos[0], event.pos[1])
73.
74.                 if (spotx, spoty) == (None, None):
75.                     # check if the user clicked on an option button
76.                     if RESET_RECT.collidepoint(event.pos):
77.                         resetAnimation(mainBoard, allMoves) # clicked on Reset button
78.                         allMoves = []
79.                     elif NEW_RECT.collidepoint(event.pos):
80.                         mainBoard, solutionSeq = generateNewPuzzle(80) # clicked on New Game button
81.                         allMoves = []
82.                     elif SOLVE_RECT.collidepoint(event.pos):
83.                         resetAnimation(mainBoard, solutionSeq + allMoves) # clicked on Solve button
```

```
84.                    allMoves = []
85.                else:
86.                    # check if the clicked tile was next to the blank spot
87.
88.                    blankx, blanky = getBlankPosition(mainBoard)
89.                    if spotx == blankx + 1 and spoty == blanky:
90.                        slideTo = LEFT
91.                    elif spotx == blankx - 1 and spoty == blanky:
92.                        slideTo = RIGHT
93.                    elif spotx == blankx and spoty == blanky + 1:
94.                        slideTo = UP
95.                    elif spotx == blankx and spoty == blanky - 1:
96.                        slideTo = DOWN
97.
98.            elif event.type == KEYUP:
99.                # check if the user pressed a key to slide a tile
100.               if event.key in (K_LEFT, K_a) and isValidMove(mainBoard, LEFT):
101.                   slideTo = LEFT
102.               elif event.key in (K_RIGHT, K_d) and isValidMove(mainBoard, RIGHT):
103.                   slideTo = RIGHT
104.               elif event.key in (K_UP, K_w) and isValidMove(mainBoard, UP):
105.                   slideTo = UP
106.               elif event.key in (K_DOWN, K_s) and isValidMove(mainBoard, DOWN):
107.                   slideTo = DOWN
108.
109.        if slideTo:
110.            slideAnimation(mainBoard, slideTo, 'Click tile or press arrow keys to slide.', 8) # show slide on screen
111.            makeMove(mainBoard, slideTo)
112.            allMoves.append(slideTo) # record the slide
113.        pygame.display.update()
114.        FPSCLOCK.tick(FPS)
115.
116.
117. def terminate():
118.     pygame.quit()
119.     sys.exit()
120.
121.
122. def checkForQuit():
123.     for event in pygame.event.get(QUIT): # get all the QUIT events
124.         terminate() # terminate if any QUIT events are present
```

第 4 章　Slide Puzzle

```
125.        for event in pygame.event.get(KEYUP): # get all the KEYUP events
126.            if event.key == K_ESCAPE:
127.                terminate() # terminate if the KEYUP event was for the Esc key
128.            pygame.event.post(event) # put the other KEYUP event objects back
129.
130.
131. def getStartingBoard():
132.     # Return a board data structure with tiles in the solved state.
133.     # For example, if BOARDWIDTH and BOARDHEIGHT are both 3, this function
134.     # returns [[1, 4, 7], [2, 5, 8], [3, 6, None]]
135.     counter = 1
136.     board = []
137.     for x in range(BOARDWIDTH):
138.         column = []
139.         for y in range(BOARDHEIGHT):
140.             column.append(counter)
141.             counter += BOARDWIDTH
142.         board.append(column)
143.         counter -= BOARDWIDTH * (BOARDHEIGHT - 1) + BOARDWIDTH - 1
144.
145.     board[BOARDWIDTH-1][BOARDHEIGHT-1] = None
146.     return board
147.
148.
149. def getBlankPosition(board):
150.     # Return the x and y of board coordinates of the blank space.
151.     for x in range(BOARDWIDTH):
152.         for y in range(BOARDHEIGHT):
153.             if board[x][y] == None:
154.                 return (x, y)
155.
156.
157. def makeMove(board, move):
158.     # This function does not check if the move is valid.
159.     blankx, blanky = getBlankPosition(board)
160.
161.     if move == UP:
162.         board[blankx][blanky], board[blankx][blanky + 1] = board[blankx][blanky + 1], board[blankx][blanky]
163.     elif move == DOWN:
164.         board[blankx][blanky], board[blankx][blanky - 1] = board[blankx][blanky - 1], board[blankx][blanky]
165.     elif move == LEFT:
166.         board[blankx][blanky], board[blankx + 1][blanky] = board[blankx + 1][blanky], board[blankx][blanky]
167.     elif move == RIGHT:
```

```
168.            board[blankx][blanky], board[blankx - 1][blanky] = board[blankx -
1][blanky], board[blankx][blanky]
169.
170.
171. def isValidMove(board, move):
172.     blankx, blanky = getBlankPosition(board)
173.     return (move == UP and blanky != len(board[0]) - 1) or \
174.            (move == DOWN and blanky != 0) or \
175.            (move == LEFT and blankx != len(board) - 1) or \
176.            (move == RIGHT and blankx != 0)
177.
178.
179. def getRandomMove(board, lastMove=None):
180.     # start with a full list of all four moves
181.     validMoves = [UP, DOWN, LEFT, RIGHT]
182.
183.     # remove moves from the list as they are disqualified
184.     if lastMove == UP or not isValidMove(board, DOWN):
185.         validMoves.remove(DOWN)
186.     if lastMove == DOWN or not isValidMove(board, UP):
187.         validMoves.remove(UP)
188.     if lastMove == LEFT or not isValidMove(board, RIGHT):
189.         validMoves.remove(RIGHT)
190.     if lastMove == RIGHT or not isValidMove(board, LEFT):
191.         validMoves.remove(LEFT)
192.
193.     # return a random move from the list of remaining moves
194.     return random.choice(validMoves)
195.
196.
197. def getLeftTopOfTile(tileX, tileY):
198.     left = XMARGIN + (tileX * TILESIZE) + (tileX - 1)
199.     top = YMARGIN + (tileY * TILESIZE) + (tileY - 1)
200.     return (left, top)
201.
202.
203. def getSpotClicked(board, x, y):
204.     # from the x & y pixel coordinates, get the x & y board coordinates
205.     for tileX in range(len(board)):
206.         for tileY in range(len(board[0])):
207.             left, top = getLeftTopOfTile(tileX, tileY)
208.             tileRect = pygame.Rect(left, top, TILESIZE, TILESIZE)
209.             if tileRect.collidepoint(x, y):
210.                 return (tileX, tileY)
211.     return (None, None)
212.
```

第 4 章　Slide Puzzle

```
213.
214. def drawTile(tilex, tiley, number, adjx=0, adjy=0):
215.     # draw a tile at board coordinates tilex and tiley, optionally a few
216.     # pixels over (determined by adjx and adjy)
217.     left, top = getLeftTopOfTile(tilex, tiley)
218.     pygame.draw.rect(DISPLAYSURF, TILECOLOR, (left + adjx, top + adjy, TILESIZE, TILESIZE))
219.     textSurf = BASICFONT.render(str(number), True, TEXTCOLOR)
220.     textRect = textSurf.get_rect()
221.     textRect.center = left + int(TILESIZE / 2) + adjx, top + int(TILESIZE / 2) + adjy
222.     DISPLAYSURF.blit(textSurf, textRect)
223.
224.
225. def makeText(text, color, bgcolor, top, left):
226.     # create the Surface and Rect objects for some text.
227.     textSurf = BASICFONT.render(text, True, color, bgcolor)
228.     textRect = textSurf.get_rect()
229.     textRect.topleft = (top, left)
230.     return (textSurf, textRect)
231.
232.
233. def drawBoard(board, message):
234.     DISPLAYSURF.fill(BGCOLOR)
235.     if message:
236.         textSurf, textRect = makeText(message, MESSAGECOLOR, BGCOLOR, 5, 5)
237.         DISPLAYSURF.blit(textSurf, textRect)
238.
239.     for tilex in range(len(board)):
240.         for tiley in range(len(board[0])):
241.             if board[tilex][tiley]:
242.                 drawTile(tilex, tiley, board[tilex][tiley])
243.
244.     left, top = getLeftTopOfTile(0, 0)
245.     width = BOARDWIDTH * TILESIZE
246.     height = BOARDHEIGHT * TILESIZE
247.     pygame.draw.rect(DISPLAYSURF, BORDERCOLOR, (left - 5, top - 5, width + 11, height + 11), 4)
248.
249.     DISPLAYSURF.blit(RESET_SURF, RESET_RECT)
250.     DISPLAYSURF.blit(NEW_SURF, NEW_RECT)
251.     DISPLAYSURF.blit(SOLVE_SURF, SOLVE_RECT)
252.
253.
254. def slideAnimation(board, direction, message, animationSpeed):
```

```
255.        # Note: This function does not check if the move is valid.
256.
257.        blankx, blanky = getBlankPosition(board)
258.        if direction == UP:
259.            movex = blankx
260.            movey = blanky + 1
261.        elif direction == DOWN:
262.            movex = blankx
263.            movey = blanky - 1
264.        elif direction == LEFT:
265.            movex = blankx + 1
266.            movey = blanky
267.        elif direction == RIGHT:
268.            movex = blankx - 1
269.            movey = blanky
270.
271.        # prepare the base surface
272.        drawBoard(board, message)
273.        baseSurf = DISPLAYSURF.copy()
274.        # draw a blank space over the moving tile on the baseSurf Surface.
275.        moveLeft, moveTop = getLeftTopOfTile(movex, movey)
276.        pygame.draw.rect(baseSurf, BGCOLOR, (moveLeft, moveTop, TILESIZE, TILESIZE))
277.
278.        for i in range(0, TILESIZE, animationSpeed):
279.            # animate the tile sliding over
280.            checkForQuit()
281.            DISPLAYSURF.blit(baseSurf, (0, 0))
282.            if direction == UP:
283.                drawTile(movex, movey, board[movex][movey], 0, -i)
284.            if direction == DOWN:
285.                drawTile(movex, movey, board[movex][movey], 0, i)
286.            if direction == LEFT:
287.                drawTile(movex, movey, board[movex][movey], -i, 0)
288.            if direction == RIGHT:
289.                drawTile(movex, movey, board[movex][movey], i, 0)
290.
291.            pygame.display.update()
292.            FPSCLOCK.tick(FPS)
293.
294.
295. def generateNewPuzzle(numSlides):
296.     # From a starting configuration, make numSlides number of moves (and
297.     # animate these moves).
298.     sequence = []
299.     board = getStartingBoard()
```

第 4 章　Slide Puzzle

```
300.        drawBoard(board, '')
301.        pygame.display.update()
302.        pygame.time.wait(500) # pause 500 milliseconds for effect
303.        lastMove = None
304.        for i in range(numSlides):
305.            move = getRandomMove(board, lastMove)
306.            slideAnimation(board, move, 'Generating new puzzle...',
int(TILESIZE / 3))
307.            makeMove(board, move)
308.            sequence.append(move)
309.            lastMove = move
310.        return (board, sequence)
311.
312.
313.    def resetAnimation(board, allMoves):
314.        # make all of the moves in allMoves in reverse.
315.        revAllMoves = allMoves[:] # gets a copy of the list
316.        revAllMoves.reverse()
317.
318.        for move in revAllMoves:
319.            if move == UP:
320.                oppositeMove = DOWN
321.            elif move == DOWN:
322.                oppositeMove = UP
323.            elif move == RIGHT:
324.                oppositeMove = LEFT
325.            elif move == LEFT:
326.                oppositeMove = RIGHT
327.            slideAnimation(board, oppositeMove, '', int(TILESIZE / 2))
328.            makeMove(board, oppositeMove)
329.
330.
331.    if __name__ == '__main__':
332.        main()
```

4.3　第二款游戏和第一款相同

Slide Puzzle 的很多代码与我们前面看到的游戏类似，尤其是在代码一开始所设置的常量。

```
1. # Slide Puzzle
2. # By Al Sweigart al@inventwithpython.com
3. # http://inventwithpython.com/pygame
4. # Creative Commons BY-NC-SA 3.0 US
5.
```

4.4 设置按钮

```
 6. import pygame, sys, random
 7. from pygame.locals import *
 8.
 9. # Create the constants (go ahead and experiment with different values)
10. BOARDWIDTH = 4  # number of columns in the board
11. BOARDHEIGHT = 4 # number of rows in the board
12. TILESIZE = 80
13. WINDOWWIDTH = 640
14. WINDOWHEIGHT = 480
15. FPS = 30
16. BLANK = None
17.
18. #                    R    G    B
19. BLACK =          (   0,   0,   0)
20. WHITE =          ( 255, 255, 255)
21. BRIGHTBLUE =     (   0,  50, 255)
22. DARKTURQUOISE =  (   3,  54,  73)
23. GREEN =          (   0, 204,   0)
24.
25. BGCOLOR = DARKTURQUOISE
26. TILECOLOR = GREEN
27. TEXTCOLOR = WHITE
28. BORDERCOLOR = BRIGHTBLUE
29. BASICFONTSIZE = 20
30.
31. BUTTONCOLOR = WHITE
32. BUTTONTEXTCOLOR = BLACK
33. MESSAGECOLOR = WHITE
34.
35. XMARGIN = int((WINDOWWIDTH - (TILESIZE * BOARDWIDTH + (BOARDWIDTH - 1))) / 2)
36. YMARGIN = int((WINDOWHEIGHT - (TILESIZE * BOARDHEIGHT + (BOARDHEIGHT - 1))) / 2)
37.
38. UP = 'up'
39. DOWN = 'down'
40. LEFT = 'left'
41. RIGHT = 'right'
```

程序最顶部的这段代码只是执行模块的基本导入并创建常量。这和第 3 章中 Memory Puzzle 游戏的开始处一样。

4.4 设置按钮

```
43. def main():
44.     global FPSCLOCK, DISPLAYSURF, BASICFONT, RESET_SURF, RESET_RECT,
```

第 4 章　Slide Puzzle

```
NEW_SURF, NEW_RECT, SOLVE_SURF, SOLVE_RECT
 45.
 46.     pygame.init()
 47.     FPSCLOCK = pygame.time.Clock()
 48.     DISPLAYSURF = pygame.display.set_mode((WINDOWWIDTH, WINDOWHEIGHT))
 49.     pygame.display.set_caption('Slide Puzzle')
 50.     BASICFONT = pygame.font.Font('freesansbold.ttf', BASICFONTSIZE)
 51.
 52.     # Store the option buttons and their rectangles in OPTIONS.
 53.     RESET_SURF, RESET_RECT = makeText('Reset',    TEXTCOLOR, TILECOLOR, WINDOWWIDTH - 120, WINDOWHEIGHT - 90)
 54.     NEW_SURF,   NEW_RECT   = makeText('New Game', TEXTCOLOR, TILECOLOR, WINDOWWIDTH - 120, WINDOWHEIGHT - 60)
 55.     SOLVE_SURF, SOLVE_RECT = makeText('Solve',    TEXTCOLOR, TILECOLOR, WINDOWWIDTH - 120, WINDOWHEIGHT - 30)
 56.
 57.     mainBoard, solutionSeq = generateNewPuzzle(80)
 58.     SOLVEDBOARD = getStartingBoard() # a solved board is the same as the board in a start state.
```

　　和第 3 章一样，从 main() 函数中调用的函数将在稍后介绍。现在，你只需要知道它们做什么以及它们返回什么值，而不需要知道它们是如何工作的。

　　main() 函数的第一部分将负责创建窗口、Clock 对象和 Font 对象。makeText() 函数将在程序的后面定义，但是现在，你只需要知道它返回一个 pygame.Surface 对象和 pygame.Rect 对象，这二者可以用来制作可点击的按钮。Slide Puzzle 游戏将有 3 个按钮：一个"Reset"按钮用来撤销玩家所做的任何移动，一个"New"按钮用来创建一个新的滑块谜题，还有一个"Solve"按钮将为玩家破解谜题。

　　对于这款游戏，我们需要两个游戏板数据结构。一个游戏板表示当前的游戏状态。另一个游戏板让其贴片处于"已求解"的状态，意味着所有的贴片都已经顺序排好。当当前游戏状态的游戏板和求解后的游戏板完全相同的时候，我们知道，玩家已经获胜了（我们不会修改第 2 个游戏板，它只是用来和当前的游戏状态游戏板进行比较的）。

　　generateNewPuzzle() 将会创建一个游戏板数据结构，它一开始就是按照顺序排列的、求解后的状态，然后在其上执行 80 次随机的滑块移动（这是由于我们给它传递了整数 80。如果想要让游戏板打得更乱，我们可以传递一个更大的整数给它）。这将会使得游戏板进入随机打乱的状态，让玩家必须求解（这个状态将会存储在名为 mainBoard 的一个变量中）。generateNewBoard() 也将返回在其上执行的所有随机移动的一个列表（这将存储在一个名为 solutionSeq 的变量中）。

4.5 使用愚笨的代码变聪明

```
59.    allMoves = [] # list of moves made from the solved configuration
```

求解一个滑块谜题可能真得颇费脑筋。我们可以编写程序让计算机来做，但是，这需要我们搞清楚解决滑块谜题的一种算法。这就很难了，要在程序中投入很多智慧和努力。

好在，有一种较为容易的方法。当创建游戏板数据结构的时候，让计算机记住它所进行的所有的随机的滑块移动，然后，只要执行相反的滑块移动，就可以求解了。由于游戏板最初是从已求解的状态开始的，撤销所有的滑块移动就应该可以返回到已求解的状态了。

例如，我们在页面的左边缘执行了一次向右滑动，这使得游戏板处于图 4-2 中右图所示的状态。

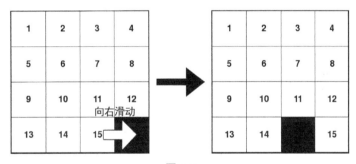

图 4-2

在向右滑动之后，如果进行相反的滑动（向左滑动），那么，游戏板将会回到最初的状态。因此，在经过了几次滑动之后，如果想要回到最初的状态，我们只需要按照相反的顺序做相反的滑动。如果先向右滑动，然后又向右滑动一次，再向下滑动一次，我们可以向上滑动一次，向左滑动一次，然后向左滑动一次，以取消之前的 3 次滑动。这比编写一个函数直接通过查看当前状态来解决谜题要容易得多。

4.6 主游戏循环

```
61.    while True: # main game loop
62.        slideTo = None # the direction, if any, a tile should slide
63.        msg = '' # contains the message to show in the upper left corner.
64.        if mainBoard == SOLVEDBOARD:
65.            msg = 'Solved!'
66.
67.        drawBoard(mainBoard, msg)
```

第 4 章　Slide Puzzle

在主游戏循环中，slideTo 变量将会记录玩家想要向哪一个方向滑动贴片（在游戏循环刚开始的时候，它是 None，并且随后会进行设置），msg 变量记录了在窗口顶部显示什么字符串。程序在第 64 行进行一次快速检查，看游戏板数据结构是否和 SOLVEDBOARD 中所存储的已求解的游戏板数据结构具有相同的值。如果是的，那么 msg 变量就修改为字符串 'Solved!'。

在调用 drawBoard()将其绘制到 DISPLAYSURF Surface 对象（这在第 67 行完成）之前以及调用 pygame.display.update()将显示 Surface 对象绘制到实际的计算机屏幕上（这在游戏循环末尾的第 291 行完成）之前，这个字符串不会出现在屏幕上。

4.7　点击按钮

```
69.         checkForQuit()
70.         for event in pygame.event.get(): # event handling loop
71.             if event.type == MOUSEBUTTONUP:
72.                 spotx, spoty = getSpotClicked(mainBoard, event.pos[0], event.pos[1])
73.
74.                 if (spotx, spoty) == (None, None):
75.                     # check if the user clicked on an option button
76.                     if RESET_RECT.collidepoint(event.pos):
77.                         resetAnimation(mainBoard, allMoves) # clicked on Reset button
78.                         allMoves = []
79.                     elif NEW_RECT.collidepoint(event.pos):
80.                         mainBoard, solutionSeq = generateNewPuzzle(80) # clicked on New Game button
81.                         allMoves = []
82.                     elif SOLVE_RECT.collidepoint(event.pos):
83.                         resetAnimation(mainBoard, solutionSeq + allMoves) # clicked on Solve button
84.                         allMoves = []
```

在进入到事件循环之前，程序在第 69 行调用 checkForQuit()，查看是否已经创建了 QUIT 事件（并且如果已经有了该事件，就终止程序）。为什么要有一个单独的函数（checkForQuit()函数）来处理 QUIT 事件，个中的原因我们将稍后说明。第 70 行的 for 循环针对自上一次调用 pygame.event.get()之后所创建的任何其他事件（或者如果 pygame.event.get()之前还没有调用过，就是从程序开始之后的所有事件）执行事件处理代码。

如果事件的类型是一个 MOUSEBUTTONUP 事件，即玩家在窗口上的某处释放了一次鼠标按钮，那么，我们将鼠标坐标传递给 getSpotClicked()函数，它将会返回游戏板上释放鼠标的位置的游戏板坐标。event.pos[0]是 X 坐标，而 event.pos[1]是 Y 坐标。

如果鼠标按钮释放并没有发生在游戏板的空格上(但是显然还是位于窗口上的某个地方，因为已经创建了一个 MOUSEBUTTONUP 事件)，那么，getSpotClicked()将返回 None。如果是这种情况，我们想要做一次额外的检查，看看玩家是否已经在 Reset、New 或 Solve 按钮上点击了（这些按钮并不在游戏板上）。

这些按钮在窗口上的坐标都存储在 pygame.Rect 对象中，而这些对象分别存储在 RESET_RECT、NEW_RECT 和 SOLVE_RECT 变量之中。我们可以将来自 Event 对象的鼠标坐标传递给 collidepoint()方法。如果鼠标的坐标位于 Rect 对象的区域之中，该方法将返回 True，否则的话，它返回 False。

4.8 用鼠标滑动贴片

```
85.             else:
86.                 # check if the clicked tile was next to the blank spot
87.
88.                 blankx, blanky = getBlankPosition(mainBoard)
89.                 if spotx == blankx + 1 and spoty == blanky:
90.                     slideTo = LEFT
91.                 elif spotx == blankx - 1 and spoty == blanky:
92.                     slideTo = RIGHT
93.                 elif spotx == blankx and spoty == blanky + 1:
94.                     slideTo = UP
95.                 elif spotx == blankx and spoty == blanky - 1:
96.                     slideTo = DOWN
```

如果 getSpotClicked()没有返回(None, None)，那么它将会返回两个整数值的一个元组，整数分别表示在游戏板上点击的位置的 X 和 Y 坐标。然后，第 89 行到第 96 行的 if 和 elif 语句，检查所点击的点是否是挨着空白位置的一个贴片（不是的话，贴片将没有可以滑动的空格）。

getBlankPosition()函数将接受游戏板数据结构，并且返回空白位置的 X 和 Y 的游戏板坐标，我们将其分别存储到变量 blankx 和 blanky 中。如果用户在其上点击的位置挨着空白位置，我们将 slideTo 变量设置为贴片应该滑动的值。

4.9 用键盘滑动贴片

```
98.         elif event.type == KEYUP:
99.             # check if the user pressed a key to slide a tile
```

第 4 章 Slide Puzzle

```
100.                if event.key in (K_LEFT, K_a) and isValidMove(mainBoard,
LEFT):
101.                    slideTo = LEFT
102.                elif event.key in (K_RIGHT, K_d) and
isValidMove(mainBoard, RIGHT):
103.                    slideTo = RIGHT
104.                elif event.key in (K_UP, K_w) and isValidMove(mainBoard,
UP):
105.                    slideTo = UP
106.                elif event.key in (K_DOWN, K_s) and isValidMove(mainBoard,
DOWN):
107.                    slideTo = DOWN
```

也可以让用户通过按下键盘按键来滑动贴片。第 100 行到第 107 行的 if 和 elif 语句允许用户通过按下箭头按键或者 WASD 键来设置 slideTo 变量（稍后介绍）。每一条 if 语句和 elif 语句还调用一次 isValidMove()，以确保贴片能够沿着该方向滑动（我们不一定必须要通过鼠标点击来进行这一调用，因为检查邻近的空白空格所做的事情是相同的）。

4.10 使用 in 操作符实现 "等于其中之一" 的技巧

表达式 event.key in (K_LEFT, K_a) 只是使得代码较为简单的一种 Python 技巧。这是表示 "如果 event.key 等于 K_LEFT 或 K_a 之一，计算为 True" 的一种方法。如下的两个表达式以完全相同的方式计算。

```
event.key in (K_LEFT, K_a)
event.key == K_LEFT or event.key == K_a
```

当你必须检查一个值是否等于多个值之中的某一个的时候，使用这个技巧真地可以节省代码录入。如下的两个表达式都以相同的方式计算。

```
spam == 'dog' or spam == 'cat' or spam == 'mouse' or spam == 'horse' or spam ==
42 or spam == 'dingo'
spam in ('dog', 'cat', 'mouse', 'horse', 42, 'dingo')
```

4.11 WASD 和箭头按键

W、A、S 和 D 按键（合在一起叫作 WASD 按键）通常在计算机游戏中用来做和箭头按键相同的事情，只不过玩家可以使用自己的左手来操作它们（因为 WASD 按键位于键盘的左边）。W 表示向上，A 表示向左，S 表示向下，而 D 表示向右。你可以很容易地记住它们，因为 WASD 按键和箭头按键具有相同的布局，如图 4-3 所示。

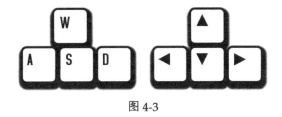

图 4-3

4.12 实际执行贴片滑动

```
109.        if slideTo:
110.            slideAnimation(mainBoard, slideTo, 'Click tile or press arrow
keys to slide.', 8) # show slide on screen
111.            makeMove(mainBoard, slideTo)
112.            allMoves.append(slideTo) # record the slide
113.        pygame.display.update()
114.        FPSCLOCK.tick(FPS)
```

既然已经处理了事件，我们应该更新游戏状态的变量并且在屏幕上显示新的状态。如果已经设置了 slideTo（不管是通过鼠标事件设置，还是用键盘事件处理代码来设置），那么，我们可以调用 slideAnimation() 来执行滑动动画。其参数是游戏板数据结构、滑动的方向、在滑动贴片的时候显示的一条消息，以及滑动的速度。

在该函数返回后，我们需要更新实际的游戏板数据结构（这通过 makeMove() 函数来实现），然后将滑动添加到目前为止所做的所有的滑动的列表 allMoves 中。这么做之后，如果玩家点击了 "Reset" 按钮，我们就知道如何撤销所有的玩家移动了。

4.13 IDLE 和终止 Pygame 程序

```
117. def terminate():
118.     pygame.quit()
119.     sys.exit()
```

通过这个函数，我们可以同时调用 pygame.quit() 和 sys.exit() 函数。这是一个语法糖，我们可以不用记住要进行两次调用，只需要调用一个函数就可以搞定了。

4.14 检查特定的事件并且将事件添加到 Pygame 的事件队列

```
122. def checkForQuit():
123.     for event in pygame.event.get(QUIT): # get all the QUIT events
124.         terminate() # terminate if any QUIT events are present
125.     for event in pygame.event.get(KEYUP): # get all the KEYUP events
126.         if event.key == K_ESCAPE:
127.             terminate() # terminate if the KEYUP event was for the Esc key
128.         pygame.event.post(event) # put the other KEYUP event objects back
```

checkForQuit()函数将会检查 QUIT 事件（或者，如果用户按下了 Esc 键），然后调用 terminate()函数。但是，这有一点技巧性，需要做一些说明。

Pygame 内部拥有自己的列表数据结构，它创建该列表并且会添加所产生的 Event 对象。这个数据结构叫作事件队列（event queue）。当不使用参数调用 pygame.event.get()函数的时候，将返回整个列表。然而，你可以给 pygame.event.get()传递诸如 QUIT 这样的一个常量，以使其只是返回内部事件队列中的 QUIT 事件（如果有的话）。剩下的事件仍然留在事件队列中，等待下一次的 pygame.event.get()调用。

应该注意，Pygame 的事件队列最多只能存储 127 个 Event 对象。如果你的程序没有足够频繁地调用 pygame.event.get()，并且队列填满了，那么，所发生的任何新的事件都不能够再添加到事件队列中。

第 123 行从 Pygame 的事件队列中提取 QUIT 事件的一个列表并返回它们。如果事件队列中有任何的 QUIT 事件，程序将终止。

第 125 行从事件队列中提取出所有的 KEYUP 事件，并且检查它们是否是针对 Esc 按键的。如果其中一个事件是针对 Esc 按键的，那么程序会终止。然而，也可能有针对 Esc 键以外的键的 KEYUP 事件。在这种情况下，我们需要将 KEYUP 事件放回到 Pygame 的事件队列中。我们使用 pygame.event.post()函数来做到这一点，该函数会将传递给它的 Event 对象添加到 Pygame 事件队列的末尾。通过这种方式，第 70 行调用 pygame.event.get()的时候，那些非 Esc 键的 KEYUP 事件将仍然在那里。否则的话，调用 checkForQuit()将会"消费"（consume）所有的 KEYUP 事件，并且那些事件也不会被处理。

如果你想要自己的程序给 Pygame 的事件队列添加 Event 对象的话，pygame.event.post()函数也很方便。

4.15 创建游戏板数据结构

```
131. def getStartingBoard():
132.     # Return a board data structure with tiles in the solved state.
133.     # For example, if BOARDWIDTH and BOARDHEIGHT are both 3, this function
134.     # returns [[1, 4, 7], [2, 5, 8], [3, 6, None]]
135.     counter = 1
136.     board = []
137.     for x in range(BOARDWIDTH):
138.         column = []
139.         for y in range(BOARDHEIGHT):
140.             column.append(counter)
141.             counter += BOARDWIDTH
142.         board.append(column)
143.         counter -= BOARDWIDTH * (BOARDHEIGHT - 1) + BOARDWIDTH - 1
144.
145.     board[BOARDWIDTH-1][BOARDHEIGHT-1] = None
146.     return board
```

getStartingBoard()将创建并返回表示"已求解"游戏板的一个数据结构,其中,所有编号的贴片都是按照顺序排列的,并且空白的贴片位于右下角。这通过嵌套的 for 循环来实现,就像在 Memory Puzzle 游戏中对游戏板数据结构所做的一样。

然而,注意第一列不会是[1, 2, 3],而是[1, 4, 7]。这是因为贴片沿着一行每次增加 1,而不是沿着一列增加 1。沿着列向下,编号会增加游戏板的宽度大小(这个值存储在 BOARDWIDTH 常量中)那么多。我们将使用 counter 变量来记录下一个贴片应该使用的编号。当一行中的贴片的编号完成的时候,我们需要将 counter 设置为下一行开始时候的编号。

4.16 不记录空白的位置

```
149. def getBlankPosition(board):
150.     # Return the x and y of board coordinates of the blank space.
151.     for x in range(BOARDWIDTH):
152.         for y in range(BOARDHEIGHT):
153.             if board[x][y] == None:
154.                 return (x, y)
```

无论何时,当代码需要得到空白空格的 XY 坐标的时候,我们可以创建一个函数来遍历整个游戏板并找到空白空格的坐标,而不是在每次移动之后记录空白空格的位置。在游戏板数据结构中,None 值用来表示空白空格。getBlankPosition()中的代码直接使用嵌套的 for

循环来找出游戏板上的哪一块空格是空白的。

4.17 通过更新游戏板数据结构来移动

```
157. def makeMove(board, move):
158.     # This function does not check if the move is valid.
159.     blankx, blanky = getBlankPosition(board)
160.
161.     if move == UP:
162.         board[blankx][blanky], board[blankx][blanky + 1] = board[blankx][blanky + 1], board[blankx][blanky]
163.     elif move == DOWN:
164.         board[blankx][blanky], board[blankx][blanky - 1] = board[blankx][blanky - 1], board[blankx][blanky]
165.     elif move == LEFT:
166.         board[blankx][blanky], board[blankx + 1][blanky] = board[blankx + 1][blanky], board[blankx][blanky]
167.     elif move == RIGHT:
168.         board[blankx][blanky], board[blankx - 1][blanky] = board[blankx - 1][blanky], board[blankx][blanky]
```

board 参数中的数据结构是一个 2D 列表，表示所有的贴片位于何处。无论何时，当玩家进行一次移动的时候，程序都需要更新这个数据结构。所发生的事情是表示贴片的值和表示空白空格的值相互交换了。

makeMove() 函数不必返回任何值，因为传递给 board 参数的是一个列表引用。这意味着在该函数中对 board 做出的任何修改，都将会反映到传递给 makeMove() 的列表的值中（你可以通过 http://invpy.com/references 了解引用的概念）。

4.18 何时不使用断言

```
171. def isValidMove(board, move):
172.     blankx, blanky = getBlankPosition(board)
173.     return (move == UP and blanky != len(board[0]) - 1) or \
174.            (move == DOWN and blanky != 0) or \
175.            (move == LEFT and blankx != len(board) - 1) or \
176.            (move == RIGHT and blankx != 0)
```

这段代码给 isValidMove() 函数传递了一个游戏板数据结构，以及玩家想要做出的一次移动。如果该移动是可能的，返回值为 True，否则的话，返回值为 False。例如，不能在同一行之中将一个贴片向左移动 100 次，因为最终空白空格将会位于边缘，并且再没有贴片能够向左移动了。

移动是有效的，还是无效的，这取决于空白空格位于何处。这个函数调用了 getBlankPosition() 来找到空白位置的 X 和 Y 坐标。第 173 行到第 176 行是带有单个表达式的一条 return 语句。前三行末尾的 \ 告诉 Python 解释器，这不是一行代码的结束（即便它确实位于一行的末尾）。这将允许我们将一行代码分多行放置，而不是只使用一个很长的、可读性很差的代码行。

由于这个表达式的括号中的各个部分，都是使用 or 操作符连接的，只要它们之中有一个为 True，那么，整个表达式就为 True。这些部分中的每一个都会检查想要移动的方向，然后看看空白空格所在的坐标是否允许该移动。

4.19 获取一次并不是那么随机的移动

```
179. def getRandomMove(board, lastMove=None):
180.     # start with a full list of all four moves
181.     validMoves = [UP, DOWN, LEFT, RIGHT]
182.
183.     # remove moves from the list as they are disqualified
184.     if lastMove == UP or not isValidMove(board, DOWN):
185.         validMoves.remove(DOWN)
186.     if lastMove == DOWN or not isValidMove(board, UP):
187.         validMoves.remove(UP)
188.     if lastMove == LEFT or not isValidMove(board, RIGHT):
189.         validMoves.remove(RIGHT)
190.     if lastMove == RIGHT or not isValidMove(board, LEFT):
191.         validMoves.remove(LEFT)
192.
193.     # return a random move from the list of remaining moves
194.     return random.choice(validMoves)
```

在游戏的开始处，我们从处于已求解的、有序状态的游戏板数据结构开始，通过随机地移动贴片来创建谜题。要确定应该向 4 个方向中的哪一个方向滑动，我们调用 getRandomMove() 函数。通常，只是使用 random.choice() 函数并且传递给它一个元组(UP, DOWN, LEFT, RIGHT)，让 Python 直接随机地选择一个方向值。但是，Sliding Puzzle 游戏有一个小小的限制，不允许我们选择一个纯粹的随机数。

如果你有一个 Sliding Puzzle 并且把一个贴片向左滑动，然后把一个贴片向右滑动，那么，你将得到和开始的时候完全相同的游戏板。在一次滑动之后，接着就向相反的方向滑动，这么做毫无意义。此外，如果空白空格在右下角，将贴片向上滑动或向左滑动也是不可能的。

getRandomMove() 中的代码将考虑到这些因素。为了防止函数选择上一次所做的移动，

第 4 章　Slide Puzzle

该函数的调用者可以给 lastMove 参数传递一个方向值。第 181 行先将 4 个方向的一个列表存储到 validMoves 变量中。lastMove 值（如果没有设置为 None 的话）会从 validMoves 中删除。根据空白空格是否位于游戏板的边缘，第 184 行到第 191 行将从 validMoves 列表中删除其他的方向值。

在 validMoves 中所剩下的值之中，有一个是通过调用 random.choice()随机选择并返回的。

4.20　将贴片坐标转换为像素坐标

```
197. def getLeftTopOfTile(tileX, tileY):
198.     left = XMARGIN + (tileX * TILESIZE) + (tileX - 1)
199.     top = YMARGIN + (tileY * TILESIZE) + (tileY - 1)
200.     return (left, top)
```

getLeftTopOfTile()函数将游戏板坐标转换为像素坐标。对于传入的游戏板 XY 坐标，该函数计算并返回游戏板空格左上角的像素的像素 XY 坐标。

4.21　将像素坐标转换为游戏板坐标

```
203. def getSpotClicked(board, x, y):
204.     # from the x & y pixel coordinates, get the x & y board coordinates
205.     for tileX in range(len(board)):
206.         for tileY in range(len(board[0])):
207.             left, top = getLeftTopOfTile(tileX, tileY)
208.             tileRect = pygame.Rect(left, top, TILESIZE, TILESIZE)
209.             if tileRect.collidepoint(x, y):
210.                 return (tileX, tileY)
211.     return (None, None)
```

getSpotClicked()所做的事情和 getLeftTopOfTile()刚好相反，它会把像素坐标转换为游戏板坐标。第 205 行到第 206 行的嵌套循环遍历了每一种可能的 XY 游戏板坐标，如果传入的像素坐标位于游戏板的该空格之内，它会返回这个游戏板坐标。由于所有的贴片都拥有 TILESIZE 常量中所设置的宽度和高度，我们可以通过获取游戏板空格的左上角的像素坐标来创建一个 Rect 对象以表示游戏板上的这一空格，然后使用 Rect 的 collidepoint()方法来查看像素坐标是否位于 Rect 对象的区域之中。

如果所传入的像素坐标没有在任何游戏板空格之上，那么，返回值(None, None)。

4.22 绘制一个贴片

```
214. def drawTile(tilex, tiley, number, adjx=0, adjy=0):
215.     # draw a tile at board coordinates tilex and tiley, optionally a few
216.     # pixels over (determined by adjx and adjy)
217.     left, top = getLeftTopOfTile(tilex, tiley)
218.     pygame.draw.rect(DISPLAYSURF, TILECOLOR, (left + adjx, top + adjy,
TILESIZE, TILESIZE))
219.     textSurf = BASICFONT.render(str(number), True, TEXTCOLOR)
220.     textRect = textSurf.get_rect()
221.     textRect.center = left + int(TILESIZE / 2) + adjx, top + int(TILESIZE
 / 2) + adjy
222.     DISPLAYSURF.blit(textSurf, textRect)
```

drawTile()函数将在游戏板上绘制单个的编号的贴片。tilex 和 tiley 参数是贴片的游戏板坐标。number 参数是一个字符串，表示贴片的编号（如'3'或'12'）。adjx 和 adjy 参数用于对贴片位置进行微调。例如，给 adjx 传递 5 将会使得贴片出现在游戏板上的 tilex 和 tiley 空格的右边 5 像素的位置。给 adjy 传递-10，将会使得贴片出现在该空格的上边的 10 像素的位置。

当我们需要在滑动过程中间绘制贴片的时候，这些调整值将会很方便。如果在调用 drawTile()的时候没有给这些参数传递值，那么，它们将默认地设置为 0。这意味着，它们将恰好处于 tilex 和 tiley 所给定的游戏板空格的位置。

Pygame 绘制函数只是使用像素坐标，因此，首先第 217 行 tilex 和 tiley 中的游戏板坐标转换为像素坐标，然后会把像素坐标存储在 left 和 top 中（因为 getLeftTopOfTile()返回了左上角的坐标）。我们通过调用一次 pygame.draw.rect()来绘制贴片的背景方块，调用函数的过程中会给 left 和 top 加上 adjx 和 adjy 值，以防止代码需要调整贴片的位置。

第 219 行到第 222 行创建了 Surface 对象，数字编号文本绘制于该对象之上。有一个 Rect 对象用于将 Surface 对象定位，然后，使用它将 Surface 对象复制到显示 Surface 上。drawTile()函数不会调用 pygame.display.update()函数，因为 drawTile()函数的调用者可能想要为剩下的游戏板绘制更多的贴片，然后才让它们出现在屏幕上。

4.23 让文本显示在屏幕上

```
225. def makeText(text, color, bgcolor, top, left):
```

第 4 章　Slide Puzzle

```
226.    # create the Surface and Rect objects for some text.
227.    textSurf = BASICFONT.render(text, True, color, bgcolor)
228.    textRect = textSurf.get_rect()
229.    textRect.topleft = (top, left)
230.    return (textSurf, textRect)
```

makeText()函数负责 Surface 和 Rect 对象的创建，以便将文本定位在屏幕上。每次想要让文本显示到屏幕上的时候，我们只是调用 makeText()，而不是执行所有这些调用。这使得程序节省了很多必须的输入量（通过 makeText()来进行对 render()和 get_rect()本身的调用，因为它通过中心点而不是左上角的点来定位文本 Surface 的对象，并且使用一种透明的背景色）。

4.24　绘制游戏板

```
233. def drawBoard(board, message):
234.    DISPLAYSURF.fill(BGCOLOR)
235.    if message:
236.        textSurf, textRect = makeText(message, MESSAGECOLOR, BGCOLOR, 5, 5)
237.        DISPLAYSURF.blit(textSurf, textRect)
238.
239.    for tilex in range(len(board)):
240.        for tiley in range(len(board[0])):
241.            if board[tilex][tiley]:
242.                drawTile(tilex, tiley, board[tilex][tiley])
```

该函数负责将整个游戏板及其所有的贴片绘制到 DISPLAYSURF 显示 Surface 对象上。第 234 行的 fill()方法完全通过绘制覆盖了之前在显示 Surface 对象上绘制的所有内容，使得我们能够重新开始。

第 235 行到第 237 行负责绘制窗口顶部的消息。我们将其用于 "Generating new puzzle..." 以及想要在窗口顶部显示的其他文本。记住，if 语句条件认为空白字符串为 False 值，因此，如果消息设置为''，那么该条件为 False，并且第 236 行和第 237 行将会跳过。接下来，嵌套的 for 循环通过调用 drawTile()将每一个贴片绘制到显示 Surface 对象。

4.25　绘制游戏板的边框

```
244.    left, top = getLeftTopOfTile(0, 0)
245.    width = BOARDWIDTH * TILESIZE
```

```
246.        height = BOARDHEIGHT * TILESIZE
247.        pygame.draw.rect(DISPLAYSURF, BORDERCOLOR, (left - 5, top - 5, width + 11, height + 11), 4)
```

第 244 行到第 247 行绘制了围绕贴片的边框。边框的左上角在游戏板坐标(0, 0)的贴片的左上角的左边 5 像素和上边 5 像素处。将游戏板的宽度和高度（分别存储在 BOARDWIDTH 和 BOARDHEIGHT 常量中），分别乘以贴片的大小（存储在 TILESIZE 常量中），从而计算出边框的宽度和高度。

在第 247 行绘制的矩形，其边框将具有 4 个像素的宽度，因此，我们将把边框从 top 和 left 变量的位置向左和向上移动 5 个像素，以使得线条的宽度不会和贴片重叠。我们还给宽度和长度增加了 11 个像素（这 11 个像素中的 5 个像素，用来作为将矩形向左上方移动的补偿）。

4.26　绘制按钮

```
249.        DISPLAYSURF.blit(RESET_SURF, RESET_RECT)
250.        DISPLAYSURF.blit(NEW_SURF, NEW_RECT)
251.        DISPLAYSURF.blit(SOLVE_SURF, SOLVE_RECT)
```

最后，我们绘制了屏幕之外的按钮。这些按钮的文本和位置不能改变，这也是为什么我们在 main() 的开始处将其保存到常量变量之中的原因。

4.27　实现贴片滑动动画

```
254. def slideAnimation(board, direction, message, animationSpeed):
255.     # Note: This function does not check if the move is valid.
256.
257.     blankx, blanky = getBlankPosition(board)
258.     if direction == UP:
259.         movex = blankx
260.         movey = blanky + 1
261.     elif direction == DOWN:
262.         movex = blankx
263.         movey = blanky - 1
264.     elif direction == LEFT:
265.         movex = blankx + 1
266.         movey = blanky
267.     elif direction == RIGHT:
268.         movex = blankx - 1
269.         movey = blanky
```

第 4 章　Slide Puzzle

　　贴片滑动代码需要做的第一件事情就是计算空白空格在哪里，以及将贴片移动到哪里。第 255 行的注释提醒我们，调用 slideAnimation()的代码应该确保传递给它的 direction 参数是可以有效移动的方向。

　　空白空格的坐标通过调用 getBlankPosition()而得到。通过这些坐标以及滑动的方向，我们可以搞清楚将要滑动的贴片的 XY 游戏板坐标。这些坐标将存储在 movex 和 movey 变量之中。

4.28　Surface 的 copy()方法

```
271.        # prepare the base surface
272.        drawBoard(board, message)
273.        baseSurf = DISPLAYSURF.copy()
274.        # draw a blank space over the moving tile on the baseSurf Surface.
275.        moveLeft, moveTop = getLeftTopOfTile(movex, movey)
276.        pygame.draw.rect(baseSurf, BGCOLOR, (moveLeft, moveTop, TILESIZE,
TILESIZE))
```

　　Surface 的 copy()方法将返回一个新的 Surface 对象，这个新的对象具有相同的需要绘制的图像。但是，它们是两个不同的 Surface 对象。在调用了 copy()方法之后，如果要使用 blit()或者 Pygame 的绘制方法在一个 Surface 对象上绘制，这将不会修改另一个 Surface 对象上的图像。在第 273 行，我们将这个副本保存到 baseSurf 变量中。接下来，我们在将要移动的贴片之上绘制了另一个空白空格。这是因为，当绘制滑动动画的每一帧的时候，我们将在 baseSurf Surface 对象的不同部分之上绘制滑动贴片。如果我们没有把移动的贴片在 baseSurf Surface 对象上绘制成空白，那么，当我们绘制移动贴片的时候，这个贴片就会还在原来的位置。在这种情况下，baseSurf 最初看上去如图 4-4 所示。

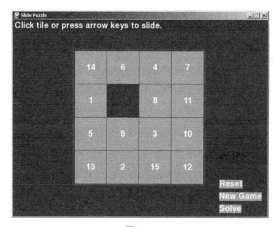

图 4-4

4.28 Surface 的 copy()方法

而当我们绘制"9"的贴片向上滑动的时候，它看上去如图 4-5 所示。

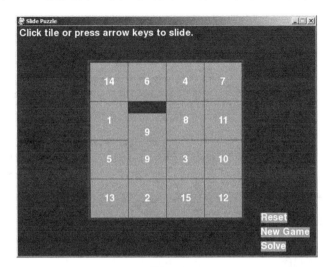

图 4-5

你可以注释掉第 276 行并运行程序，自己来看看这种效果。

```
278.    for i in range(0, TILESIZE, animationSpeed):
279.        # animate the tile sliding over
280.        checkForQuit()
281.        DISPLAYSURF.blit(baseSurf, (0, 0))
282.        if direction == UP:
283.            drawTile(movex, movey, board[movex][movey], 0, -i)
284.        if direction == DOWN:
285.            drawTile(movex, movey, board[movex][movey], 0, i)
286.        if direction == LEFT:
287.            drawTile(movex, movey, board[movex][movey], -i, 0)
288.        if direction == RIGHT:
289.            drawTile(movex, movey, board[movex][movey], i, 0)
290.
291.        pygame.display.update()
292.        FPSCLOCK.tick(FPS)
```

为了绘制滑动动画的帧，我们必须把 baseSurf Surface 绘制到显示 Surface 上，然后，在动画的每一帧之上，绘制滑动的贴片，使其距离其最终位置（也就是空白空格的最初位置）越来越近。两个相邻接的贴片之间的空格，保持和单个贴片相同的大小，我们将这个大小存储在 TILESIZE 中。这段代码使用一个从 0 到 TILESIZE 运行的 for 循环。

通常，这意味着，我们把贴片向上绘制 0 个像素，然后在下一帧中，将贴片向上绘制 1 个像素，2 个像素，然后是 3 个像素，依次类推。每一帧将花 1/30 秒。如果将 TILESIZE 设

97

置为 80（就像本书中的程序在第 12 行所做的那样），那么，滑动一个贴片将要花 2.5 秒，这实际上有点慢。

因此，我们让 for 循环从 0 到 TILESIZE 循环迭代的时候，每帧差异为几个像素。跳过的像素的数目存储在 animationSpeed 之中，这在调用 slideAnimation() 的时候传入。例如，如果 animationSpeed 设置为 8，并且常量 TILESIZE 设置为 80，那么，for 循环和 range(0, TILESIZE, animationSpeed) 将会把 i 变量设置为 0、8、16、24、32、40、48、56、64 和 72（并不包括 80，因为 range() 不超过第二个参数，但是不包括第二个参数）。这意味着整个滑动动画将在 10 帧中完成，即游戏以 30 帧/秒运行的话，动画会在 1/3 秒中完成。第 282 行到第 289 行确保了我们绘制贴片朝着正确的方向滑动（根据 direction 变量中的值）。在动画完成之后，该函数返回。注意，当动画进行的时候，用户创建的任何事件都不会处理。这些事件将会在下一次执行到达 main() 函数中的第 70 行的时候处理，或者在执行 checkForQuit() 函数中的代码的时候处理。

4.29 创建新的谜题

```
295. def generateNewPuzzle(numSlides):
296.     # From a starting configuration, make numSlides number of moves (and
297.     # animate these moves).
298.     sequence = []
299.     board = getStartingBoard()
300.     drawBoard(board, '')
301.     pygame.display.update()
302.     pygame.time.wait(500) # pause 500 milliseconds for effect
```

在每次新游戏开始的时候，将会调用 generateNewPuzzle() 函数。它将通过调用 getStartingBoard() 创建一个新的游戏板数据结构并随机地打乱它们。generateNewPuzzle() 的第一行获取该游戏板，然后将其绘制到屏幕上（冻结半秒钟的时间，以便让玩家看一会儿新的游戏板）。

```
303.     lastMove = None
304.     for i in range(numSlides):
305.         move = getRandomMove(board, lastMove)
306.         slideAnimation(board, move, 'Generating new puzzle...', int(TILESIZE / 3))
307.         makeMove(board, move)
308.         sequence.append(move)
309.         lastMove = move
310.     return (board, sequence)
```

numSlides 参数将告诉该函数做多少次随机移动。执行随机移动的代码是第 305 行的

getRandomMove()调用，随后调用slideAnimation()在屏幕上执行动画。由于执行滑动动画并没有真正地更新游戏板数据结构，我们通过在307行调用makeMove()来更新游戏板。

我们需要记录所做的每一次随机移动，以便用户随后可以点击"Solve"按钮，并且让程序撤销所有的这些随机的移动（4.5节将会介绍我们为什么这么做，以及如何做到这一点）。因此，在第308行，移动会添加到sequence中的移动列表之中。

然后，我们将这些随机的移动存储到一个名为 lastMove 的变量中，它将会在下一次迭代的时候传递给 getRandomMove()。这防止了下一次的随机移动撤销我们刚刚执行过的随机移动。

所有这些都需要进行numSlides多次，因此，我们将第305行到第309行的代码放入到一个for循环中。当游戏板打乱以后，我们返回游戏板数据结构以及对其所做的随机移动的列表。

4.30　实现游戏板重置动画

```
313. def resetAnimation(board, allMoves):
314.     # make all of the moves in allMoves in reverse.
315.     revAllMoves = allMoves[:] # gets a copy of the list
316.     revAllMoves.reverse()
317.
318.     for move in revAllMoves:
319.         if move == UP:
320.             oppositeMove = DOWN
321.         elif move == DOWN:
322.             oppositeMove = UP
323.         elif move == RIGHT:
324.             oppositeMove = LEFT
325.         elif move == LEFT:
326.             oppositeMove = RIGHT
327.         slideAnimation(board, oppositeMove, '', int(TILESIZE / 2))
328.         makeMove(board, oppositeMove)
```

当玩家在"Reset"或"Solve"上点击的时候，Slide Puzzle 游戏程序需要撤销对游戏板所做的所有移动。移动的方向值的列表将作为allMoves参数的实参传递。

第315行使用列表分片创建了allMoves列表的一个副本。还记得吧，如果你没有在:之前指定一个数字，Python 假设该分片应该从列表的最前面开始。并且，如果你没有在:之后指定一个数字，Python 假设分片应该持续到列表的最末尾。因此，allMoves[:]创建了整个allMoves列表的一个列表分片。这实际上会生成列表的一个副本，并存储到 revAllMoves 之中，而不只是创建列表引用的一个副本（参见 http://invpy.com/references 了解更多细节）。

要撤销allMoves中的所有移动，我们需要执行和allMoves中的移动相反的移动，并且

第 4 章　Slide Puzzle

以相反的顺序来执行。有一个名为 reverse() 的列表方法，可以将列表中的项的顺序颠倒过来。第 318 行的 for 循环遍历了方向值的列表。记住，我们想要相反的移动，因此，第 319 行到第 326 行的 if 和 elif 语句在 oppositeMove 变量中设置了正确的方向值。然后，我们调用 slideAnimation() 来执行动画，并调用 makeMove() 来更新游戏板数据结构。

```
331. if __name__ == '__main__':
332.     main()
```

就像在 Memory Puzzle 游戏中一样，当所有的 def 语句执行完并创建了所有函数之后，我们调用 main() 函数开始了程序的实际内容。

这就是 Slide Puzzle 程序的所有内容。但是，让我们来看看这款游戏中遇到的一些通用的编程概念。

4.31　时间 vs.内存的权衡

当然，还有一些不同的方法来编写 Slide Puzzle 游戏，以使得即便代码不同，但其外观和行为都完全相同。可以有很多种不同的方式来编写一个程序以执行任务。最为常见的区别是要在执行时间和内存使用之前求得平衡。

通常，程序运行得越快越好。对于那些需要很多计算的程序来说尤其如此，不管是科学的天气模拟程序，还是需要绘制大量详细的 3D 图形的游戏。尽可能地使用最少的内存也是很好的。程序使用的变量越多，使用的列表越大，它所占用的内存也会越大（可以通过 http://invpy.com/profiling 了解如何度量计算机内存的使用和执行时间）。

到目前为止，本书中的程序都不大也不复杂，还不必担心节约内存或者优化执行时间。但是，随着你成为一名更加熟练的程序员，可能要考虑这些问题。

例如，考虑一下 getBlankPosition() 函数。该函数需要花些时间来运行，因为它遍历所有可能的游戏板坐标以找到空白空格在何处。相反，我们也可以用一个 blankspacex 和 blankspacey 变量来记录这些 XY 坐标，从而不必在每次想要知道空白空格位于何处的时候遍历整个游戏板（但是这样一来，无论何时完成移动，我们还需要更新 blankspacex 和 blankspacey 变量。负责更新这两个变量的代码可以放到 makeMove() 之中）。使用这些变量会占用更多的内存，但是，确实节省了执行时间，从而可以让程序运行得更快。

另一个例子是我们将处于已解决状态的一个游戏板数据结构保存在 SOLVEDBOARD 变量中，以便可以将当前的游戏板和 SOLVEDBOARD 相比较，看看玩家是否已经解决了谜题。每次想要执行这一检查的时候，我们也可以调用 getStartingBoard() 函数，并将其返回值和当前的游戏板进行比较。这样一来，我们就不需要 SOLVEDBOARD 变量了。这么做会为我们节省一些内存，但是，程序将会花费更长一点儿的时间运行，因为每次我们执行这一检查的时候，它都需要重新创建已解决状态的游戏板数据结构。

然而，有一点必须牢记。编写可阅读的代码是一项非常重要的技能。代码是"可读的"，就是说代码是容易理解的，特别是对那些没有编写该段代码的程序员来说。如果另一个程序员看到你的程序的源代码，并且很容易就能够搞清楚它在做什么，那么，这个程序的可读性很好。可读性很重要，是因为当你想要给自己的程序修改 Bug 或添加新的功能的时候（并且，修改 Bug 和添加新功能的想法总是会出现），具有可读性的程序会使得这些任务很容易。

4.32 没人在乎几个字节

此外，还有一点看上去似乎很傻，但是必须在本书中讲清楚，因为它看似显而易见，但很多人颇为怀疑。你应该知道这一点，即使用诸如 x 或 num 这样的较短的变量名称，而不是诸如 blankx 或 numSlides 这样更具有描述性的变量名称，在程序真正运行的时候，这并不会节省任何内存。使用那些较长一点的名字要更好一些，因为它们使得你的程序更为可读。

你可能也会提出一些聪明的技巧，它们确实能在这里或那里节省几个字节的内存。一个技巧是当你不再需要一个变量的时候，可以复用这个变量名以用于另一个不同的用途，而不是直接使用两个不同名称的变量。

请尽量避免这种尝试。通常，这些技巧会降低代码的可读性，并且使得很难调试你的程序。现代的计算机有着数以百万字节计的内存，在这里或那里节省区区几个字节，而让代码变得更容易令人类程序员混淆，实在是得不偿失。

4.33 没人在乎几百万个纳秒

类似的，有的时候，你想要以某种方式重新组织代码而使其运行得略微快上几个纳秒。这些技巧通常也会令代码难以阅读。想想吧，当你需要花费几百万个纳秒去阅读这些句子的时候，在程序中节省下来的区区几个纳秒，玩家丝毫也不会留意。

4.34 本章小结

本章没有介绍任何在 Memory Puzzle 中没有用到的、新的 Pygame 编程概念，除了使用 Surface 对象的 copy() 方法之外。要知道，些许不同的概念可能会使得你编写出完全不同的游戏。作为练习，你可以从 http://invpy.com/buggy/slidepuzzle 下载 Sliding Puzzle 程序的带有 Bug 的版本。

第 5 章 Simulate

5.1 如何玩 Simulate 游戏

Simulate 游戏是 Simon 的一个翻版。屏幕上有 4 个彩色的按钮，如图 5-1 所示。按钮以某种随机的模式点亮，然后，玩家必须通过以正确的顺序按下按钮来重复这一模式。每次玩家成功地模拟了模式，模式会变得更长。玩家努力尽可能地去匹配模式。

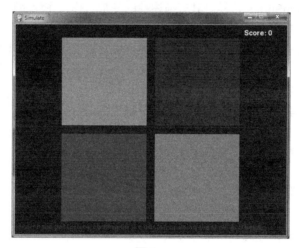

图 5-1

5.2 Simulate 的源代码

可以从 http://invpy.com/simulate.py 下载源代码。如果你得到了任何错误的消息，看一下错误消息中提到的行号，并且检查你的代码是否有任何的录入错误。你可以将代码复制并粘贴到位于 http://invpy.com/diff/simulate 的 Web 表单中，看看你的代码和本书中的代码是否有差别。

可以通过如下地址下载该程序中要用到的 4 个声音文件。

- http://invpy.com/beep1.ogg
- http://invpy.com/beep2.ogg
- http://invpy.com/beep3.ogg
- http://invpy.com/beep4.ogg

5.2 Simulate 的源代码

```
 1. # Simulate (a Simon clone)
 2. # By Al Sweigart al@inventwithpython.com
 3. # http://inventwithpython.com/pygame
 4. # Creative Commons BY-NC-SA 3.0 US
 5.
 6. import random, sys, time, pygame
 7. from pygame.locals import *
 8.
 9. FPS = 30
10. WINDOWWIDTH = 640
11. WINDOWHEIGHT = 480
12. FLASHSPEED = 500 # in milliseconds
13. FLASHDELAY = 200 # in milliseconds
14. BUTTONSIZE = 200
15. BUTTONGAPSIZE = 20
16. TIMEOUT = 4 # seconds before game over if no button is pushed.
17.
18. #                 R    G    B
19. WHITE        = (255, 255, 255)
20. BLACK        = (  0,   0,   0)
21. BRIGHTRED    = (255,   0,   0)
22. RED          = (155,   0,   0)
23. BRIGHTGREEN  = (  0, 255,   0)
24. GREEN        = (  0, 155,   0)
25. BRIGHTBLUE   = (  0,   0, 255)
26. BLUE         = (  0,   0, 155)
27. BRIGHTYELLOW = (255, 255,   0)
28. YELLOW       = (155, 155,   0)
29. DARKGRAY     = ( 40,  40,  40)
30. bgColor = BLACK
31.
32. XMARGIN = int((WINDOWWIDTH - (2 * BUTTONSIZE) - BUTTONGAPSIZE) / 2)
33. YMARGIN = int((WINDOWHEIGHT - (2 * BUTTONSIZE) - BUTTONGAPSIZE) / 2)
34.
35. # Rect objects for each of the four buttons
36. YELLOWRECT = pygame.Rect(XMARGIN, YMARGIN, BUTTONSIZE, BUTTONSIZE)
37. BLUERECT   = pygame.Rect(XMARGIN + BUTTONSIZE + BUTTONGAPSIZE, YMARGIN, BUTTONSIZE, BUTTONSIZE)
38. REDRECT    = pygame.Rect(XMARGIN, YMARGIN + BUTTONSIZE + BUTTONGAPSIZE, BUTTONSIZE, BUTTONSIZE)
39. GREENRECT  = pygame.Rect(XMARGIN + BUTTONSIZE + BUTTONGAPSIZE, YMARGIN + BUTTONSIZE + BUTTONGAPSIZE, BUTTONSIZE, BUTTONSIZE)
40.
41. def main():
42.     global FPSCLOCK, DISPLAYSURF, BASICFONT, BEEP1, BEEP2, BEEP3, BEEP4
43.
```

第 5 章　Simulate

```
 44.    pygame.init()
 45.    FPSCLOCK = pygame.time.Clock()
 46.    DISPLAYSURF = pygame.display.set_mode((WINDOWWIDTH, WINDOWHEIGHT))
 47.    pygame.display.set_caption('Simulate')
 48.
 49.    BASICFONT = pygame.font.Font('freesansbold.ttf', 16)
 50.
 51.    infoSurf = BASICFONT.render('Match the pattern by clicking on the button or using the Q, W, A, S keys.', 1, DARKGRAY)
 52.    infoRect = infoSurf.get_rect()
 53.    infoRect.topleft = (10, WINDOWHEIGHT - 25)
 54.    # load the sound files
 55.    BEEP1 = pygame.mixer.Sound('beep1.ogg')
 56.    BEEP2 = pygame.mixer.Sound('beep2.ogg')
 57.    BEEP3 = pygame.mixer.Sound('beep3.ogg')
 58.    BEEP4 = pygame.mixer.Sound('beep4.ogg')
 59.
 60.    # Initialize some variables for a new game
 61.    pattern = [] # stores the pattern of colors
 62.    currentStep = 0 # the color the player must push next
 63.    lastClickTime = 0 # timestamp of the player's last button push
 64.    score = 0
 65.    # when False, the pattern is playing. when True, waiting for the player to click a colored button:
 66.    waitingForInput = False
 67.
 68.    while True: # main game loop
 69.        clickedButton = None # button that was clicked (set to YELLOW, RED, GREEN, or BLUE)
 70.        DISPLAYSURF.fill(bgColor)
 71.        drawButtons()
 72.
 73.        scoreSurf = BASICFONT.render('Score: ' + str(score), 1, WHITE)
 74.        scoreRect = scoreSurf.get_rect()
 75.        scoreRect.topleft = (WINDOWWIDTH - 100, 10)
 76.        DISPLAYSURF.blit(scoreSurf, scoreRect)
 77.
 78.        DISPLAYSURF.blit(infoSurf, infoRect)
 79.
 80.        checkForQuit()
 81.        for event in pygame.event.get(): # event handling loop
 82.            if event.type == MOUSEBUTTONUP:
 83.                mousex, mousey = event.pos
 84.                clickedButton = getButtonClicked(mousex, mousey)
 85.            elif event.type == KEYDOWN:
 86.                if event.key == K_q:
```

```
87.                    clickedButton = YELLOW
88.                elif event.key == K_w:
89.                    clickedButton = BLUE
90.                elif event.key == K_a:
91.                    clickedButton = RED
92.                elif event.key == K_s:
93.                    clickedButton = GREEN
94.
95.
96.
97.        if not waitingForInput:
98.            # play the pattern
99.            pygame.display.update()
100.            pygame.time.wait(1000)
101.            pattern.append(random.choice((YELLOW, BLUE, RED, GREEN)))
102.            for button in pattern:
103.                flashButtonAnimation(button)
104.                pygame.time.wait(FLASHDELAY)
105.            waitingForInput = True
106.        else:
107.            # wait for the player to enter buttons
108.            if clickedButton and clickedButton == pattern[currentStep]:
109.                # pushed the correct button
110.                flashButtonAnimation(clickedButton)
111.                currentStep += 1
112.                lastClickTime = time.time()
113.
114.                if currentStep == len(pattern):
115.                    # pushed the last button in the pattern
116.                    changeBackgroundAnimation()
117.                    score += 1
118.                    waitingForInput = False
119.                    currentStep = 0 # reset back to first step
120.
121.            elif (clickedButton and clickedButton != pattern[currentStep]) or (currentStep != 0 and time.time() - TIMEOUT > lastClickTime):
122.                # pushed the incorrect button, or has timed out
123.                gameOverAnimation()
124.                # reset the variables for a new game:
125.                pattern = []
126.                currentStep = 0
127.                waitingForInput = False
128.                score = 0
129.                pygame.time.wait(1000)
130.                changeBackgroundAnimation()
131.
```

第 5 章 Simulate

```
132.            pygame.display.update()
133.            FPSCLOCK.tick(FPS)
134.
135.
136. def terminate():
137.     pygame.quit()
138.     sys.exit()
139.
140.
141. def checkForQuit():
142.     for event in pygame.event.get(QUIT): # get all the QUIT events
143.         terminate() # terminate if any QUIT events are present
144.     for event in pygame.event.get(KEYUP): # get all the KEYUP events
145.         if event.key == K_ESCAPE:
146.             terminate() # terminate if the KEYUP event was for the Esc key
147.         pygame.event.post(event) # put the other KEYUP event objects back
148.
149.
150. def flashButtonAnimation(color, animationSpeed=50):
151.     if color == YELLOW:
152.         sound = BEEP1
153.         flashColor = BRIGHTYELLOW
154.         rectangle = YELLOWRECT
155.     elif color == BLUE:
156.         sound = BEEP2
157.         flashColor = BRIGHTBLUE
158.         rectangle = BLUERECT
159.     elif color == RED:
160.         sound = BEEP3
161.         flashColor = BRIGHTRED
162.         rectangle = REDRECT
163.     elif color == GREEN:
164.         sound = BEEP4
165.         flashColor = BRIGHTGREEN
166.         rectangle = GREENRECT
167.
168.     origSurf = DISPLAYSURF.copy()
169.     flashSurf = pygame.Surface((BUTTONSIZE, BUTTONSIZE))
170.     flashSurf = flashSurf.convert_alpha()
171.     r, g, b = flashColor
172.     sound.play()
173.     for start, end, step in ((0, 255, 1), (255, 0, -1)): # animation loop
174.         for alpha in range(start, end, animationSpeed * step):
175.             checkForQuit()
176.             DISPLAYSURF.blit(origSurf, (0, 0))
177.             flashSurf.fill((r, g, b, alpha))
```

```
178.              DISPLAYSURF.blit(flashSurf, rectangle.topleft)
179.              pygame.display.update()
180.              FPSCLOCK.tick(FPS)
181.     DISPLAYSURF.blit(origSurf, (0, 0))
182.
183.
184. def drawButtons():
185.     pygame.draw.rect(DISPLAYSURF, YELLOW, YELLOWRECT)
186.     pygame.draw.rect(DISPLAYSURF, BLUE,   BLUERECT)
187.     pygame.draw.rect(DISPLAYSURF, RED,    REDRECT)
188.     pygame.draw.rect(DISPLAYSURF, GREEN,  GREENRECT)
189.
190.
191. def changeBackgroundAnimation(animationSpeed=40):
192.     global bgColor
193.     newBgColor = (random.randint(0, 255), random.randint(0, 255), random.randint(0, 255))
194.
195.     newBgSurf = pygame.Surface((WINDOWWIDTH, WINDOWHEIGHT))
196.     newBgSurf = newBgSurf.convert_alpha()
197.     r, g, b = newBgColor
198.     for alpha in range(0, 255, animationSpeed): # animation loop
199.         checkForQuit()
200.         DISPLAYSURF.fill(bgColor)
201.
202.         newBgSurf.fill((r, g, b, alpha))
203.         DISPLAYSURF.blit(newBgSurf, (0, 0))
204.
205.         drawButtons() # redraw the buttons on top of the tint
206.
207.         pygame.display.update()
208.         FPSCLOCK.tick(FPS)
209.     bgColor = newBgColor
210.
211.
212. def gameOverAnimation(color=WHITE, animationSpeed=50):
213.     # play all beeps at once, then flash the background
214.     origSurf = DISPLAYSURF.copy()
215.     flashSurf = pygame.Surface(DISPLAYSURF.get_size())
216.     flashSurf = flashSurf.convert_alpha()
217.     BEEP1.play() # play all four beeps at the same time, roughly.
218.     BEEP2.play()
219.     BEEP3.play()
220.     BEEP4.play()
221.     r, g, b = color
222.     for i in range(3): # do the flash 3 times
```

```
223.          for start, end, step in ((0, 255, 1), (255, 0, -1)):
224.              # The first iteration in this loop sets the following for loop
225.              # to go from 0 to 255, the second from 255 to 0.
226.              for alpha in range(start, end, animationSpeed * step): # animation loop
227.                  # alpha means transparency. 255 is opaque, 0 is invisible
228.                  checkForQuit()
229.                  flashSurf.fill((r, g, b, alpha))
230.                  DISPLAYSURF.blit(origSurf, (0, 0))
231.                  DISPLAYSURF.blit(flashSurf, (0, 0))
232.                  drawButtons()
233.                  pygame.display.update()
234.                  FPSCLOCK.tick(FPS)
235.
236.
237.
238. def getButtonClicked(x, y):
239.     if YELLOWRECT.collidepoint( (x, y) ):
240.         return YELLOW
241.     elif BLUERECT.collidepoint( (x, y) ):
242.         return BLUE
243.     elif REDRECT.collidepoint( (x, y) ):
244.         return RED
245.     elif GREENRECT.collidepoint( (x, y) ):
246.         return GREEN
247.     return None
248.
249.
250. if __name__ == '__main__':
251.     main()
```

5.3 常用初始内容

```
1. # Simulate (a Simon clone)
2. # By Al Sweigart al@inventwithpython.com
3. # http://inventwithpython.com/pygame
4. # Creative Commons BY-NC-SA 3.0 US
5.
6. import random, sys, time, pygame
7. from pygame.locals import *
8.
9. FPS = 30
10. WINDOWWIDTH = 640
11. WINDOWHEIGHT = 480
```

```
12. FLASHSPEED = 500 # in milliseconds
13. FLASHDELAY = 200 # in milliseconds
14. BUTTONSIZE = 200
15. BUTTONGAPSIZE = 20
16. TIMEOUT = 4 # seconds before game over if no button is pushed.
17.
18. #                    R    G    B
19. WHITE         = (255, 255, 255)
20. BLACK         = (  0,   0,   0)
21. BRIGHTRED     = (255,   0,   0)
22. RED           = (155,   0,   0)
23. BRIGHTGREEN   = (  0, 255,   0)
24. GREEN         = (  0, 155,   0)
25. BRIGHTBLUE    = (  0,   0, 255)
26. BLUE          = (  0,   0, 155)
27. BRIGHTYELLOW  = (255, 255,   0)
28. YELLOW        = (155, 155,   0)
29. DARKGRAY      = ( 40,  40,  40)
30. bgColor = BLACK
31.
32. XMARGIN = int((WINDOWWIDTH - (2 * BUTTONSIZE) - BUTTONGAPSIZE) / 2)
33. YMARGIN = int((WINDOWHEIGHT - (2 * BUTTONSIZE) - BUTTONGAPSIZE) / 2)
```

在这里，我们设置了一些常用的常量，可能随后我们想要修改这些常量，它们包括：4个按钮的大小、用于按钮的阴影的颜色（当按钮点亮的时候，我们使用明亮的颜色），以及在游戏超时之前玩家需要按下序列中的下一个按钮的时间量。

5.4 设置按钮

```
35. # Rect objects for each of the four buttons
36. YELLOWRECT = pygame.Rect(XMARGIN, YMARGIN, BUTTONSIZE, BUTTONSIZE)
37. BLUERECT   = pygame.Rect(XMARGIN + BUTTONSIZE + BUTTONGAPSIZE, YMARGIN, BUTTONSIZE, BUTTONSIZE)
38. REDRECT    = pygame.Rect(XMARGIN, YMARGIN + BUTTONSIZE + BUTTONGAPSIZE, BUTTONSIZE, BUTTONSIZE)
39. GREENRECT  = pygame.Rect(XMARGIN + BUTTONSIZE + BUTTONGAPSIZE, YMARGIN + BUTTONSIZE + BUTTONGAPSIZE, BUTTONSIZE, BUTTONSIZE)
```

就像Sliding Puzzle游戏中有针对"Reset"、"Solve"和"New Game"的按钮一样，Simulate游戏有4个矩形的区域，以及当用户点击这些区域的时候负责进行处理的代码。程序需要针对这4个按钮使用Rect对象，以便可以在其上调用collidepoint()方法。第36行到第39行使用相应的坐标和大小设置了这些Rect对象。

第 5 章　Simulate

5.5　main()函数

```
41. def main():
42.     global FPSCLOCK, DISPLAYSURF, BASICFONT, BEEP1, BEEP2, BEEP3, BEEP4
43.
44.     pygame.init()
45.     FPSCLOCK = pygame.time.Clock()
46.     DISPLAYSURF = pygame.display.set_mode((WINDOWWIDTH, WINDOWHEIGHT))
47.     pygame.display.set_caption('Simulate')
48.
49.     BASICFONT = pygame.font.Font('freesansbold.ttf', 16)
50.
51.     infoSurf = BASICFONT.render('Match the pattern by clicking on the
button or using the Q, W, A, S keys.', 1, DARKGRAY)
52.     infoRect = infoSurf.get_rect()
53.     infoRect.topleft = (10, WINDOWHEIGHT - 25)
54.     # load the sound files
55.     BEEP1 = pygame.mixer.Sound('beep1.ogg')
56.     BEEP2 = pygame.mixer.Sound('beep2.ogg')
57.     BEEP3 = pygame.mixer.Sound('beep3.ogg')
58.     BEEP4 = pygame.mixer.Sound('beep4.ogg')
```

　　main()实现了大部分的程序，并且在需要的时候调用其他的函数。其调用了 Pygame 的设置函数来初始化该库，创建了一个 Clock 对象，创建了一个窗口，设置了标题，并且创建了一个 Font 对象以用来在窗口上显示分数和指令。这些函数调用所创建的对象将存储在全局变量中，以便能够在其他的函数中使用它们。但是，它们的值不会改变，因此它们基本上是常量。

　　第 55 行到第 58 行加载了声音文件，以便当玩家点击每一个按钮的时候，Simulate 能够播放声音效果。pygame.mixer.Sound()构造函数将返回一个 Sound 对象，我们将其存储在 BEEP1 到 BEEP4 的变量之中，这些是在第 42 行所创建的全局变量。

5.6　程序中用到的一些局部变量

```
60.     # Initialize some variables for a new game
61.     pattern = [] # stores the pattern of colors
62.     currentStep = 0 # the color the player must push next
63.     lastClickTime = 0 # timestamp of the player's last button push
64.     score = 0
65.     # when False, the pattern is playing. when True, waiting for the
```

110

```
player to click a colored button:
66.            waitingForInput = False
```

Pattern 变量是颜色值（YELLOW、RED、BLUE 或 GREEN）的一个列表，这些颜色值记录了玩家必须记住的模式。例如，如果模式的值是[RED, RED, YELLOW, RED, BLUE, BLUE, RED, GREEN]，那么玩家必须首先点击红色的按钮两次，然后点击黄色按钮，再点击红色按钮，依次类推，直到最后点击绿色按钮。当玩家完成每一个回合，都会有一个新的随机颜色添加到这个列表的末尾。

currentStep 变量将记录玩家接下来必须点击模式列表中的哪一种颜色。如果 currentStep 是 0，并且 pattern 是[GREEN, RED, RED, YELLOW]，那么玩家必须点击绿色按钮。如果他们点击了任何其他的按钮，代码将会使得游戏结束。

TIMEOUT 变量使得玩家在一定的秒数之内点击模式中的下一个按钮，否则的话，代码将使得游戏结束。为了检查从上一次按钮点击开始是否经过了足够的时间，lastClickTime 变量需要记录玩家上一次点击按钮的时间（Python 有一个名为 time 的模块以及一个返回当前时间的 time.time()函数。稍后会介绍这些）。

Score 变量用来记录分数，这有点让人难以相信。同样难以置信的是该程序也有两种模式。其中一种是程序为玩家播放按钮顺序（在这种情况下，waitingForInput 设置为 False）。另一种模式是程序完成了按钮顺序的播放，等待玩家以正确的顺序点击按钮（在这种情况下，waitingForInput 设置为 True）。

5.7 绘制游戏板并处理输入

```
68.    while True: # main game loop
69.        clickedButton = None # button that was clicked (set to YELLOW,
RED, GREEN, or BLUE)
70.        DISPLAYSURF.fill(bgColor)
71.        drawButtons()
72.
73.        scoreSurf = BASICFONT.render('Score: ' + str(score), 1, WHITE)
74.        scoreRect = scoreSurf.get_rect()
75.        scoreRect.topleft = (WINDOWWIDTH - 100, 10)
76.        DISPLAYSURF.blit(scoreSurf, scoreRect)
77.
78.        DISPLAYSURF.blit(infoSurf, infoRect)
```

第 68 行是主游戏循环的开始。在每次迭代的开始，clickedButton 将重新设置为 None。如果在该迭代中点击了一个按钮，clickedButton 将设置为与该按钮匹配的一种颜色值（YELLOW、RED、GREEN 或 BLUE）。第 70 行调用了 fill()方法来重新绘制整个显示 Surface，

第 5 章　Simulate

以便我们能够从头开始绘制。通过调用 drawButtons()（稍后介绍）来绘制 4 个彩色按钮。第 73 行到第 76 行创建了表示分数的文本。

这里还是用文本来告诉玩家其当前的分数。和第 51 行调用 render()方法显示指示性文本不同，分数的文本是要修改的。它最开始为'Score: 0'，然后变成'Score: 1'，随后是'Score: 2'，依次类推。这就是为什么通过在游戏循环中第 73 行调用 render()方法来创建新的 Surface 对象的原因。由于指示性文本（"Match the pattern by…"）不会修改，我们只需要在游戏循环之外的第 50 行调用 render()一次就可以了。

5.8　检查鼠标点击

```
80.         checkForQuit()
81.         for event in pygame.event.get(): # event handling loop
82.             if event.type == MOUSEBUTTONUP:
83.                 mousex, mousey = event.pos
84.                 clickedButton = getButtonClicked(mousex, mousey)
```

第 80 行快速检查了 QUIT 事件，第 81 行是事件处理循环的开始。任何鼠标点击的 XY 坐标都将存储到 mousex 和 mousey 变量中。如果鼠标点击位于 4 个按钮中的某一个之上，getButtonClicked()函数将返回所点击的按钮的一个 Color 对象（否则的话，它返回 None）。

5.9　检查键盘按下

```
85.             elif event.type == KEYDOWN:
86.                 if event.key == K_q:
87.                     clickedButton = YELLOW
88.                 elif event.key == K_w:
89.                     clickedButton = BLUE
90.                 elif event.key == K_a:
91.                     clickedButton = RED
92.                 elif event.key == K_s:
93.                     clickedButton = GREEN
```

第 85 行到第 93 行检查任何的 KEYDOWN 事件（当用户按下键盘上的一个按键的时候，会创建该事件）。Q、W、A 和 S 键对应到各个按钮，因为它们在键盘上也构成一个方形排列。

Q 位于 4 个键盘按键的左上方，就像是黄色按钮位于屏幕的左上方一样，因此，我们认为按下 Q 键就等同于点击了黄色按钮。可以通过将 clickedButton 变量设置为常量变量 YELLOW 中的值，从而做到这一点。我们可以对其他 3 个按键做同样的事情。通过这种方

112

式，用户可以使用鼠标或者键盘来玩 Simulate。

5.10 游戏循环的两种状态

```
97.        if not waitingForInput:
98.            # play the pattern
99.            pygame.display.update()
100.           pygame.time.wait(1000)
101.           pattern.append(random.choice((YELLOW, BLUE, RED, GREEN)))
102.           for button in pattern:
103.               flashButtonAnimation(button)
104.               pygame.time.wait(FLASHDELAY)
105.           waitingForInput = True
```

　　该程序可以处于两种不同的模式（或状态）。当 waitingForInput 为 False 的时候，程序将会显示按钮模式的动画。当 waitingForInput 为 True 的时候，程序将会等待用户选择按钮。第 97 行到第 105 行将涵盖程序显示模式动画的情况。由于这是在游戏的开始处进行的，或者是在玩家完成一个模式的时候进行的，第 101 行将向模式列表添加一个随机的颜色，以使得该模式更长一步。第 102 行到第 104 行遍历了模式列表中的每一个值，并且调用 flashButtonAnimation() 将按钮点亮。当它点亮了模式列表中的所有按钮之后，程序将 waitingForInput 变量设置为 True。

5.11 搞清楚玩家是否按下了正确的按钮

```
106.        else:
107.            # wait for the player to enter buttons
108.            if clickedButton and clickedButton == pattern[currentStep]:
109.                # pushed the correct button
110.                flashButtonAnimation(clickedButton)
111.                currentStep += 1
112.                lastClickTime = time.time()
```

　　如果 waitingForInput 为 True，那么第 106 行的 else 语句将会执行。第 108 行检查玩家是否在游戏循环的迭代之中点击了一个按钮，以及这个按钮是否是正确的按钮。currentStep 变量记录了模式列表中玩家应该点击的下一个按钮的索引。

　　例如，如果模式设置为[YELLOW, RED, RED]，并且 currentStep 变量设置为 0（就像玩家初次启动游戏时候一样），那么，玩家需要点击的正确按钮是 pattern[0]（黄色按钮）。

　　如果玩家已经点击了正确的按钮，我们想要通过调用 flashButtonAnimation() 让玩家点

第 5 章 Simulate

击的按钮闪烁一下，然后，增加 currentStep 以进行下一步，将 lastClickTime 变量更新为当前时间（time.time()函数返回一个浮点数,表示从 1970 年 1 月 1 日开始的秒数,因此，我们可以使用它来记录时间)。

```
114.            if currentStep == len(pattern):
115.                # pushed the last button in the pattern
116.                changeBackgroundAnimation()
117.                score += 1
118.                waitingForInput = False
119.                currentStep = 0 # reset back to first step
```

第 114 行到第 199 行位于 else 语句中，该 else 语句从第 106 行开始。如果执行位于这条 else 语句中，我们知道玩家点击了正确的按钮。第 114 行通过检查 currentStep 中存储的整数是否等于模式列表中的值的数目，从而判断这是否是模式列表中的最后一个正确的按钮。

如果检查结果为 True，那么，我们通过调用 changeBackgroundAnimation()来修改背景颜色。这是让玩家知道他们已经输入了正确的模式的一种简单方法。分数增加，将 currentStep 重新设置为 0，将 waitingForInput 变量设置为 False，以便在游戏循环的下一次迭代的时候，代码会给模式列表添加一个新的 Color 值并且闪烁按钮。

```
121.            elif (clickedButton and clickedButton != pattern[currentStep])
or (currentStep != 0 and time.time() - TIMEOUT > lastClickTime):
```

如果玩家没有点击正确的按钮，第 121 行的 elif 语句将处理这种情况，要么玩家点击了错误的按钮，要么玩家等了太长时间而没有点击按钮。不管是哪一种情况，我们需要显示"游戏结束"动画，并且开始新的游戏。

这条 elif 语句条件检查的（clickedButton and clickedButton != pattern[currentStep]）部分，检查是否点击了一个按钮以及是否点击了错误的按钮。你可以将其与第 108 行的 if 语句的条件 clickedButton and clickedButton == pattern[currentStep]比较，如果玩家点击了按钮并且点击了正确的按钮的话，后面这个条件的结果为 True。

第 121 行的 elif 条件的另一个部分是（currentStep != 0 and time.time() - TIMEOUT > lastClickTime）。它负责确保玩家没有超时。注意，该部分的条件是由一个 and 关键字连接起来的两个表达式。这意味着，and 关键字两边的条件都需要为 True。

要达到"超时"，这必须不是玩家的第一次按钮点击。但是，一旦玩家开始点击按钮，他必须保持足够快速地点击按钮，直到完成了整个模式（或者点击了错误的模式按钮并导致游戏结束）。如果 currentStep != 0 为 True，那么，我们知道玩家已经开始按钮了。

5.12 新纪元时间

此外，为了实现超时，当前时间（由 time.time()返回）减去 4 秒后（因为 TIMEOUT 中存储的是 4），必须比点击按钮的最后时间要大（存储在 lastClickTime 中）。time.time() - TIMEOUT > lastClickTime 之所以有效，和新纪元时间的工作方式有关系。新纪元时间（Epoch time，也叫作 UNIX 新纪元时间）是从 1970 年 1 月 1 日开始经过的秒数。1970 年 1 月 1 日这个日期叫作 UNIX 新纪元（UNIX Epoch）。

例如，当从交互式 shell 运行 time.time()的时候（别忘了先导入 time 模块），如下所示：

```
>>> import time
>>> time.time()
1320460242.118
```

这个数字的含义是调用 time.time()的时刻从 1970 年 1 月 1 日午夜开始经过了 1320460242 秒的时候（这可以转换为 2011 年 11 月 4 日的晚上的 7:30:42。你可以通过 http://invpy.com/epochtime 了解如何将 UNIX 新纪元时间转换为常规的时间）。

如果在几秒钟之后从交互式 shell 调用 time.time()，如下所示：

```
>>> time.time()
1320460261.315
```

UNIX 新纪元时间之后的 1320460261.315 秒是 2011 年 11 月 4 日晚上的 7:31:01(实际上，如果要更精确的话，是 7:31 又 0.315 秒)。

如果必须处理字符串的话，那么处理时间的时候会很困难。如果只有字符串值'7:30:42 PM'和'7:31:01 PM'进行比较的话，很难分辨出已经经过了 19 秒。但是，有了新纪元时间，这只需要做数字的减法，1320460261.315 - 1320460242.118 会得到 19.197000026702881。这个值就是两个时间之间相差的秒数（剩下的 0.000026702881 是我们在做浮点数数学计算的时候遇到的很小的一个含入错误。你可以通过 http://invpy.com/roundingerrors 了解浮点含入错误的更多细节）。

回到第 121 行，如果 time.time() - TIMEOUT > lastClickTime 结果为 True，那么，从 time.time()调用后又经过 4 秒，并且这个时间存储到了 lastClickTime。如果该条件为假，那么，经过的时间还不到 4 秒。

```
122.            # pushed the incorrect button, or has timed out
123.            gameOverAnimation()
124.            # reset the variables for a new game:
125.            pattern = []
```

第 5 章　Simulate

```
126.            currentStep = 0
127.            waitingForInput = False
128.            score = 0
129.            pygame.time.wait(1000)
130.            changeBackgroundAnimation()
```

如果玩家点击了错误的按钮或者超时了，程序应该会播放"游戏结束"动画，然后为了新游戏重新设置变量。这涉及将 pattern 列表设置为一个空白列表，将 currentStep 设置为 0，将 waitingForInput 设置为 False，以及将 score 设置为 0。会暂停一小会儿，然后将设置一个新的背景颜色，以告诉玩家开始一次新的游戏，这将会开始游戏循环的下一次迭代。

5.13　将游戏板绘制到屏幕

```
132.        pygame.display.update()
133.        FPSCLOCK.tick(FPS)
```

就像其他的游戏程序一样，游戏循环中所做的最后一件事情是将显示 Surface 对象绘制到屏幕上并调用 tick()方法。

5.14　相同的旧的 terminate()函数

```
136. def terminate():
137.     pygame.quit()
138.     sys.exit()
139.
140.
141. def checkForQuit():
142.     for event in pygame.event.get(QUIT): # get all the QUIT events
143.         terminate() # terminate if any QUIT events are present
144.     for event in pygame.event.get(KEYUP): # get all the KEYUP events
145.         if event.key == K_ESCAPE:
146.             terminate() # terminate if the KEYUP event was for the Esc key
147.         pygame.event.post(event) # put the other KEYUP event objects back
```

terminate()和 checkForQuit()函数在第 4 章中用到过并介绍过，因此，我们在这里不再重复介绍。

5.15 复用常量变量

```
150. def flashButtonAnimation(color, animationSpeed=50):
151.     if color == YELLOW:
152.         sound = BEEP1
153.         flashColor = BRIGHTYELLOW
154.         rectangle = YELLOWRECT
155.     elif color == BLUE:
156.         sound = BEEP2
157.         flashColor = BRIGHTBLUE
158.         rectangle = BLUERECT
159.     elif color == RED:
160.         sound = BEEP3
161.         flashColor = BRIGHTRED
162.         rectangle = REDRECT
163.     elif color == GREEN:
164.         sound = BEEP4
165.         flashColor = BRIGHTGREEN
166.         rectangle = GREENRECT
```

根据传递给 color 参数的实参是哪一个 Color 值，闪烁时候的声音、闪烁颜色以及闪烁的矩形区域都将是不同的。第 151 行到第 166 行，根据 color 参数中的值，以不同方式设置了这 3 个局域变量：sound、flashColor 和 rectangle。

5.16 实现按钮闪烁动画

```
168.     origSurf = DISPLAYSURF.copy()
169.     flashSurf = pygame.Surface((BUTTONSIZE, BUTTONSIZE))
170.     flashSurf = flashSurf.convert_alpha()
171.     r, g, b = flashColor
172.     sound.play()
```

按钮闪烁动画的过程很简单：在动画的每一帧上，绘制常规的游戏板，然后在其上，将闪烁的明亮颜色的按钮版本绘制于按钮之上。从动画的第 1 帧开始，明亮颜色的 alpha 值为 0，随后，在每一帧之后 alpha 值都会缓慢地增加，直到其完全不透明并且明亮颜色的版本完全地绘制于常规按钮颜色之上。这会使得按钮看上去像是慢慢变亮的。

点亮按钮只是动画的前一半。后一半是让按钮变淡。这也通过相同的代码来完成，只不过 alpha 值不是在每一帧中增加，而是减少。随着 alpha 值变得越来越小，在顶部绘制的明亮颜色会变得越来越不可见，直到只能看到最初的、颜色暗淡游戏板。

第 5 章　Simulate

为了在代码中做到这一点，第 168 行创建了显示 Surface 对象的一个版本，并将其存储在 origSurf 中。第 169 行创建了一个新的 Surface 对象，其大小为单个按钮那么大，并且将其存储到 flashSurf 中。在 flashSurf 上调用 convert_alpha() 方法，以便该 Surface 对象能够具有一个绘制于其上的透明色（否则的话，将会忽略 Color 对象中的 alpha 值并自动假设其为 255）。在你自己的游戏程序中，如果在让颜色透明度生效的时候遇到麻烦，确保你想要使用透明颜色绘制的任何 Suface 对象上调用了 convert_alpha() 方法。

第 171 行创建了名为 r、g 和 b 的单个的局部变量，用来存储单个的 RGB 值，这些 RGB 值构成的元组，存储在 flashColor 中。这只是一些语法糖，使得该函数中的剩余的代码更容易阅读。在开始实现按钮闪烁动画之前，第 172 行将播放按钮的声音效果。在声音效果开始播放之后，程序执行继续进行，因此，在按钮闪烁动画的过程中，声音也将播放。

```
173.    for start, end, step in ((0, 255, 1), (255, 0, -1)): # animation loop
174.        for alpha in range(start, end, animationSpeed * step):
175.            checkForQuit()
176.            DISPLAYSURF.blit(origSurf, (0, 0))
177.            flashSurf.fill((r, g, b, alpha))
178.            DISPLAYSURF.blit(flashSurf, rectangle.topleft)
179.            pygame.display.update()
180.            FPSCLOCK.tick(FPS)
181.    DISPLAYSURF.blit(origSurf, (0, 0))
```

记住，要执行动画，我们首先要使用 alpha 值从 0 到 255 增加的颜色来绘制 flashSurf，以实现动画中逐渐增亮的部分。然后，执行变暗的过程，我们想要让 alpha 值从 255 到 0。可以使用如下的代码来做到这些。

```
for alpha in range(0, 255, animationSpeed): # brightening
    checkForQuit()
    DISPLAYSURF.blit(origSurf, (0, 0))
    flashSurf.fill((r, g, b, alpha))
    DISPLAYSURF.blit(flashSurf, rectangle.topleft)
    pygame.display.update()
    FPSCLOCK.tick(FPS)
for alpha in range(255, 0, -animationSpeed): # dimming
    checkForQuit()
    DISPLAYSURF.blit(origSurf, (0, 0))
    flashSurf.fill((r, g, b, alpha))
    DISPLAYSURF.blit(flashSurf, rectangle.topleft)
    pygame.display.update()
    FPSCLOCK.tick(FPS)
```

但是要注意，for 循环中的代码处理了帧的绘制，并且彼此是相同的。如果我们编写了类似上面的代码，那么，第一个 for 循环将处理动画的变亮部分（其中 alpha 值从 0 到 255），

5.16 实现按钮闪烁动画

第二个 for 循环将处理动画的变暗部分（其中 alpha 值从 255 到 0）。注意，对于第二个 for 循环，range()调用的第三个参数是一个负数。

无论何时，当我们拥有诸如这样的相同的代码的时候，可能能够缩短代码，而不必要重复它。这正是我们在第 173 行对该 for 循环所做的事情，在第 174 行，只是为 range()提供了不同的值。

```
173.    for start, end, step in ((0, 255, 1), (255, 0, -1)): # animation loop
174.        for alpha in range(start, end, animationSpeed * step):
```

第 173 行的 for 循环的第一次迭代中，start 设置为 0，end 设置为 255，而 step 设置为 1。通过这种方式，当第 174 行的 for 循环执行的时候，它将调用 range(0, 255, animationSpeed)（注意，animationSpeed * 1 和 animationSpeed 相同。将一个数字乘以 1，会得到相同的数字）。

然后，第 174 行的 for 循环执行，并且执行增亮的动画。

在第 173 行的 for 循环的第 2 次迭代中（内部的 for 循环总是有，并且仅有两次迭代），start 设置为 255，end 设置为 0，并且 step 设置为-1。当第 174 行的 for 循环执行的时候，它调用 range(255, 0, -animationSpeed)（注意，animationSpeed * -1 等于-animationSpeed，因为将一个数字乘以-1，会返回该数字的负数形式）。

通过这种方式，我们不必使用两个分开的 for 循环并重复其中的所有代码。这里再次给出第 174 行的 for 循环中的代码。

```
175.            checkForQuit()
176.            DISPLAYSURF.blit(origSurf, (0, 0))
177.            flashSurf.fill((r, g, b, alpha))
178.            DISPLAYSURF.blit(flashSurf, rectangle.topleft)
179.            pygame.display.update()
180.            FPSCLOCK.tick(FPS)
181.    DISPLAYSURF.blit(origSurf, (0, 0))
```

检查任何的 QUIT 事件（防止用户试图在动画过程中关闭程序），然后，将 origSurf Suface 复制到显示 Surface。然后，通过调用 fill()绘制 flashSurf Surface（提供在第 171 行所得到的颜色的 r、g、b 值以及该 for 循环在 alpha 变量中设置的 alpha 值）。然后，将 flashSurf Surface 复制到显示 Surface。

为了让显示 Surface 出现在屏幕上，在第 179 行调用 pygame.display.update()。为了确保动画不会像计算机绘制它那么快地播放，我们通过调用 tick()方法增加了短暂的暂停（如果想要看看闪烁动画很慢地播放的样子，用一个诸如 1 或 2 这样的较低的值作为 tick()的参数，而不是使用 FPS）。

第 5 章 Simulate

5.17 绘制按钮

```
184. def drawButtons():
185.     pygame.draw.rect(DISPLAYSURF, YELLOW, YELLOWRECT)
186.     pygame.draw.rect(DISPLAYSURF, BLUE,   BLUERECT)
187.     pygame.draw.rect(DISPLAYSURF, RED,    REDRECT)
188.     pygame.draw.rect(DISPLAYSURF, GREEN,  GREENRECT)
```

由于每一个按钮只是在特定位置具有特定颜色的一个矩形，我们只是调用 pygame.draw.rect()4 次，就可以将按钮绘制到显示 Surface 上。我们用来定位按钮的 Color 对象和 Rect 对象并没有变化，这就是为什么要将其存储在诸如 YELLOW 和 YELLOWRECT 这样的常量变量中的原因。

5.18 实现背景颜色改变的动画

```
191. def changeBackgroundAnimation(animationSpeed=40):
192.     global bgColor
193.     newBgColor = (random.randint(0, 255), random.randint(0, 255), random.randint(0, 255))
194.
195.     newBgSurf = pygame.Surface((WINDOWWIDTH, WINDOWHEIGHT))
196.     newBgSurf = newBgSurf.convert_alpha()
197.     r, g, b = newBgColor
198.     for alpha in range(0, 255, animationSpeed): # animation loop
199.         checkForQuit()
200.         DISPLAYSURF.fill(bgColor)
201.
202.         newBgSurf.fill((r, g, b, alpha))
203.         DISPLAYSURF.blit(newBgSurf, (0, 0))
204.
205.         drawButtons() # redraw the buttons on top of the tint
206.
207.         pygame.display.update()
208.         FPSCLOCK.tick(FPS)
209.     bgColor = newBgColor
```

只要玩家正确地完成了整个模式的输入，背景颜色改变动画就会播放。在第 198 行开始的这个循环的每一次迭代中，都会绘制整个显示 Surface（用一个较小的透明度的新的背景颜色来混合，直到该背景被新的颜色完全覆盖）。循环的每一次迭代所完成的步骤如下。

- 第 200 行使用旧的背景色（存储在 bgColor 中）填充了整个显示 Surface（存储在

- 第 202 行使用新的背景色的 RGB 值（以及在每次迭代中变化的 alpha 透明度值，第 198 行的 for 循环正是从改变 alpha 值开始的）来填充一个不同的 Surface 对象（存储在 newBgSurf 中）。
- 第 203 行将 newBgSurf Surface 绘制到 DISPLAYSURF 中的显示 Surface。没有在一开始的时候就将半透明的新的背景色绘制到 DISPLAYSURF 上，这是因为 fill() 方法只是替换了 Surface 上的颜色，而 blit() 方法将混合颜色。
- 现在，我们有了想要的背景方式，将通过在第 205 行调用一次 drawButtons()，在其上绘制按钮。
- 第 207 行和第 208 行只是将显示 Surface 绘制到屏幕上并添加一个暂停。

在 changeBackgroundAnimation() 函数的开始处有一条 global 语句，这是用于 bgColor 变量的，因为该函数使用第 209 行的一条赋值语句修改了该变量的内容。任何函数都能够读取没有使用 global 语句指定的一个全局变量的值。如果该函数给没有 globla 语句指定的一个全局变量赋值，Python 则会认为该变量是局部变量，而只不过刚好拥有和全局变量相同的名称而已。main() 函数使用了 bgColor 变量，但是并不需要对它使用一条 global 语句，因为它只是读取了 bgColor 的内容，而不会给 bgColor 赋一个新的值。http://invpy.com/global 更为详细地介绍了这一概念。

5.19 游戏结束动画

```
212. def gameOverAnimation(color=WHITE, animationSpeed=50):
213.     # play all beeps at once, then flash the background
214.     origSurf = DISPLAYSURF.copy()
215.     flashSurf = pygame.Surface(DISPLAYSURF.get_size())
216.     flashSurf = flashSurf.convert_alpha()
217.     BEEP1.play() # play all four beeps at the same time, roughly.
218.     BEEP2.play()
219.     BEEP3.play()
220.     BEEP4.play()
221.     r, g, b = color
222.     for i in range(3): # do the flash 3 times
```

下一行（第 223 行以下）的 for 循环的每一次迭代都将执行一次闪烁。要执行 3 次闪烁，将所有代码放到拥有 3 次迭代的一个 for 循环中。如果想要更多或者更少的闪烁，那么，修改在第 222 行传递给 range() 的整数就行了。

```
223.         for start, end, step in ((0, 255, 1), (255, 0, -1)):
```

第 5 章 Simulate

第 223 行的 for 循环和第 173 行的 for 循环完全相同。第 224 行的 for 循环,将使用 start、end 和 step 变量来控制如何改变 alpha 变量。如果你需要回忆这些循环是如何工作的,请再次阅读 5.16 小节。

```
224.            # The first iteration in this loop sets the following for loop
225.            # to go from 0 to 255, the second from 255 to 0.
226.            for alpha in range(start, end, animationSpeed * step): # animation loop
227.                # alpha means transparency. 255 is opaque, 0 is invisible
228.                checkForQuit()
229.                flashSurf.fill((r, g, b, alpha))
230.                DISPLAYSURF.blit(origSurf, (0, 0))
231.                DISPLAYSURF.blit(flashSurf, (0, 0))
232.                drawButtons()
233.                pygame.display.update()
234.                FPSCLOCK.tick(FPS)
```

这个动画循环和前面 5.16 小节中的闪烁动画代码工作方式相同。存储在 origSurf 中的最初的 Surface 对象的副本,绘制到了显示 Surface 上,然后将 flashSurf(新的闪烁颜色已经绘制在其上)复制到显示 Surface 的上面。在设置好背景颜色之后,在第 232 行,将按钮绘制于其上。最后,通过调用 pygame.display.update(),将显示 Surface 绘制到屏幕上。

第 226 行的 for 循环,为在动画中的每一帧所使用的颜色调整 alpha 值(先是增加,然后递减)。

5.20 将像素坐标转换为按钮

```
238. def getButtonClicked(x, y):
239.     if YELLOWRECT.collidepoint( (x, y) ):
240.         return YELLOW
241.     elif BLUERECT.collidepoint( (x, y) ):
242.         return BLUE
243.     elif REDRECT.collidepoint( (x, y) ):
244.         return RED
245.     elif GREENRECT.collidepoint( (x, y) ):
246.         return GREEN
247.     return None
248.
249.
250. if __name__ == '__main__':
251.     main()
```

getButtonClicked() 函数直接接受 XY 坐标,并且如果点击的是某一个按钮的话,它会返回 YELLOW、BLUE、RED 或 REEN，如果 XY 坐标不在 4 个按钮中的任何一个之上，返回 None。

5.21 显式比隐式好

你可能注意到了，getButtonClicked() 的代码最终在第 247 行带有一条 return None 的语句。将这条语句写出来似乎是一件奇怪的事情，因为所有的函数如果根本没有任何 return 语句的话，它们就会返回 None。我们即使完全漏掉第 247 行，程序还是会以完全相同的方式工作。那为什么要多此一举呢？

通常，当函数到达末尾并且隐式地返回 None（也就是说，没有一条 return 语句明确地表示它要返回 None），调用它的代码根本不会在意返回值。所有的函数调用必须返回一个值（以便它们能够得到某个结果并且成为表达式的一部分），但是，我们的代码并不总是用到返回值。例如，考虑一下 print() 函数。从技术上讲，这个函数返回 None 值，但是，我们并不在意它。

```
>>> spam = print('Hello')
Hello
>>> spam == None
True
>>>
```

然而，当 getButtonClicked() 返回 None，这意味着传递给它的坐标并不在 4 个按钮的任何一个之上。为了清楚地表达 getButtonClicked() 返回 None 这种情况，我们在该函数的末尾使用了 return None 这行代码。

为了让代码更具有可读性，让代码显式（也就是说，即便是显而易见的，也清楚地表明）比隐式（留待阅读代码的人自行搞清楚它是如何工作的，而不彻底地告诉他们）好。实际上，"显式总是比隐式好"，这是 Python 的信条之一。

这个信条是关于如何编写好的代码的一组名言。这是 Python 交互式 shell 中的一个复活节彩蛋，即一个隐藏起来的小小惊喜，如果你试图导入一个名为 this 的模块，就会显示出这组 "The Zen of Python" 信条。在交互式 shell 中尝试一下。

```
>>> import this
The Zen of Python, by Tim Peters

Beautiful is better than ugly.
Explicit is better than implicit.
Simple is better than complex.
Complex is better than complicated.
```

第 5 章 Simulate

```
Flat is better than nested.
Sparse is better than dense.
Readability counts.
Special cases aren't special enough to break the rules.
Although practicality beats purity.
Errors should never pass silently.
Unless explicitly silenced.
In the face of ambiguity, refuse the temptation to guess.
There should be one-- and preferably only one --obvious way to do it.
Although that way may not be obvious at first unless you're Dutch.
Now is better than never.
Although never is often better than *right* now.
If the implementation is hard to explain, it's a bad idea.
If the implementation is easy to explain, it may be a good idea.
Namespaces are one honking great idea -- let's do more of those!
```

如果你想要更详细地了解这些信条中的每一条的含义,请访问 http://invpy.com/zen。

第 6 章   Wormy

6.1   Wormy 游戏的玩法

Wormy 是 Nibbles 游戏的翻版。玩家一开始控制一条短短的、不断地在屏幕上移动的虫子。玩家无法让虫子停下来或者减慢速度,但是,他们可以控制虫子转向哪个方向,如图 6-1 所示。一个红色苹果会随机地出现在屏幕上,玩家必须移动虫子让其吃到苹果。每次虫子吃到一个苹果,虫子都会长长一段,并且一个新的苹果会随机地出现在屏幕上。如果虫子碰到自己或者碰到屏幕的边缘,游戏结束。

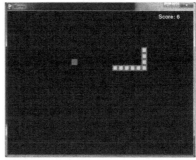

图 6-1

6.2   Wormy 的源代码

可以从 http://invpy.com/wormy.py 下载 Wormy 的源代码。如果得到任何的错误消息,查看一下错误消息中提到的代码行,并且检查代码是否有任何录入错误。可以将代码复制并粘贴到位于 http://invpy.com/diff/wormy 的 Web 表单中,看看你的代码和本书中的代码有何不同。

```
1. # Wormy (a Nibbles clone)
2. # By Al Sweigart al@inventwithpython.com
3. # http://inventwithpython.com/pygame
4. # Creative Commons BY-NC-SA 3.0 US
5.
6. import random, pygame, sys
```

第6章　Wormy

```
 7. from pygame.locals import *
 8.
 9. FPS = 15
10. WINDOWWIDTH = 640
11. WINDOWHEIGHT = 480
12. CELLSIZE = 20
13. assert WINDOWWIDTH % CELLSIZE == 0, "Window width must be a multiple of cell size."
14. assert WINDOWHEIGHT % CELLSIZE == 0, "Window height must be a multiple of cell size."
15. CELLWIDTH = int(WINDOWWIDTH / CELLSIZE)
16. CELLHEIGHT = int(WINDOWHEIGHT / CELLSIZE)
17.
18. #             R    G    B
19. WHITE     = (255, 255, 255)
20. BLACK     = (  0,   0,   0)
21. RED       = (255,   0,   0)
22. GREEN     = (  0, 255,   0)
23. DARKGREEN = (  0, 155,   0)
24. DARKGRAY  = ( 40,  40,  40)
25. BGCOLOR = BLACK
26.
27. UP = 'up'
28. DOWN = 'down'
29. LEFT = 'left'
30. RIGHT = 'right'
31.
32. HEAD = 0 # syntactic sugar: index of the worm's head
33.
34. def main():
35.     global FPSCLOCK, DISPLAYSURF, BASICFONT
36.
37.     pygame.init()
38.     FPSCLOCK = pygame.time.Clock()
39.     DISPLAYSURF = pygame.display.set_mode((WINDOWWIDTH, WINDOWHEIGHT))
40.     BASICFONT = pygame.font.Font('freesansbold.ttf', 18)
41.     pygame.display.set_caption('Wormy')
42.
43.     showStartScreen()
44.     while True:
45.         runGame()
46.         showGameOverScreen()
47.
48.
49. def runGame():
50.     # Set a random start point.
```

```
51.     startx = random.randint(5, CELLWIDTH - 6)
52.     starty = random.randint(5, CELLHEIGHT - 6)
53.     wormCoords = [{'x': startx,     'y': starty},
54.                   {'x': startx - 1, 'y': starty},
55.                   {'x': startx - 2, 'y': starty}]
56.     direction = RIGHT
57.
58.     # Start the apple in a random place.
59.     apple = getRandomLocation()
60.
61.     while True: # main game loop
62.         for event in pygame.event.get(): # event handling loop
63.             if event.type == QUIT:
64.                 terminate()
65.             elif event.type == KEYDOWN:
66.                 if (event.key == K_LEFT or event.key == K_a) and direction != RIGHT:
67.                     direction = LEFT
68.                 elif (event.key == K_RIGHT or event.key == K_d) and direction != LEFT:
69.                     direction = RIGHT
70.                 elif (event.key == K_UP or event.key == K_w) and direction != DOWN:
71.                     direction = UP
72.                 elif (event.key == K_DOWN or event.key == K_s) and direction != UP:
73.                     direction = DOWN
74.                 elif event.key == K_ESCAPE:
75.                     terminate()
76.
77.         # check if the worm has hit itself or the edge
78.         if wormCoords[HEAD]['x'] == -1 or wormCoords[HEAD]['x'] == CELLWIDTH or wormCoords[HEAD]['y'] == -1 or wormCoords[HEAD]['y'] == CELLHEIGHT:
79.             return # game over
80.         for wormBody in wormCoords[1:]:
81.             if wormBody['x'] == wormCoords[HEAD]['x'] and wormBody['y'] == wormCoords[HEAD]['y']:
82.                 return # game over
83.
84.         # check if worm has eaten an apply
85.         if wormCoords[HEAD]['x'] == apple['x'] and wormCoords[HEAD]['y'] == apple['y']:
86.             # don't remove worm's tail segment
87.             apple = getRandomLocation() # set a new apple somewhere
```

第6章 Wormy

```
88.         else:
89.             del wormCoords[-1] # remove worm's tail segment
90.
91.         # move the worm by adding a segment in the direction it is moving
92.         if direction == UP:
93.             newHead = {'x': wormCoords[HEAD]['x'], 'y': wormCoords[HEAD]['y'] - 1}
94.         elif direction == DOWN:
95.             newHead = {'x': wormCoords[HEAD]['x'], 'y': wormCoords[HEAD]['y'] + 1}
96.         elif direction == LEFT:
97.             newHead = {'x': wormCoords[HEAD]['x'] - 1, 'y': wormCoords[HEAD]['y']}
98.         elif direction == RIGHT:
99.             newHead = {'x': wormCoords[HEAD]['x'] + 1, 'y': wormCoords[HEAD]['y']}
100.        wormCoords.insert(0, newHead)
101.        DISPLAYSURF.fill(BGCOLOR)
102.        drawGrid()
103.        drawWorm(wormCoords)
104.        drawApple(apple)
105.        drawScore(len(wormCoords) - 3)
106.        pygame.display.update()
107.        FPSCLOCK.tick(FPS)
108.
109. def drawPressKeyMsg():
110.     pressKeySurf = BASICFONT.render('Press a key to play.', True, DARKGRAY)
111.     pressKeyRect = pressKeySurf.get_rect()
112.     pressKeyRect.topleft = (WINDOWWIDTH - 200, WINDOWHEIGHT - 30)
113.     DISPLAYSURF.blit(pressKeySurf, pressKeyRect)
114.
115.
116. def checkForKeyPress():
117.     if len(pygame.event.get(QUIT)) > 0:
118.         terminate()
119.
120.     keyUpEvents = pygame.event.get(KEYUP)
121.     if len(keyUpEvents) == 0:
122.         return None
123.     if keyUpEvents[0].key == K_ESCAPE:
124.         terminate()
125.     return keyUpEvents[0].key
126.
127.
128. def showStartScreen():
```

```
129.        titleFont = pygame.font.Font('freesansbold.ttf', 100)
130.        titleSurf1 = titleFont.render('Wormy!', True, WHITE, DARKGREEN)
131.        titleSurf2 = titleFont.render('Wormy!', True, GREEN)
132.
133.        degrees1 = 0
134.        degrees2 = 0
135.        while True:
136.            DISPLAYSURF.fill(BGCOLOR)
137.            rotatedSurf1 = pygame.transform.rotate(titleSurf1, degrees1)
138.            rotatedRect1 = rotatedSurf1.get_rect()
139.            rotatedRect1.center = (WINDOWWIDTH / 2, WINDOWHEIGHT / 2)
140.            DISPLAYSURF.blit(rotatedSurf1, rotatedRect1)
141.
142.            rotatedSurf2 = pygame.transform.rotate(titleSurf2, degrees2)
143.            rotatedRect2 = rotatedSurf2.get_rect()
144.            rotatedRect2.center = (WINDOWWIDTH / 2, WINDOWHEIGHT / 2)
145.            DISPLAYSURF.blit(rotatedSurf2, rotatedRect2)
146.
147.            drawPressKeyMsg()
148.
149.            if checkForKeyPress():
150.                pygame.event.get() # clear event queue
151.                return
152.            pygame.display.update()
153.            FPSCLOCK.tick(FPS)
154.            degrees1 += 3 # rotate by 3 degrees each frame
155.            degrees2 += 7 # rotate by 7 degrees each frame
156.
157.
158. def terminate():
159.     pygame.quit()
160.     sys.exit()
161.
162.
163. def getRandomLocation():
164.     return {'x': random.randint(0, CELLWIDTH - 1), 'y': random.randint(0, CELLHEIGHT - 1)}
165.
166.
167. def showGameOverScreen():
168.     gameOverFont = pygame.font.Font('freesansbold.ttf', 150)
169.     gameSurf = gameOverFont.render('Game', True, WHITE)
170.     overSurf = gameOverFont.render('Over', True, WHITE)
171.     gameRect = gameSurf.get_rect()
172.     overRect = overSurf.get_rect()
173.     gameRect.midtop = (WINDOWWIDTH / 2, 10)
```

第6章　Wormy

```
174.        overRect.midtop = (WINDOWWIDTH / 2, gameRect.height + 10 + 25)
175.
176.        DISPLAYSURF.blit(gameSurf, gameRect)
177.        DISPLAYSURF.blit(overSurf, overRect)
178.        drawPressKeyMsg()
179.        pygame.display.update()
180.        pygame.time.wait(500)
181.        checkForKeyPress() # clear out any key presses in the event queue
182.
183.        while True:
184.            if checkForKeyPress():
185.                pygame.event.get() # clear event queue
186.                return
187.
188.    def drawScore(score):
189.        scoreSurf = BASICFONT.render('Score: %s' % (score), True, WHITE)
190.        scoreRect = scoreSurf.get_rect()
191.        scoreRect.topleft = (WINDOWWIDTH - 120, 10)
192.        DISPLAYSURF.blit(scoreSurf, scoreRect)
193.
194.
195.    def drawWorm(wormCoords):
196.        for coord in wormCoords:
197.            x = coord['x'] * CELLSIZE
198.            y = coord['y'] * CELLSIZE
199.            wormSegmentRect = pygame.Rect(x, y, CELLSIZE, CELLSIZE)
200.            pygame.draw.rect(DISPLAYSURF, DARKGREEN, wormSegmentRect)
201.            wormInnerSegmentRect = pygame.Rect(x + 4, y + 4, CELLSIZE - 8, CELLSIZE - 8)
202.            pygame.draw.rect(DISPLAYSURF, GREEN, wormInnerSegmentRect)
203.
204.
205.    def drawApple(coord):
206.        x = coord['x'] * CELLSIZE
207.        y = coord['y'] * CELLSIZE
208.        appleRect = pygame.Rect(x, y, CELLSIZE, CELLSIZE)
209.        pygame.draw.rect(DISPLAYSURF, RED, appleRect)
210.
211.
212.    def drawGrid():
213.        for x in range(0, WINDOWWIDTH, CELLSIZE): # draw vertical lines
214.            pygame.draw.line(DISPLAYSURF, DARKGRAY, (x, 0), (x, WINDOWHEIGHT))
215.        for y in range(0, WINDOWHEIGHT, CELLSIZE): # draw horizontal lines
216.            pygame.draw.line(DISPLAYSURF, DARKGRAY, (0, y), (WINDOWWIDTH, y))
217.
218.
219.    if __name__ == '__main__':
220.        main()
```

6.3 栅格

如果你玩一会儿游戏就会注意到，苹果和虫子身体的各个段总是会刚好和线条组成的栅格对齐，如图 6-2 所示。我们把这个栅格中的每一个方块叫作一个单元格（并不是所有的栅格中都这么称呼这一空格，这只是我所使用的一个名字）。单元格有它们自己的笛卡尔坐标系统，(0, 0)是左上角的单元格，而(31, 23)是右小角的单元格。

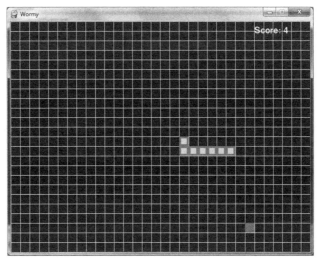

图 6-2

6.4 设置代码

```
 1. # Wormy (a Nibbles clone)
 2. # By Al Sweigart al@inventwithpython.com
 3. # http://inventwithpython.com/pygame
 4. # Creative Commons BY-NC-SA 3.0 US
 5.
 6. import random, pygame, sys
 7. from pygame.locals import *
 8.
 9. FPS = 15
10. WINDOWWIDTH = 640
11. WINDOWHEIGHT = 480
12. CELLSIZE = 20
```

第 6 章　Wormy

```
13. assert WINDOWWIDTH % CELLSIZE == 0, "Window width must be a multiple of
cell size."
14. assert WINDOWHEIGHT % CELLSIZE == 0, "Window height must be a multiple of
cell size."
15. CELLWIDTH = int(WINDOWWIDTH / CELLSIZE)
16. CELLHEIGHT = int(WINDOWHEIGHT / CELLSIZE)
```

在程序开始处的代码只是设置了游戏中用到的一些常量。单元格的宽度和高度存储在 CELLSIZE 中。第 13 行和第 14 行的 assert 语句保证了单元格和窗口很好地匹配。例如，如果 CELLSIZE 是 10，而 WINDOWWIDTH 或 WINDOWHEIGHT 常量设置为 15，那么，只有 1.5 个单元格才能匹配。assert 语句确保了窗口中的单元格的数目刚好是整数。

```
18. #              R    G    B
19. WHITE     = (255, 255, 255)
20. BLACK     = (  0,   0,   0)
21. RED       = (255,   0,   0)
22. GREEN     = (  0, 255,   0)
23. DARKGREEN = (  0, 155,   0)
24. DARKGRAY  = ( 40,  40,  40)
25. BGCOLOR = BLACK
26.
27. UP = 'up'
28. DOWN = 'down'
29. LEFT = 'left'
30. RIGHT = 'right'
31.
32. HEAD = 0 # syntactic sugar: index of the worm's head
```

第 19 行到第 32 行设置了另一些常量。HEAD 常量将在本章后面介绍。

6.5　main()函数

```
34. def main():
35.     global FPSCLOCK, DISPLAYSURF, BASICFONT
36.
37.     pygame.init()
38.     FPSCLOCK = pygame.time.Clock()
39.     DISPLAYSURF = pygame.display.set_mode((WINDOWWIDTH, WINDOWHEIGHT))
40.     BASICFONT = pygame.font.Font('freesansbold.ttf', 18)
41.     pygame.display.set_caption('Wormy')
42.
43.     showStartScreen()
44.     while True:
```

```
45.        runGame()
46.        showGameOverScreen()
```

在Wormy游戏程序中，我们已经把代码的主要部分放到了一个名为runGame()的函数中。这是因为，我们只想要在程序开始的时候（通过调用showStartScreen()函数）显示"初始屏幕"（带有旋转的"Wormy"文本的动画）一次。然后，我们想要调用runGame()，它将启动Wormy游戏。当玩家的虫子碰到了一面墙或者碰到自己从而导致游戏结束的时候，runGame()函数将会返回。此时，我们将通过调用showGameOverScreen()在屏幕上显示游戏结束。当showGameOverScreen()函数调用返回的时候，循环回到了起始处并再次调用runGame()。第44行的while循环将不断地进行，直到程序终止。

6.6 单独的runGame()函数

```
49. def runGame():
50.     # Set a random start point.
51.     startx = random.randint(5, CELLWIDTH - 6)
52.     starty = random.randint(5, CELLHEIGHT - 6)
53.     wormCoords = [{'x': startx,     'y': starty},
54.                   {'x': startx - 1, 'y': starty},
55.                   {'x': startx - 2, 'y': starty}]
56.     direction = RIGHT
57.
58.     # Start the apple in a random place.
59.     apple = getRandomLocation()
```

在游戏开始处，我们想要让虫子从一个随机的位置开始（但是不要太接近游戏板的边缘），因此，我们在startx和starty中存储了一个随机的坐标（记住，CELLWIDTH和CELLHEIGHT是窗口宽度和高度所能容纳的单元格数，而不是宽度和高度的像素数）。

虫子的身体存储在字典值的一个列表中。虫子的每一个身体段，都有一个字典值。该字典将拥有键'x'和'y'，表示身体段的XY坐标。身体的头部位于startx和starty。另外两个身体段分别位于头部左边的1个和2个单元格处。

虫子的头部总是位于wormCoords[0]的身体部分。为了让这段代码更具可读性，我们在第32行将HEAD常量设置为0，以便能够使用wormCoords[HEAD]，而不是wormCoords[0]。

第 6 章　Wormy

6.7　事件处理循环

```
61.     while True: # main game loop
62.         for event in pygame.event.get(): # event handling loop
63.             if event.type == QUIT:
64.                 terminate()
65.             elif event.type == KEYDOWN:
66.                 if (event.key == K_LEFT or event.key == K_a) and direction != RIGHT:
67.                     direction = LEFT
68.                 elif (event.key == K_RIGHT or event.key == K_d) and direction != LEFT:
69.                     direction = RIGHT
70.                 elif (event.key == K_UP or event.key == K_w) and direction != DOWN:
71.                     direction = UP
72.                 elif (event.key == K_DOWN or event.key == K_s) and direction != UP:
73.                     direction = DOWN
74.                 elif event.key == K_ESCAPE:
75.                     terminate()
```

第 61 行开始了主游戏循环，第 62 行开始了事件处理循环。如果事件是 QUIT 事件，那么，我们调用 terminate()（其定义和前面的游戏程序中的 terminate() 函数相同）。

否则，如果事件是一个 KEYDOWN 事件，那么，我们检查按下的键是否是一个箭头键或 WASD 键。我们还想要进行一次额外的检查，以保证虫子不会转身碰到自己。例如，如果虫子向左移动，那么，如果玩家偶然按下了向右键头键，虫子将会立即开始向右移动而碰到自己。这就是为什么要检查 direction 变量的当前值的原因。通过这种方式，如果玩家偶然按下了一个箭头键并可能会导致虫子立即碰到自己，我们可以直接忽略该按键事件。

6.8　碰撞检测

```
77.     # check if the worm has hit itself or the edge
78.     if wormCoords[HEAD]['x'] == -1 or wormCoords[HEAD]['x'] == CELLWIDTH or wormCoords[HEAD]['y'] == -1 or wormCoords[HEAD]['y'] == CELLHEIGHT:
79.         return # game over
80.     for wormBody in wormCoords[1:]:
81.         if wormBody['x'] == wormCoords[HEAD]['x'] and wormBody['y'] == wormCoords[HEAD]['y']:
82.             return # game over
```

当虫子的头部移动到栅格的边缘之外,或者当其头部移动到已经由身体的另一段所占据的单元格的时候,虫子就会被碰死。

我们可以通过查看头部的 X 坐标(存储在 wormCoords[HEAD]['x'])是否是-1(这表示超过了栅格的左边缘)或等于 CELLWIDTH(这表示超过了右边缘,因为最右边的单元格的 X 坐标比 CELLWIDTH 小 1),从而检测头部是否超出了单元格的边界。

如果头部的 Y 坐标(存储在 wormCoords[HEAD]['y'])为-1(这表示超过了栅格的上边缘)或 CELLHEIGHT(这表示超过了下边缘),那么,头部也超出了栅格。

在当前游戏的末尾,我们所要做的只是从 runGame()返回。当 runGame()返回到 main()中的函数调用处,runGame()调用之后的一行(第 46 行)是调用 showGameOver Screen(),它使得一个较大的"Game Over"文本显示出来。这就是为什么在第 79 行使用 return 语句的原因。

第 80 行遍历了 wormCoords 中头部(其索引为 0)以后的每一个身体段(这就是为什么 for 循环遍历了 wormCoords[1:],而不只是 wormCoords)。如果身体段的'x'和'y'值都和头部的'x'和'y'值相同,那么,我们也通过从 runGame()函数返回而结束游戏。

6.9 检测和苹果的碰撞

```
84.        # check if worm has eaten an apply
85.        if wormCoords[HEAD]['x'] == apple['x'] and wormCoords[HEAD]['y'] == apple['y']:
86.            # don't remove worm's tail segment
87.            apple = getRandomLocation() # set a new apple somewhere
88.        else:
89.            del wormCoords[-1] # remove worm's tail segment
```

我们在虫子的头部和苹果的 XY 坐标之间做类似的碰撞检测。如果它们匹配,我们将苹果的坐标设置为一个随机的新的位置(这个位置通过 getRandomLocation()的返回值而获得)。

如果头部没有和苹果碰撞,那么,我们删除掉 wormCoords 列表中最后的身体段。记住,负整数表示索引是从列表的尾部开始计算的。因此,0 是列表中第一项的索引,1 是第二项的索引,而-1 则是列表中的最后一项,-2 是列表中的倒数第二项。

第 91 行到第 100 行的代码(将在 6.10 节中介绍)将把一个新的身体段(作为头部)添加到虫子前进的方向上。这将使得虫子长长了一个身体段。当虫子吃了一个苹果的时候,我们并没有删除最后的身体段,虫子整体的长度增加 1。但是,当第 89 行删除了最后的身体段时,由于随后就添加了一个新的头部段,虫子大小仍然保持相同。

6.10 移动虫子

```
91.        # move the worm by adding a segment in the direction it is moving
92.        if direction == UP:
93.            newHead = {'x': wormCoords[HEAD]['x'], 'y': wormCoords[HEAD]['y'] - 1}
94.        elif direction == DOWN:
95.            newHead = {'x': wormCoords[HEAD]['x'], 'y': wormCoords[HEAD]['y'] + 1}
96.        elif direction == LEFT:
97.            newHead = {'x': wormCoords[HEAD]['x'] - 1, 'y': wormCoords[HEAD]['y']}
98.        elif direction == RIGHT:
99.            newHead = {'x': wormCoords[HEAD]['x'] + 1, 'y': wormCoords[HEAD]['y']}
100.       wormCoords.insert(0, newHead)
```

要移动虫子，因此向 wormCoords 列表的开始处添加了一个新的身体段。由于这个身体段添加到了列表的开始，它将变成新的头部。新的头部的坐标刚好挨着旧的头部的坐标。是 X 坐标，还是 Y 坐标，是添加 1，还是减去 1，这都取决于虫子移动的方向。

在第 100 行，新的头部段通过 insert() 列表方法添加到 wormCoords。

6.11 insert()列表方法

和只能够向列表末尾添加项的 append() 列表方法不同，insert() 列表方法可以向列表中的任何地方添加项。insert() 的第一个参数是项应该位于何处的一个索引（最初位于该索引的项以及在其之后所有的项，其索引都增加 1）。如果传递给第一个参数的值比列表的长度还要大，该项直接添加到列表的末尾（就像 append() 的做法一样）。insert() 的第二个参数是要添加的项的值。在交互式 shell 中输入如下的内容，看看 insert() 是如何工作的。

```
>>> spam = ['cat', 'dog', 'bat']
>>> spam.insert(0, 'frog')
>>> spam
['frog', 'cat', 'dog', 'bat']
>>> spam.insert(10, 42)
>>> spam
['frog', 'cat', 'dog', 'bat', 42]
>>> spam.insert(2, 'horse')
>>> spam
['frog', 'cat', 'horse', 'dog', 'bat', 42]
>>>
```

6.12 绘制屏幕

```
101.        DISPLAYSURF.fill(BGCOLOR)
102.        drawGrid()
103.        drawWorm(wormCoords)
104.        drawApple(apple)
105.        drawScore(len(wormCoords) - 3)
106.        pygame.display.update()
107.        FPSCLOCK.tick(FPS)
```

runGame()函数中绘制屏幕的代码相当简单。第 101 行使用背景颜色填充了整个显示 Surface。第 102 行到第 105 行将栅格、虫子、苹果以及分数绘制到显示 Surface。然后，调用 pygame.display.update()把显示 Surface 绘制到实际的计算机屏幕上。

6.13 在屏幕上绘制"Press a key"文本

```
109. def drawPressKeyMsg():
110.     pressKeySurf = BASICFONT.render('Press a key to play.', True, DARKGRAY)
111.     pressKeyRect = pressKeySurf.get_rect()
112.     pressKeyRect.topleft = (WINDOWWIDTH - 200, WINDOWHEIGHT - 30)
113.     DISPLAYSURF.blit(pressKeySurf, pressKeyRect)
```

尽管初始屏幕动画在播放，或者游戏结束屏幕在显示，但是在屏幕右下角还是有一些较小的文本显示"Press a key to play"。我们将这些文本显示放在一个单独的函数中，并且直接通过 showStartScreen() 和 showGameOverScreen() 调用该函数，而不是在 showStartScreen()和 showGameOverScreen()中输入显示文本的代码。

6.14 checkForKeyPress()函数

```
116. def checkForKeyPress():
117.     if len(pygame.event.get(QUIT)) > 0:
118.         terminate()
119.
120.     keyUpEvents = pygame.event.get(KEYUP)
121.     if len(keyUpEvents) == 0:
122.         return None
```

第 6 章 Wormy

```
123.        if keyUpEvents[0].key == K_ESCAPE:
124.            terminate()
125.        return keyUpEvents[0].key
```

这个函数首先检查事件队列中是否有任何 QUIT 事件。第 117 行对 pygame.event.get() 的调用会返回事件队列中的所有 QUIT 事件的一个列表（因为我们传递了 QUIT 作为参数）。如果事件队列中没有 QUIT 事件，那么 pygame.event.get() 返回的列表将是一个空列表：[]。

如果 pygame.event.get() 返回一个空列表，第 117 行对 len() 的调用将返回 0。如果 pygame.event.get() 返回的列表中有 0 个以上的项（并且记住，该列表中任何的项将只能是 QUIT 事件，因为我们传递了 QUIT 作为 pygame.event.get() 的参数），那么在第 118 行将会调用 terminate() 函数，并且程序终止。

此后，对 pygame.event.get() 的调用将会得到事件队列中的任何 KEYUP 事件的一个列表。如果这个按键事件是针对 Esc 键的，那么，程序在此情况下也会终止。否则，pygame.event.get() 所返回的列表中的第一个按键事件对象，将由这个 checkFor KeyPress() 函数返回。

6.15 初始屏幕

```
128. def showStartScreen():
129.     titleFont = pygame.font.Font('freesansbold.ttf', 100)
130.     titleSurf1 = titleFont.render('Wormy!', True, WHITE, DARKGREEN)
131.     titleSurf2 = titleFont.render('Wormy!', True, GREEN)
132.
133.     degrees1 = 0
134.     degrees2 = 0
135.     while True:
136.         DISPLAYSURF.fill(BGCOLOR)
```

当 Wormy 游戏程序第一次开始运行的时候，玩家不会自动开始玩游戏。相反，会出现一个初始屏幕，告诉玩家他们在运行什么程序。初始屏幕还给玩家一次机会为开始玩游戏而做好准备（否则，玩家可能因为没有准备而在第一次游戏的时候就把虫子给碰死了）。

Wormy 初始屏幕需要两个 Surface 对象，"Wormy" 文本绘制在它们之上。这就是第 130 行和第 131 行的 render() 方法调用所创建的内容。文本将会比较大，第 129 行的 Font() 构造函数创建了一个 Font 对象，它的大小为 100 点。第一个"Wormy"文本将会是白色的字体显示在暗绿色的背景之上，另一个将会是绿色文本带有透明的背景。

第 135 行开始了初始屏幕的动画循环。在这个动画中，两段文本将会旋转，并且绘制到显示 Surface 对象上。

6.16 旋转初始屏幕文本

```
137.        rotatedSurf1 = pygame.transform.rotate(titleSurf1, degrees1)
138.        rotatedRect1 = rotatedSurf1.get_rect()
139.        rotatedRect1.center = (WINDOWWIDTH / 2, WINDOWHEIGHT / 2)
140.        DISPLAYSURF.blit(rotatedSurf1, rotatedRect1)
141.
142.        rotatedSurf2 = pygame.transform.rotate(titleSurf2, degrees2)
143.        rotatedRect2 = rotatedSurf2.get_rect()
144.        rotatedRect2.center = (WINDOWWIDTH / 2, WINDOWHEIGHT / 2)
145.        DISPLAYSURF.blit(rotatedSurf2, rotatedRect2)
146.
147.        drawPressKeyMsg()
148.
149.        if checkForKeyPress():
150.            pygame.event.get() # clear event queue
151.            return
152.        pygame.display.update()
153.        FPSCLOCK.tick(FPS)
```

showStartScreen()函数将旋转 Surface 对象上的图像,而"Wormy!"文本正是写在该 Surface 对象上的。第一个参数是要制作其旋转的副本的 Surface 对象。第二个参数是旋转 Surface 的角度数。pygame.transform.rotate()并不会改变传递给它的 Surface 对象,而是返回一个新的 Surface 对象,旋转后的图形绘制于新的 Surface 对象之上。

注意,新的 Surface 对象可能会比最初的那个对象要大,因为所有的 Surface 对象表示一个矩形区域,而旋转的 Surface 对象的角就可能会超出之前最初的 Surface 对象的宽度和高度。图 6-3 所示是一个黑色的矩形,以及其自身的一个略微旋转的版本。为了让一个 Surface 对象能够位于旋转的矩形之中

图 6-3

(图 6-3 中用灰色表示的矩形),它必须要比最初的黑色矩形的 Surface 对象大。

旋转的程度用角度给出,这是度量旋转的方法。一个圆有 360 度。根本不旋转是 0 度。沿着逆时针方向旋转四分之一个圆是 90 度。要顺时针方向旋转,传递一个负的整数。旋转 360 度,就是将图像完全旋转一圈,这意味着最终得到的图像和旋转 0 度的图像是相同的。实际上,如果你传递给 pygame.transform.rotate()的旋转参数是 360 或者更大,Pygame 会自动地用它减去 360 直到得到一个小于 360 的数字。图 6-4 展示了几个不同旋转度数的例子。

第 6 章　Wormy

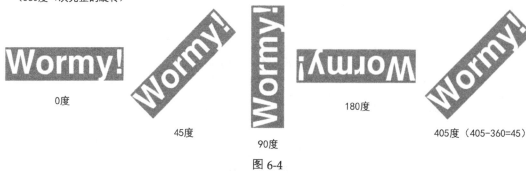

图 6-4

在第 140 行到第 145 行，在动画循环的每一帧上，两次旋转后的"Wormy!" Surface 对象复制到了显示 Surface 上。

第 147 行的 drawPressKeyMsg()函数调用，将"Press a key to play"文本绘制到了显示 Surface 对象的右下角。这个动画循环将保持循环，直到 checkForKeyPress()返回一个非 None 的值，如果玩家按下一个键的话，就会发生这种情况。在返回之前，调用 pygame.event.get()直接清除在初始屏幕显示的过程中累积于事件队列中的任何其他的事件。

6.17　旋转并不完美

你可能会问，为什么将旋转的 Surface 存储在一个单独的变量中，而不是直接覆盖 titleSurf1 和 titleSurf2 变量？原因有两点。

首先，旋转一幅 2D 图像并不完全完美。旋转图像总是近似的。如果你将一幅图像逆时针方向旋转 10 度，然后按照顺时针方向旋转回去 10 度，图像和开始的时候并不完全一致。可以这样考虑，做一次影印，拿着第一次的影印件再做一次影印，然后再做另外一次影印。如果持续这么做，图像会变得越来越糟糕，因为细小的扭曲会叠加在一起。

这一点的唯一例外情况是如果将图像旋转 90 度的倍数，例如，0 度、90 度、180 度、270 度或 360 度，在这种情况下，像素会旋转而不会有任何扭曲。

其次，如果你旋转一幅 2D 图像，那么，旋转后的图像会比最初的图像略微大一点。如果你旋转了已经旋转后的图像，那么，下一次旋转后的图像还将会略微大一点。如果你持续这么做，最终图像将会变得太大，以至于 Pygame 无法处理它，程序将会崩溃并得到一条错误消息：pygame.error: Width or height is too large。

```
154.        degrees1 += 3 # rotate by 3 degrees each frame
155.        degrees2 += 7 # rotate by 7 degrees each frame
```

我们旋转两个"Wormy!"文本 Surface 对象的量，分别存储在 degrees1 和 degrees2 中。

在动画循环的每一次迭代中，我们都将 degree1 中的数字增加 3，而将 degree2 中的数字增加 7。这意味着，在动画循环的下一次迭代中，白色"Wormy!"文本的 Surface 对象将会多旋转 3 度，而绿色"Wormy!"文本的 Surface 对象将会多旋转 7 度。这就是为什么一个 Surface 对象比另外一个旋转得慢的原因。

```
158. def terminate():
159.     pygame.quit()
160.     sys.exit()
```

terminate()函数调用 pygame.quit()和 sys.exit()，以便游戏能够正确地退出。这和之前的游戏程序中的 terminate()函数相同。

6.18 决定苹果出现在哪里

```
163. def getRandomLocation():
164.     return {'x': random.randint(0, CELLWIDTH - 1), 'y': random.randint(0, CELLHEIGHT - 1)}
```

当需要苹果的新的坐标的时候，调用 getRandomLocation()函数。该函数返回一个字典，它带有键'x'和'y'，其值用来设置随机的 XY 坐标。

6.19 游戏结束屏幕

```
167. def showGameOverScreen():
168.     gameOverFont = pygame.font.Font('freesansbold.ttf', 150)
169.     gameSurf = gameOverFont.render('Game', True, WHITE)
170.     overSurf = gameOverFont.render('Over', True, WHITE)
171.     gameRect = gameSurf.get_rect()
172.     overRect = overSurf.get_rect()
173.     gameRect.midtop = (WINDOWWIDTH / 2, 10)
174.     overRect.midtop = (WINDOWWIDTH / 2, gameRect.height + 10 + 25)
175.
176.     DISPLAYSURF.blit(gameSurf, gameRect)
177.     DISPLAYSURF.blit(overSurf, overRect)
178.     drawPressKeyMsg()
179.     pygame.display.update()
```

游戏结束屏幕类似于初始屏幕，只不过它不是动画的。单词"Game"和"Over"渲染到两个 Surface 对象上，然后绘制到屏幕上。

第 6 章　Wormy

```
180.        pygame.time.wait(500)
181.        checkForKeyPress() # clear out any key presses in the event queue
182.
183.        while True:
184.            if checkForKeyPress():
185.                pygame.event.get() # clear event queue
186.                return
```

Game Over 文本将会停留在屏幕上，直到玩家按下一个键。为了确保玩家不会太快地偶然按下一个键，我们将在第 180 行调用 pygame.time.wait()以暂停半秒钟（参数 500 表示暂停 500 毫秒，也就是半秒钟的时间）。

然后，调用 checkForKeyPress()，以便将 showGameOverScreen()函数之后发生的任何的按键事件都忽略。暂停以及忽略按键事件都是为了防止出现如下的情况：假设玩家试图在最后一分钟从屏幕边缘走开，但是，按键太迟了以至于碰到了游戏板的边界。如果发生了这种情况，那么，按键按下将会发生在 showGameOverScreen()之后，并且按键按下会导致游戏结束屏幕立即消失。

此后，下一次游戏会立即开始，这可能会让玩家感到惊讶。因此，添加一个暂停会帮助游戏变得更加用户友好一些。

6.20　绘制函数

绘制分数、虫子、苹果和栅格的代码都放在各自的绘制函数中。

```
188. def drawScore(score):
189.     scoreSurf = BASICFONT.render('Score: %s' % (score), True, WHITE)
190.     scoreRect = scoreSurf.get_rect()
191.     scoreRect.topleft = (WINDOWWIDTH - 120, 10)
192.     DISPLAYSURF.blit(scoreSurf, scoreRect)
```

drawScore()函数直接读取传递给其参数 score 的分数文本，并将其绘制到显示 Surface 对象。

```
195. def drawWorm(wormCoords):
196.     for coord in wormCoords:
197.         x = coord['x'] * CELLSIZE
198.         y = coord['y'] * CELLSIZE
199.         wormSegmentRect = pygame.Rect(x, y, CELLSIZE, CELLSIZE)
200.         pygame.draw.rect(DISPLAYSURF, DARKGREEN, wormSegmentRect)
201.         wormInnerSegmentRect = pygame.Rect(x + 4, y + 4, CELLSIZE - 8,
```

```
CELLSIZE - 8)
202.            pygame.draw.rect(DISPLAYSURF, GREEN, wormInnerSegmentRect)
```

drawWorm()函数将为虫子的每一个身体段绘制一个绿色的矩形。身体段传入到 wormCoords 参数中，这是一个字典的列表，字典中带有一个'x'和一个'y'键。第 196 行的 for 循环遍历了 wormCoords 中的每一个字典值。

由于栅格坐标占据了整个窗口，并且也是从 0,0 像素开始的，将栅格坐标转换为像素坐标相当容易。第 197 行和第 198 行直接将 coord['x']和 coord['y']坐标乘以 CELLSIZE。

第 199 行创建了一个 Rect 对象表示虫子的身体段，它将在第 200 行传递给 pygame.draw.rect()函数。记住，栅格中的每一个单元格的宽度和高度都是 CELLSIZE，因此，身体段的 Rectangular 对象也应该是这个大小。第 200 行绘制了一个暗绿色的矩形表示身体段。然后，在其上，绘制了一个小的亮绿色的矩形。这使得虫子看上去好看一些。

内部的亮绿色的矩形从单元格的左上角再分别向下和向右 4 个像素的位置开始。这个矩形的宽度和高度比单元格的大小要小 8 个像素，因此，单元格的右边和下边都将留下 4 个像素的边距。

```
205. def drawApple(coord):
206.     x = coord['x'] * CELLSIZE
207.     y = coord['y'] * CELLSIZE
208.     appleRect = pygame.Rect(x, y, CELLSIZE, CELLSIZE)
209.     pygame.draw.rect(DISPLAYSURF, RED, appleRect)
```

drawApple()函数和 drawWorm()函数很相似，只不过红色的苹果是一个单个的矩形，它填满了单元格，该函数所需要做的只是转换为像素坐标（这在第 206 行和第 207 行执行），用苹果的位置和大小创建 Rect 对象（在第 208 行），然后，将这个 Rect 对象传递给 pygame.draw.rect()函数。

```
212. def drawGrid():
213.     for x in range(0, WINDOWWIDTH, CELLSIZE): # draw vertical lines
214.         pygame.draw.line(DISPLAYSURF, DARKGRAY, (x, 0), (x, WINDOWHEIGHT))
215.     for y in range(0, WINDOWHEIGHT, CELLSIZE): # draw horizontal lines
216.         pygame.draw.line(DISPLAYSURF, DARKGRAY, (0, y), (WINDOWWIDTH, y))
```

为了让单元格的栅格更为可见，我们调用了 pygame.draw.line()绘制出栅格的每一个垂直的和水平的线条。

通常，为了绘制所需的 32 条垂直线条，我们需要使用如下的坐标调用 pygame.draw.line()32 次。

第 6 章 Wormy

```
pygame.draw.line(DISPLAYSURF, DARKGRAY, (0, 0), (0, WINDOWHEIGHT))
pygame.draw.line(DISPLAYSURF, DARKGRAY, (20, 0), (20, WINDOWHEIGHT))
pygame.draw.line(DISPLAYSURF, DARKGRAY, (40, 0), (40, WINDOWHEIGHT))
pygame.draw.line(DISPLAYSURF, DARKGRAY, (60, 0), (60, WINDOWHEIGHT))
...skipped for brevity...
pygame.draw.line(DISPLAYSURF, DARKGRAY, (560, 0), (560, WINDOWHEIGHT))
pygame.draw.line(DISPLAYSURF, DARKGRAY, (580, 0), (580, WINDOWHEIGHT))
pygame.draw.line(DISPLAYSURF, DARKGRAY, (600, 0), (600, WINDOWHEIGHT))
pygame.draw.line(DISPLAYSURF, DARKGRAY, (620, 0), (620, WINDOWHEIGHT))
```

我们可以在 for 循环中只使用一行代码，而不是输入所有这些代码行。

注意，垂直线条的模式是开始点和结束点的 X 坐标从 0 开始，一直到 620，每次增加 20 个像素。起始点的 Y 坐标总是为 0，而结束点的 Y 坐标总是为 WINDOWHEIGHT。这意味着 for 循环应该按照 range(0, 640, 20)来迭代。这就是为什么第 213 行的 for 循环迭代为 range(0, WINDOWWIDTH, CELLSIZE)的原因。

对于水平线条，其坐标是：

```
pygame.draw.line(DISPLAYSURF, DARKGRAY, (0, 0), (WINDOWWIDTH, 0))
pygame.draw.line(DISPLAYSURF, DARKGRAY, (0, 20), (WINDOWWIDTH, 20))
pygame.draw.line(DISPLAYSURF, DARKGRAY, (0, 40), (WINDOWWIDTH, 40))
pygame.draw.line(DISPLAYSURF, DARKGRAY, (0, 60), (WINDOWWIDTH, 60))
...skipped for brevity...
pygame.draw.line(DISPLAYSURF, DARKGRAY, (0, 400), (WINDOWWIDTH, 400))
pygame.draw.line(DISPLAYSURF, DARKGRAY, (0, 420), (WINDOWWIDTH, 420))
pygame.draw.line(DISPLAYSURF, DARKGRAY, (0, 440), (WINDOWWIDTH, 440))
pygame.draw.line(DISPLAYSURF, DARKGRAY, (0, 460), (WINDOWWIDTH, 460))
```

Y 坐标的范围从 0 到 460，每次增加 20 个像素。起始点的 X 坐标总是为 0，而结束点的 X 坐标总是为 WINDOWWIDTH。这里还是使用 for 循环，从而不必录入所有这些 pygame.draw.line()调用。

注意调用所需的常规模式并且恰当地使用循环，这是节省很多录入的一种聪明的编程技巧。如果我们输入所有的 56 次 pygame.draw.line()调用，程序也能同样地工作。但是，稍微聪明一点，我们就可以节省很多的工作。

```
219. if __name__ == '__main__':
220.     main()
```

在所有的函数、常量以及全局变量都定义和创建之后，调用 main()函数以开始游戏。

6.21 不要复用变量名

再次看一下 drawWorm()函数中的如下几行代码。

6.21 不要复用变量名

```
199.            wormSegmentRect = pygame.Rect(x, y, CELLSIZE, CELLSIZE)
200.            pygame.draw.rect(DISPLAYSURF, DARKGREEN, wormSegmentRect)
201.            wormInnerSegmentRect = pygame.Rect(x + 4, y + 4, CELLSIZE - 8,
CELLSIZE - 8)
202.            pygame.draw.rect(DISPLAYSURF, GREEN, wormInnerSegmentRect)
```

注意，在第 199 行和第 201 行创建了两个不同的 Rect 对象。第 199 行创建的 Rect 对象存储在局部变量 wormSegmentRect 中，并且在第 200 行传递给了 pygame.draw.rect() 函数。第 201 行创建的 Rect 对象存储在局部变量 wormInnerSegmentRect 中，并且在 202 行传递给 pygame.draw.rect() 函数。

每次创建一个变量，它都会占用少量的计算机内存。你可能认为对于两个 Rect 复用 wormSegmentRect 变量会更聪明一些，就像下面这样。

```
199.            wormSegmentRect = pygame.Rect(x, y, CELLSIZE, CELLSIZE)
200.            pygame.draw.rect(DISPLAYSURF, DARKGREEN, wormSegmentRect)
201.            wormSegmentRect = pygame.Rect(x + 4, y + 4, CELLSIZE - 8, CELLSIZE
- 8)
202.            pygame.draw.rect(DISPLAYSURF, GREEN, wormInnerSegmentRect)
```

由于在第 199 行由 pygame.Rect() 返回的 Rect 对象在第 200 行之后就不需要了，我们可以覆盖这个值，并且在第 201 行复用该变量来存储 pygame.Rect() 所返回的 Rect 对象。因此，我们现在可以使用较少的变量从而节省了内存，对吗？

尽管从技术上讲是这样的，但你真地只能节省几个字节。现代计算机拥有数以百万字节计的内存。因此，这点节约微不足道。同时，复用变量减少了代码的可读性。如果一名程序员在编写完这段代码之后再阅读它，他将会看到在第 200 行和第 202 行，wormSegmentRect 变量传递给了 pygame.draw.rect() 调用。如果他们试图找到 wormSegmentRect 变量第一次赋值的地方，应该会看到第 199 行的 pygame.Rect() 调用。他们可能不会意识到，第 199 行的 pygame.Rect() 调用所返回的 Rectangular 对象和第 202 行传递给 pygame.draw.rect() 的 Rect 对象是不同的。

像这样的小事情，会使得人们很难理解程序到底是如何工作的。并不仅是那些查看你的代码的其他程序员会搞混淆，当你在编写完代码的数周之后，你可能都很难记得它到底是如何工作的了。代码可读性比在这里或那里节省几个字节的内存要重要得多。

要进行额外的编程实践，你可以从 http://invpy.com/buggy/wormy 下载 Wormy 的带 Bug 的版本，并尝试搞清楚如何修正 Bug。

第 7 章 Tetromino

Tetromino 是 Tetris 的翻版。不同的形状的砖块（每一个砖块都是由 4 个方块组成的）从屏幕的顶部落下来，玩家必须引导它们下落，以便形成完整的、中间没有间隙的一行，如图 7-1 所示。一旦形成了这样的完整一行，这一行就会消失，并且它上面的每一行都会向下移动一行。玩家试图不断地形成完整的行，直到屏幕填满并且一个新落下来的砖块无法容纳在屏幕之中。

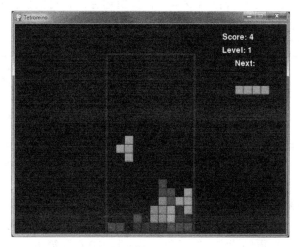

图 7-1

7.1 一些 Tetromino 术语

在本章中，对于游戏中的不同内容，我会使用一组术语。

- 游戏板（Board）——游戏板由 10×20 个空格组成，砖块会落下来并堆积到这些空格中。
- 方块（Box）——一个方块是游戏板上的、单个的实心的正方形空格。
- 砖块（Piece）——从游戏板顶部落下的东西，玩家可以旋转和放置它们。每个砖块拥有一个形状，并且由 4 个方块组成。
- 形状（Shape）——形状是游戏中的砖块的不同类型。形状的名称是 T、S、Z、J、L、I 和 O，如图 7-2 所示。
- 模板（Template）——这是表示形状的所有可能的旋转的数据结构的一个列表。
- 着陆（Landed）——当砖块到达了游戏板的底部，或者碰到了游戏板上的一个方块，我们就说砖块着陆了。此时，下一个砖块开始下落。

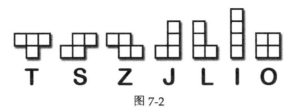

图 7-2

7.2 Tetromino 的源代码

可以从 http://invpy.com/tetromino.py 下载 Wormy 的源代码。如果得到任何的错误消息，查看一下错误消息中提到的代码行，并且检查代码是否有任何录入错误。可以将代码复制并粘贴到位于 http://invpy.com/diff/tetromino 的 Web 表单中，看看你的代码和本书中的代码是否有区别。

还需要将背景音乐文件和 *tetromino.py* 文件放到同一文件夹下。你也可以从这里下载这些文件：

- http://invpy.com/tetrisb.mid
- http://invpy.com/tetrisc.mid

```
1. # Tetromino (a Tetris clone)
2. # By Al Sweigart al@inventwithpython.com
3. # http://inventwithpython.com/pygame
4. # Creative Commons BY-NC-SA 3.0 US
5.
6. import random, time, pygame, sys
7. from pygame.locals import *
8.
9. FPS = 25
10. WINDOWWIDTH = 640
11. WINDOWHEIGHT = 480
12. BOXSIZE = 20
13. BOARDWIDTH = 10
14. BOARDHEIGHT = 20
15. BLANK = '.'
16.
17. MOVESIDEWAYSFREQ = 0.15
18. MOVEDOWNFREQ = 0.1
19.
20. XMARGIN = int((WINDOWWIDTH - BOARDWIDTH * BOXSIZE) / 2)
21. TOPMARGIN = WINDOWHEIGHT - (BOARDHEIGHT * BOXSIZE) - 5
22.
23. #               R    G    B
```

第 7 章　Tetromino

```
24. WHITE       = (255, 255, 255)
25. GRAY        = (185, 185, 185)
26. BLACK       = (  0,   0,   0)
27. RED         = (155,   0,   0)
28. LIGHTRED    = (175,  20,  20)
29. GREEN       = (  0, 155,   0)
30. LIGHTGREEN  = ( 20, 175,  20)
31. BLUE        = (  0,   0, 155)
32. LIGHTBLUE   = ( 20,  20, 175)
33. YELLOW      = (155, 155,   0)
34. LIGHTYELLOW = (175, 175,  20)
35.
36. BORDERCOLOR = BLUE
37. BGCOLOR = BLACK
38. TEXTCOLOR = WHITE
39. TEXTSHADOWCOLOR = GRAY
40. COLORS      = (     BLUE,      GREEN,      RED,       YELLOW)
41. LIGHTCOLORS = (LIGHTBLUE, LIGHTGREEN, LIGHTRED, LIGHTYELLOW)
42. assert len(COLORS) == len(LIGHTCOLORS) # each color must have light color
43.
44. TEMPLATEWIDTH = 5
45. TEMPLATEHEIGHT = 5
46.
47. S_SHAPE_TEMPLATE = [['.....',
48.                      '.....',
49.                      '..OO.',
50.                      '.OO..',
51.                      '.....'],
52.                     ['.....',
53.                      '..O..',
54.                      '..OO.',
55.                      '...O.',
56.                      '.....']]
57.
58. Z_SHAPE_TEMPLATE = [['.....',
59.                      '.....',
60.                      '.OO..',
61.                      '..OO.',
62.                      '.....'],
63.                     ['.....',
64.                      '..O..',
65.                      '.OO..',
66.                      '.O...',
67.                      '.....']]
68.
69. I_SHAPE_TEMPLATE = [['..O..',
```

```
70.                        '..O..',
71.                        '..O..',
72.                        '..O..',
73.                        '.....'],
74.                       ['.....',
75.                        '.....',
76.                        'OOOO.',
77.                        '.....',
78.                        '.....']]
79.
80. O_SHAPE_TEMPLATE = [['.....',
81.                     '.....',
82.                     '.OO..',
83.                     '.OO..',
84.                     '.....']]
85.
86. J_SHAPE_TEMPLATE = [['.....',
87.                     '.O...',
88.                     '.OOO.',
89.                     '.....',
90.                     '.....'],
91.                    ['.....',
92.                     '..OO.',
93.                     '..O..',
94.                     '..O..',
95.                     '.....'],
96.                    ['.....',
97.                     '.....',
98.                     '.OOO.',
99.                     '...O.',
100.                    '.....'],
101.                   ['.....',
102.                    '..O..',
103.                    '..O..',
104.                    '.OO..',
105.                    '.....']]
106.
107. L_SHAPE_TEMPLATE = [['.....',
108.                     '...O.',
109.                     '.OOO.',
110.                     '.....',
111.                     '.....'],
112.                    ['.....',
113.                     '..O..',
114.                     '..O..',
115.                     '..OO.',
```

```
116.                        '.....'],
117.                       ['.....',
118.                        '.....',
119.                        '.000.',
120.                        '.0...',
121.                        '.....']],
122.                       ['.....',
123.                        '.00..',
124.                        '..0..',
125.                        '..0..',
126.                        '.....']]

127.
128. T_SHAPE_TEMPLATE = [['.....',
129.                     '..0..',
130.                     '.000.',
131.                     '.....',
132.                     '.....'],
133.                    ['.....',
134.                     '..0..',
135.                     '..00.',
136.                     '..0..',
137.                     '.....'],
138.                    ['.....',
139.                     '.....',
140.                     '.000.',
141.                     '..0..',
142.                     '.....'],
143.                    ['.....',
144.                     '..0..',
145.                     '.00..',
146.                     '..0..',
147.                     '.....']]
148.
149. SHAPES = {'S': S_SHAPE_TEMPLATE,
150.           'Z': Z_SHAPE_TEMPLATE,
151.           'J': J_SHAPE_TEMPLATE,
152.           'L': L_SHAPE_TEMPLATE,
153.           'I': I_SHAPE_TEMPLATE,
154.           'O': O_SHAPE_TEMPLATE,
155.           'T': T_SHAPE_TEMPLATE}
156.
157.
158. def main():
159.     global FPSCLOCK, DISPLAYSURF, BASICFONT, BIGFONT
160.     pygame.init()
161.     FPSCLOCK = pygame.time.Clock()
```

```
162.        DISPLAYSURF = pygame.display.set_mode((WINDOWWIDTH, WINDOWHEIGHT))
163.        BASICFONT = pygame.font.Font('freesansbold.ttf', 18)
164.        BIGFONT = pygame.font.Font('freesansbold.ttf', 100)
165.        pygame.display.set_caption('Tetromino')
166.
167.        showTextScreen('Tetromino')
168.        while True: # game loop
169.            if random.randint(0, 1) == 0:
170.                pygame.mixer.music.load('tetrisb.mid')
171.            else:
172.                pygame.mixer.music.load('tetrisc.mid')
173.            pygame.mixer.music.play(-1, 0.0)
174.            runGame()
175.            pygame.mixer.music.stop()
176.            showTextScreen('Game Over')
177.
178.
179.    def runGame():
180.        # setup variables for the start of the game
181.        board = getBlankBoard()
182.        lastMoveDownTime = time.time()
183.        lastMoveSidewaysTime = time.time()
184.        lastFallTime = time.time()
185.        movingDown = False # note: there is no movingUp variable
186.        movingLeft = False
187.        movingRight = False
188.        score = 0
189.        level, fallFreq = calculateLevelAndFallFreq(score)
190.
191.        fallingPiece = getNewPiece()
192.        nextPiece = getNewPiece()
193.
194.        while True: # main game loop
195.            if fallingPiece == None:
196.                # No falling piece in play, so start a new piece at the top
197.                fallingPiece = nextPiece
198.                nextPiece = getNewPiece()
199.                lastFallTime = time.time() # reset lastFallTime
200.
201.                if not isValidPosition(board, fallingPiece):
202.                    return # can't fit a new piece on the board, so game over
203.
204.            checkForQuit()
205.            for event in pygame.event.get(): # event handling loop
206.                if event.type == KEYUP:
207.                    if (event.key == K_p):
```

第7章　Tetromino

```
208.                    # Pausing the game
209.                    DISPLAYSURF.fill(BGCOLOR)
210.                    pygame.mixer.music.stop()
211.                    showTextScreen('Paused') # pause until a key press
212.                    pygame.mixer.music.play(-1, 0.0)
213.                    lastFallTime = time.time()
214.                    lastMoveDownTime = time.time()
215.                    lastMoveSidewaysTime = time.time()
216.                elif (event.key == K_LEFT or event.key == K_a):
217.                    movingLeft = False
218.                elif (event.key == K_RIGHT or event.key == K_d):
219.                    movingRight = False
220.                elif (event.key == K_DOWN or event.key == K_s):
221.                    movingDown = False
222.
223.            elif event.type == KEYDOWN:
224.                # moving the block sideways
225.                if (event.key == K_LEFT or event.key == K_a) and isValidPosition(board, fallingPiece, adjX=-1):
226.                    fallingPiece['x'] -= 1
227.                    movingLeft = True
228.                    movingRight = False
229.                    lastMoveSidewaysTime = time.time()
230.
231.                elif (event.key == K_RIGHT or event.key == K_d) and isValidPosition(board, fallingPiece, adjX=1):
232.                    fallingPiece['x'] += 1
233.                    movingRight = True
234.                    movingLeft = False
235.                    lastMoveSidewaysTime = time.time()
236.
237.                # rotating the block (if there is room to rotate)
238.                elif (event.key == K_UP or event.key == K_w):
239.                    fallingPiece['rotation'] = (fallingPiece['rotation'] + 1) % len(SHAPES[fallingPiece['shape']])
240.                    if not isValidPosition(board, fallingPiece):
241.                        fallingPiece['rotation'] = (fallingPiece['rotation'] - 1) % len(SHAPES[fallingPiece['shape']])
242.                elif (event.key == K_q): # rotate the other direction
243.                    fallingPiece['rotation'] = (fallingPiece['rotation'] - 1) % len(SHAPES[fallingPiece['shape']])
244.                    if not isValidPosition(board, fallingPiece):
245.                        fallingPiece['rotation'] = (fallingPiece['rotation'] + 1) % len(SHAPES[fallingPiece['shape']])
246.
247.                # making the block fall faster with the down key
```

```
248.                elif (event.key == K_DOWN or event.key == K_s):
249.                    movingDown = True
250.                    if isValidPosition(board, fallingPiece, adjY=1):
251.                        fallingPiece['y'] += 1
252.                    lastMoveDownTime = time.time()
253.
254.                # move the current block all the way down
255.                elif event.key == K_SPACE:
256.                    movingDown = False
257.                    movingLeft = False
258.                    movingRight = False
259.                    for i in range(1, BOARDHEIGHT):
260.                        if not isValidPosition(board, fallingPiece, adjY=i):
261.                            break
262.                    fallingPiece['y'] += i - 1
263.
264.        # handle moving the block because of user input
265.        if (movingLeft or movingRight) and time.time() - lastMoveSidewaysTime > MOVESIDEWAYSFREQ:
266.            if movingLeft and isValidPosition(board, fallingPiece, adjX=-1):
267.                fallingPiece['x'] -= 1
268.            elif movingRight and isValidPosition(board, fallingPiece, adjX=1):
269.                fallingPiece['x'] += 1
270.            lastMoveSidewaysTime = time.time()
271.
272.        if movingDown and time.time() - lastMoveDownTime > MOVEDOWNFREQ and isValidPosition(board, fallingPiece, adjY=1):
273.            fallingPiece['y'] += 1
274.            lastMoveDownTime = time.time()
275.
276.        # let the piece fall if it is time to fall
277.        if time.time() - lastFallTime > fallFreq:
278.            # see if the piece has landed
279.            if not isValidPosition(board, fallingPiece, adjY=1):
280.                # falling piece has landed, set it on the board
281.                addToBoard(board, fallingPiece)
282.                score += removeCompleteLines(board)
283.                level, fallFreq = calculateLevelAndFallFreq(score)
284.                fallingPiece = None
285.            else:
286.                # piece did not land, just move the block down
287.                fallingPiece['y'] += 1
288.                lastFallTime = time.time()
```

第7章　Tetromino

```
289.
290.         # drawing everything on the screen
291.         DISPLAYSURF.fill(BGCOLOR)
292.         drawBoard(board)
293.         drawStatus(score, level)
294.         drawNextPiece(nextPiece)
295.         if fallingPiece != None:
296.             drawPiece(fallingPiece)
297.
298.         pygame.display.update()
299.         FPSCLOCK.tick(FPS)
300.
301.
302. def makeTextObjs(text, font, color):
303.     surf = font.render(text, True, color)
304.     return surf, surf.get_rect()
305.
306.
307. def terminate():
308.     pygame.quit()
309.     sys.exit()
310.
311.
312. def checkForKeyPress():
313.     # Go through event queue looking for a KEYUP event.
314.     # Grab KEYDOWN events to remove them from the event queue.
315.     checkForQuit()
316.
317.     for event in pygame.event.get([KEYDOWN, KEYUP]):
318.         if event.type == KEYDOWN:
319.             continue
320.         return event.key
321.     return None
322.
323.
324. def showTextScreen(text):
325.     # This function displays large text in the
326.     # center of the screen until a key is pressed.
327.     # Draw the text drop shadow
328.     titleSurf, titleRect = makeTextObjs(text, BIGFONT, TEXTSHADOWCOLOR)
329.     titleRect.center = (int(WINDOWWIDTH / 2), int(WINDOWHEIGHT / 2))
330.     DISPLAYSURF.blit(titleSurf, titleRect)
331.
332.     # Draw the text
333.     titleSurf, titleRect = makeTextObjs(text, BIGFONT, TEXTCOLOR)
```

```
334.        titleRect.center = (int(WINDOWWIDTH / 2) - 3, int(WINDOWHEIGHT / 2) -
3)
335.        DISPLAYSURF.blit(titleSurf, titleRect)
336.
337.        # Draw the additional "Press a key to play." text.
338.        pressKeySurf, pressKeyRect = makeTextObjs('Press a key to play.',
BASICFONT, TEXTCOLOR)
339.        pressKeyRect.center = (int(WINDOWWIDTH / 2), int(WINDOWHEIGHT / 2) +
100)
340.        DISPLAYSURF.blit(pressKeySurf, pressKeyRect)
341.
342.        while checkForKeyPress() == None:
343.            pygame.display.update()
344.            FPSCLOCK.tick()
345.
346.
347.    def checkForQuit():
348.        for event in pygame.event.get(QUIT): # get all the QUIT events
349.            terminate() # terminate if any QUIT events are present
350.        for event in pygame.event.get(KEYUP): # get all the KEYUP events
351.            if event.key == K_ESCAPE:
352.                terminate() # terminate if the KEYUP event was for the Esc key
353.            pygame.event.post(event) # put the other KEYUP event objects back
354.
355.
356.    def calculateLevelAndFallFreq(score):
357.        # Based on the score, return the level the player is on and
358.        # how many seconds pass until a falling piece falls one space.
359.        level = int(score / 10) + 1
360.        fallFreq = 0.27 - (level * 0.02)
361.        return level, fallFreq
362.
363.    def getNewPiece():
364.        # return a random new piece in a random rotation and color
365.        shape = random.choice(list(SHAPES.keys()))
366.        newPiece = {'shape': shape,
367.                    'rotation': random.randint(0, len(SHAPES[shape]) - 1),
368.                    'x': int(BOARDWIDTH / 2) - int(TEMPLATEWIDTH / 2),
369.                    'y': -2, # start it above the board (i.e. less than 0)
370.                    'color': random.randint(0, len(COLORS)-1)}
371.        return newPiece
372.
373.
374.    def addToBoard(board, piece):
375.        # fill in the board based on piece's location, shape, and rotation
376.        for x in range(TEMPLATEWIDTH):
```

第 7 章　Tetromino

```
377.            for y in range(TEMPLATEHEIGHT):
378.                if SHAPES[piece['shape']][piece['rotation']][y][x] != BLANK:
379.                    board[x + piece['x']][y + piece['y']] = piece['color']
380.
381.
382. def getBlankBoard():
383.     # create and return a new blank board data structure
384.     board = []
385.     for i in range(BOARDWIDTH):
386.         board.append([BLANK] * BOARDHEIGHT)
387.     return board
388.
389.
390. def isOnBoard(x, y):
391.     return x >= 0 and x < BOARDWIDTH and y < BOARDHEIGHT
392.
393.
394. def isValidPosition(board, piece, adjX=0, adjY=0):
395.     # Return True if the piece is within the board and not colliding
396.     for x in range(TEMPLATEWIDTH):
397.         for y in range(TEMPLATEHEIGHT):
398.             isAboveBoard = y + piece['y'] + adjY < 0
399.             if isAboveBoard or SHAPES[piece['shape']][piece['rotation']][y][x] == BLANK:
400.                 continue
401.             if not isOnBoard(x + piece['x'] + adjX, y + piece['y'] + adjY):
402.                 return False
403.             if board[x + piece['x'] + adjX][y + piece['y'] + adjY] != BLANK:
404.                 return False
405.     return True
406.
407. def isCompleteLine(board, y):
408.     # Return True if the line filled with boxes with no gaps.
409.     for x in range(BOARDWIDTH):
410.         if board[x][y] == BLANK:
411.             return False
412.     return True
413.
414.
415. def removeCompleteLines(board):
416.     # Remove any completed lines on the board, move everything above them down, and return the number of complete lines.
417.     numLinesRemoved = 0
418.     y = BOARDHEIGHT - 1 # start y at the bottom of the board
```

```
419.        while y >= 0:
420.            if isCompleteLine(board, y):
421.                # Remove the line and pull boxes down by one line.
422.                for pullDownY in range(y, 0, -1):
423.                    for x in range(BOARDWIDTH):
424.                        board[x][pullDownY] = board[x][pullDownY-1]
425.                # Set very top line to blank.
426.                for x in range(BOARDWIDTH):
427.                    board[x][0] = BLANK
428.                numLinesRemoved += 1
429.                # Note on the next iteration of the loop, y is the same.
430.                # This is so that if the line that was pulled down is also
431.                # complete, it will be removed.
432.            else:
433.                y -= 1 # move on to check next row up
434.        return numLinesRemoved
435.
436.
437. def convertToPixelCoords(boxx, boxy):
438.     # Convert the given xy coordinates of the board to xy
439.     # coordinates of the location on the screen.
440.     return (XMARGIN + (boxx * BOXSIZE)), (TOPMARGIN + (boxy * BOXSIZE))
441.
442.
443. def drawBox(boxx, boxy, color, pixelx=None, pixely=None):
444.     # draw a single box (each tetromino piece has four boxes)
445.     # at xy coordinates on the board. Or, if pixelx & pixely
446.     # are specified, draw to the pixel coordinates stored in
447.     # pixelx & pixely (this is used for the "Next" piece).
448.     if color == BLANK:
449.         return
450.     if pixelx == None and pixely == None:
451.         pixelx, pixely = convertToPixelCoords(boxx, boxy)
452.     pygame.draw.rect(DISPLAYSURF, COLORS[color], (pixelx + 1, pixely + 1, BOXSIZE - 1, BOXSIZE - 1))
453.     pygame.draw.rect(DISPLAYSURF, LIGHTCOLORS[color], (pixelx + 1, pixely + 1, BOXSIZE - 4, BOXSIZE - 4))
454.
455.
456. def drawBoard(board):
457.     # draw the border around the board
458.     pygame.draw.rect(DISPLAYSURF, BORDERCOLOR, (XMARGIN - 3, TOPMARGIN - 7, (BOARDWIDTH * BOXSIZE) + 8, (BOARDHEIGHT * BOXSIZE) + 8), 5)
459.
460.     # fill the background of the board
```

第 7 章　Tetromino

```
461.        pygame.draw.rect(DISPLAYSURF, BGCOLOR, (XMARGIN, TOPMARGIN, BOXSIZE *
BOARDWIDTH, BOXSIZE * BOARDHEIGHT))
462.        # draw the individual boxes on the board
463.        for x in range(BOARDWIDTH):
464.            for y in range(BOARDHEIGHT):
465.                drawBox(x, y, board[x][y])
466.
467.
468. def drawStatus(score, level):
469.        # draw the score text
470.        scoreSurf = BASICFONT.render('Score: %s' % score, True, TEXTCOLOR)
471.        scoreRect = scoreSurf.get_rect()
472.        scoreRect.topleft = (WINDOWWIDTH - 150, 20)
473.        DISPLAYSURF.blit(scoreSurf, scoreRect)
474.
475.        # draw the level text
476.        levelSurf = BASICFONT.render('Level: %s' % level, True, TEXTCOLOR)
477.        levelRect = levelSurf.get_rect()
478.        levelRect.topleft = (WINDOWWIDTH - 150, 50)
479.        DISPLAYSURF.blit(levelSurf, levelRect)
480.
481.
482. def drawPiece(piece, pixelx=None, pixely=None):
483.        shapeToDraw = SHAPES[piece['shape']][piece['rotation']]
484.        if pixelx == None and pixely == None:
485.            # if pixelx & pixely hasn't been specified, use the location
stored in the piece data structure
486.            pixelx, pixely = convertToPixelCoords(piece['x'], piece['y'])
487.
488.        # draw each of the blocks that make up the piece
489.        for x in range(TEMPLATEWIDTH):
490.            for y in range(TEMPLATEHEIGHT):
491.                if shapeToDraw[y][x] != BLANK:
492.                    drawBox(None, None, piece['color'], pixelx + (x *
BOXSIZE), pixely + (y * BOXSIZE))
493.
494.
495. def drawNextPiece(piece):
496.        # draw the "next" text
497.        nextSurf = BASICFONT.render('Next:', True, TEXTCOLOR)
498.        nextRect = nextSurf.get_rect()
499.        nextRect.topleft = (WINDOWWIDTH - 120, 80)
500.        DISPLAYSURF.blit(nextSurf, nextRect)
501.        # draw the "next" piece
502.        drawPiece(piece, pixelx=WINDOWWIDTH-120, pixely=100)
503.
```

```
504.
505. if __name__ == '__main__':
506.     main()
```

7.3 常用设置代码

```
 1. # Tetromino (a Tetris clone)
 2. # By Al Sweigart al@inventwithpython.com
 3. # http://inventwithpython.com/pygame
 4. # Creative Commons BY-NC-SA 3.0 US
 5.
 6. import random, time, pygame, sys
 7. from pygame.locals import *
 8.
 9. FPS = 25
10. WINDOWWIDTH = 640
11. WINDOWHEIGHT = 480
12. BOXSIZE = 20
13. BOARDWIDTH = 10
14. BOARDHEIGHT = 20
15. BLANK = '.'
```

这些是 Tetromino 游戏所使用的常量。每一个方块都是宽度和高度为 20 个像素的一个正方形。游戏板本身拥有 10 个方块的宽度和 20 个方块的高度。BLANK 常量用作表示游戏板数据结构中的空白空格的一个值。

7.4 设置按下键的定时常量

```
17. MOVESIDEWAYSFREQ = 0.15
18. MOVEDOWNFREQ = 0.1
```

每次玩家按下向左箭头或向右箭头键的时候，下落的砖块都应该分别向左或向右移动一个方块。然而，玩家也可能保持按住了向左箭头或向右箭头键，以使得下落的砖块持续移动。设置 MOVESIDEWAYSFREQ 常量，以便向左箭头或向右箭头键每次持续按下超过 0.15 秒的时候，砖块相应地移动一个空格。MOVEDOWNFREQ 常量也是一样的，只不过它告诉我们当玩家按住向下箭头的时候，砖块需要多么频繁地向下移动一个方块。

7.5 更多的设置代码

```
20. XMARGIN = int((WINDOWWIDTH - BOARDWIDTH * BOXSIZE) / 2)
21. TOPMARGIN = WINDOWHEIGHT - (BOARDHEIGHT * BOXSIZE) - 5
```

　　程序需要计算游戏板的左边和右边有多少个像素，以便在程序的后面使用。WINDOWWIDTH 是整个窗口宽度的总的像素数目。游戏板的总宽度拥有 BOARDWIDTH 多个方块，每个方块的宽度为 BOXSIZE 多个像素。如果我们从游戏板的宽度中减去所有方块的宽度（后者是 BOARDWIDTH * BOXSIZE），我们就得到了游戏板左边和右边的边距。如果将其除以 2，那么，就有了其中一个边距的大小。由于边距的大小是相同的，我们将 XMARGIN 用作左边距或右边距都可以。

　　我们可以用类似的方法来计算游戏板顶部和窗口顶部之间的空间的大小，如图 7-3 所示。游戏板将绘制于窗口底部之上的 5 个像素处，因此从 topmargin 减去 5 以考虑到这一点。

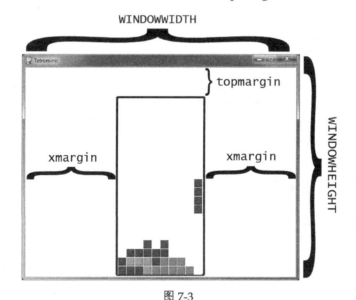

图 7-3

```
23. #                   R    G    B
24. WHITE       = (255, 255, 255)
25. GRAY        = (185, 185, 185)
26. BLACK       = (  0,   0,   0)
27. RED         = (155,   0,   0)
28. LIGHTRED    = (175,  20,  20)
```

```
29. GREEN       = (  0, 155,   0)
30. LIGHTGREEN  = ( 20, 175,  20)
31. BLUE        = (  0,   0, 155)
32. LIGHTBLUE   = ( 20,  20, 175)
33. YELLOW      = (155, 155,   0)
34. LIGHTYELLOW = (175, 175,  20)
35.
36. BORDERCOLOR = BLUE
37. BGCOLOR = BLACK
38. TEXTCOLOR = WHITE
39. TEXTSHADOWCOLOR = GRAY
40. COLORS      = (     BLUE,      GREEN,      RED,      YELLOW)
41. LIGHTCOLORS = (LIGHTBLUE, LIGHTGREEN, LIGHTRED, LIGHTYELLOW)
42. assert len(COLORS) == len(LIGHTCOLORS) # each color must have light color
```

方块将显示为 4 种颜色：蓝色、绿色、红色和黄色。当我们绘制方块的时候，方块上会有一点点浅颜色的高亮。因此，这意味着我们需要创建浅蓝色、浅绿色、浅红色和浅黄色。这 4 种颜色中的每一种都将存储到名为 COLORS（用于常规颜色）和 LIGHTCOLORS（用于浅颜色）的元组之中。

7.6 设置砖块模式

```
44. TEMPLATEWIDTH = 5
45. TEMPLATEHEIGHT = 5
46.
47. S_SHAPE_TEMPLATE = [['.....',
48.                     '.....',
49.                     '..OO.',
50.                     '.OO..',
51.                     '.....'],
52.                    ['.....',
53.                     '..O..',
54.                     '..OO.',
55.                     '...O.',
56.                     '.....']]
57.
58. Z_SHAPE_TEMPLATE = [['.....',
59.                     '.....',
60.                     '.OO..',
61.                     '..OO.',
62.                     '.....'],
63.                    ['.....',
64.                     '..O..',
```

```
 65.                        '.OO..',
 66.                        '.O...',
 67.                        '.....']]
 68.
 69. I_SHAPE_TEMPLATE = [['..O..',
 70.                     '..O..',
 71.                     '..O..',
 72.                     '..O..',
 73.                     '.....'],
 74.                    ['.....',
 75.                     '.....',
 76.                     'OOOO.',
 77.                     '.....',
 78.                     '.....']]
 79.
 80. O_SHAPE_TEMPLATE = [['.....',
 81.                      '.....',
 82.                      '.OO..',
 83.                      '.OO..',
 84.                      '.....']]
 85.
 86. J_SHAPE_TEMPLATE = [['.....',
 87.                      '.O...',
 88.                      '.OOO.',
 89.                      '.....',
 90.                      '.....'],
 91.                     ['.....',
 92.                      '..OO.',
 93.                      '..O..',
 94.                      '..O..',
 95.                      '.....'],
 96.                     ['.....',
 97.                      '.....',
 98.                      '.OOO.',
 99.                      '...O.',
100.                      '.....'],
101.                     ['.....',
102.                      '..O..',
103.                      '..O..',
104.                      '.OO..',
105.                      '.....']]
106.
107. L_SHAPE_TEMPLATE = [['.....',
108.                      '...O.',
109.                      '.OOO.',
110.                      '.....',
```

```
111.                          '.....'],
112.                         ['.....',
113.                          '...O..',
114.                          '...O..',
115.                          '...OO.',
116.                          '.....'],
117.                         ['.....',
118.                          '.....',
119.                          '.OOO.',
120.                          '...O..',
121.                          '.....'],
122.                         ['.....',
123.                          '.OO..',
124.                          '..O..',
125.                          '..O..',
126.                          '.....']]
127.
128. T_SHAPE_TEMPLATE = [['.....',
129.                      '..O..',
130.                      '.OOO.',
131.                      '.....',
132.                      '.....'],
133.                     ['.....',
134.                      '..O..',
135.                      '..OO.',
136.                      '..O..',
137.                      '.....'],
138.                     ['.....',
139.                      '.....',
140.                      '.OOO.',
141.                      '..O..',
142.                      '.....'],
143.                     ['.....',
144.                      '..O..',
145.                      '.OO..',
146.                      '..O..',
147.                      '.....']]
```

游戏程序需要知道每种形状是如何形成的，包括该形状所有可能的旋转。为了做到这一点，我们将创建字符串列表的列表。字符串的内部列表将表示一种形状的单个的一次旋转，如下：

```
['.....',
 '.....',
 '..OO.',
 '.OO..',
 '.....']
```

第 7 章 Tetromino

我们将编写剩下的代码,以便它能够解释上面这样的一个字符串列表,其中点号表示的是空白空格,O 表示的是一个方块,如图 7-4 所示。

图 7-4

7.7 将"一行代码"分隔到多行

你可以看到,这个列表在文件编辑器中跨了很多行。这绝对是有效的 Python 语句,因为 Python 解释器直到看到结束方括号]的时候,才意识到列表结束了。缩进无关紧要,因为 Python 知道你不想对位于列表中间的一个新的语句块使用不同的缩进。如下这段代码也能很好地工作。

```
spam = ['hello', 3.14, 'world', 42, 10, 'fuzz']
eggs = ['hello', 3.14,
   'world'
       , 42,
      10, 'fuzz']
```

尽管如此,当然,如果我们将 eggs 列表中的所有的项对齐,并且像 spam 那样放到一行中,该列表的代码的可读性要好很多。

通常,在文件编辑器中将一行代码分隔到多行,需要在每行的末尾放置一个\字符。这个\字符告诉 Python,"代码从下一行开始继续"(这个斜杠第一次在 Sliding Puzzle 游戏中的 isValidMove()函数中用到)。

我们将创建字符串的这些列表的一个列表,并且将其存储到诸如 S_SHAPE_TEMPLATE 的变量中,从而生成形状的"模板"数据结构。通过这种方式,len(S_SHAPE_TEMPLATE) 将表示 S 形状有多少种可能的旋转,而 S_SHAPE_TEMPLATE[0]将表示 S 形状的第一个可能的旋转。第 47 行到第 147 行将为每一种形状创建"模板"数据结构。

想象一下,每种可能的砖块都位于一个小小的 5×5 的空白空格的小板子中,板子上的一些空格由方块填满。如下的表达式 S_SHAPE_TEMPLATE[0]所表示的第一种可能的旋转。

```
S_SHAPE_TEMPLATE[0][2][2] == 'O'
S_SHAPE_TEMPLATE[0][2][3] == 'O'
S_SHAPE_TEMPLATE[0][3][1] == 'O'
S_SHAPE_TEMPLATE[0][3][2] == 'O'
```

如果我们将这个形状画到纸面上,会看到如图 7-5 所示的内容。

7.8 main()函数

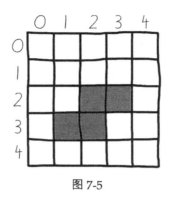

图 7-5

这就是我们如何将 Tetromino 砖块这样的内容表示为字符串和列表这样的 Python 值。TEMPLATEWIDTH 和 TEMPLATEHEIGHT 常量只是设置每一个形状的旋转的每一行和每一列有多大（模板将总是 5×5 的大小）。

```
149. SHAPES = {'S': S_SHAPE_TEMPLATE,
150.           'Z': Z_SHAPE_TEMPLATE,
151.           'J': J_SHAPE_TEMPLATE,
152.           'L': L_SHAPE_TEMPLATE,
153.           'I': I_SHAPE_TEMPLATE,
154.           'O': O_SHAPE_TEMPLATE,
155.           'T': T_SHAPE_TEMPLATE}
```

SHAPES 变量将是一个字典，它存储了所有不同的模板。由于每个模板都拥有一个单个的形状的所有可能的旋转，这意味着，SHAPES 变量包含了每一种可能的形状的所有可能的旋转。这将会是包含了游戏中的所有形状数据的一个数据结构。

7.8　main()函数

```
158. def main():
159.     global FPSCLOCK, DISPLAYSURF, BASICFONT, BIGFONT
160.     pygame.init()
161.     FPSCLOCK = pygame.time.Clock()
162.     DISPLAYSURF = pygame.display.set_mode((WINDOWWIDTH, WINDOWHEIGHT))
163.     BASICFONT = pygame.font.Font('freesansbold.ttf', 18)
164.     BIGFONT = pygame.font.Font('freesansbold.ttf', 100)
165.     pygame.display.set_caption('Tetromino')
166.
167.     showTextScreen('Tetromino')
```

main()函数还负责创建了一些其他的全局常量，并且显示了在游戏运行时候出现的初始屏幕。

```
168.    while True: # game loop
169.        if random.randint(0, 1) == 0:
170.            pygame.mixer.music.load('tetrisb.mid')
171.        else:
172.            pygame.mixer.music.load('tetrisc.mid')
173.        pygame.mixer.music.play(-1, 0.0)
174.        runGame()
175.        pygame.mixer.music.stop()
176.        showTextScreen('Game Over')
```

实际的游戏代码都在 runGame() 之中。这里的 main() 函数直接随机地确定了开始播放哪一个背景音乐（要么是 *tetrisb.mid*，要么是 *tetrisc.mid* MIDI 音乐文件），然后，调用 runGame() 开始游戏。当玩家失败的时候，runGame() 将返回 main()，然后 main() 会停止背景音乐，并且显示游戏结束屏幕。

当玩家按下一个键，显示游戏结束屏幕的 showTextScreen() 函数将返回。游戏循环将回过去从第 169 行开始，并再次启动游戏。

7.9 开始新的游戏

```
179. def runGame():
180.     # setup variables for the start of the game
181.     board = getBlankBoard()
182.     lastMoveDownTime = time.time()
183.     lastMoveSidewaysTime = time.time()
184.     lastFallTime = time.time()
185.     movingDown = False # note: there is no movingUp variable
186.     movingLeft = False
187.     movingRight = False
188.     score = 0
189.     level, fallFreq = calculateLevelAndFallFreq(score)
190.
191.     fallingPiece = getNewPiece()
192.     nextPiece = getNewPiece()
```

在游戏开始并且砖块开始下落之前，我们需要将一些变量初始化为游戏开始时候的值。在第 191 行，fallingPiece 变量将设置为能够由玩家旋转的当前下落的砖块。在第 192 行，nextPiece 变量将设置为在屏幕的"Next"部分出现的砖块，以便玩家知道在放置了正在下落的砖块之后，将要落下来的是哪一种砖块。

7.10 游戏循环

```
194.    while True: # main game loop
195.        if fallingPiece == None:
196.            # No falling piece in play, so start a new piece at the top
197.            fallingPiece = nextPiece
198.            nextPiece = getNewPiece()
199.            lastFallTime = time.time() # reset lastFallTime
200.
201.            if not isValidPosition(board, fallingPiece):
202.                return # can't fit a new piece on the board, so game over
203.
204.        checkForQuit()
```

游戏的主循环从第 194 行开始，它负责在砖块落向底部的过程中，游戏的主要部分的代码。在下落的砖块已经着陆之后，fallingPiece 变量设置为 None。这意味着 nextPiece 中的砖块将会复制到 fallingPiece 变量中，并且，会将一个随机的新的砖块放入到 nextPiece 中。新的砖块可以通过 getNewPiece()函数生成。lastFallTime 变量也重新设置为当前时间，以便砖块能够在 fallFreq 中所设置的那么多秒之内下落。

getNewPiece()生成的砖块放置于游戏板上方一点的位置，通常这个砖块的一部分已经在游戏板区域之内了。但是，如果由于游戏板已经填满了（在这种情况下，在第 201 行调用 isValidPosition()将会返回 False），而导致这是一个无效的位置，那么，我们知道游戏板已经满了，玩家失败了。当发生这种情况的时候，runGame()函数将会返回。

7.11 事件处理循环

```
205.    for event in pygame.event.get(): # event handling loop
206.        if event.type == KEYUP:
```

事件处理循环负责玩家旋转下落的砖块、移动下落的砖块或者暂停游戏。

7.12 暂停游戏

```
207.            if (event.key == K_p):
208.                # Pausing the game
209.                DISPLAYSURF.fill(BGCOLOR)
```

第 7 章　Tetromino

```
210.                    pygame.mixer.music.stop()
211.                    showTextScreen('Paused') # pause until a key press
212.                    pygame.mixer.music.play(-1, 0.0)
213.                    lastFallTime = time.time()
214.                    lastMoveDownTime = time.time()
215.                    lastMoveSidewaysTime = time.time()
```

如果玩家按下了 P 键，游戏应该暂停。我们需要向玩家隐藏游戏板（否则，玩家可以通过暂停游戏并花时间来决定向哪里移动砖块，从而作弊）。代码通过调用 DISPLAYSURF.fill (BGCOLOR)使得显示 Surface 成为空白，并且停止音乐。调用 showTextScreen()函数来显示"Paused"文本，以等待玩家按下一个键以继续游戏。

一旦玩家按下了一个键，showTextScreen()将返回。第 212 行将重新设置背景音乐。此外，由于自玩家暂停游戏之后，可能经过了很长的时间，lastFallTime、lastMoveDownTime 和 lastMoveSidewaysTime 变量应该都重新设置为当前时间(这在第 213 行到第 215 行完成)。

7.13　使用移动变量来处理用户输入

```
216.                    elif (event.key == K_LEFT or event.key == K_a):
217.                        movingLeft = False
218.                    elif (event.key == K_RIGHT or event.key == K_d):
219.                        movingRight = False
220.                    elif (event.key == K_DOWN or event.key == K_s):
221.                        movingDown = False
```

松开一个箭头键（或者 WSAD 键）将会把 movingLeft、movingRight 或 movingDown 变量设置为 False，表示玩家不再想要让砖块朝着该方向移动。随后的代码，将根据这些"moving"变量中的 Boolean 值来确定做什么。注意，向上箭头键和 W 键用于旋转砖块，而不是向上移动砖块。这就是为什么没有 movingUp 变量的原因。

7.14　检查移动或旋转是否有效

```
223.                elif event.type == KEYDOWN:
224.                    # moving the block sideways
225.                    if (event.key == K_LEFT or event.key == K_a) and isValidPosition(board, fallingPiece, adjX=-1):
226.                        fallingPiece['x'] -= 1
227.                        movingLeft = True
```

```
228.            movingRight = False
229.            lastMoveSidewaysTime = time.time()
```

当按下向左箭头键的时候（并且向左移动下落的砖块是有效的，通过调用 isValid Position()来确定是否有效），那么，我们应该通过将 fallingPiece['x']的值减去 1，从而把砖块的位置向左移动一个空格。isValidPosition()函数还有名为 adjX 和 adjY 的可选参数。通常，isValidPosition()函数检查作为第二个参数传递的砖块对象所提供的位置数据。然而，有时候我们不想要检查砖块的当前位置，而是要检查该位置之上的一些空格。

如果给 adjX("adjusted X"的缩写)传递-1，那么它不会检查砖块的数据结构中的位置的有效性，而是检查砖块所处的位置的左边的一个空格是否是有效的。给 adjX 传递 1 的话，将会检查砖块右边的一个空格。还有一个可选的 adjY 参数。传递-1 给 adjY，将会检查砖块当前所在位置的上面的一个空格，而给 adjY 传递值 3，将会检查砖块所在位置下面的 3 个空格。

movingLeft 变量设置为 True，并且为了确保落下的砖块不会既向左又向右移动，在第 228 行将 movingRight 变量设置为 False。在第 229 行，lastMoveSidewaysTime 变量将更新为当前时间。

设置了这些变量，以便玩家能够只是按住方向箭头键以保持砖块移动。如果 movingLeft 变量设置为 True，程序就知道已经按下了向左箭头键（或 A 键）并且没有松开它。如果自 lastMoveSidewaysTime 中存储的时间之后经过了 0.15 秒（这个数字存储在 MOVESIDEWAYSFREQ 中），那么，程序是时候将下落的砖块再次向左移动了。

lastMoveSidewaysTime 的作用，就像是 lastClickTime 变量在第 5 章的 Simulate 程序中的作用一样。

```
231.            elif (event.key == K_RIGHT or event.key == K_d) and
isValidPosition(board, fallingPiece, adjX=1):
232.                fallingPiece['x'] += 1
233.                movingRight = True
234.                movingLeft = False
235.                lastMoveSidewaysTime = time.time()
```

第 231 行到第 235 行的代码几乎与第 225 行到第 229 行的代码相同，只不过它处理的是在按下向右箭头键（或 D 键）的时候，将下落的砖块向右移动。

```
237.            # rotating the block (if there is room to rotate)
238.            elif (event.key == K_UP or event.key == K_w):
```

第 7 章 Tetromino

```
239.                    fallingPiece['rotation'] = (fallingPiece['rotation'] +
1) % len(SHAPES[fallingPiece['shape']])
```

向上箭头键（或 W 键）将会把砖块旋转为其下一个旋转状态。代码所需要做的只是将 fallingPiece 字典中的'rotation'键的值增加 1。然而，如果增加'rotation'键的值大于旋转的总数目，那么，用该形状可能的旋转的总数目（这就是 len(SHAPES[falling Piece['shape']]的含义）来模除它，然后，这个值将回滚到从 0 开始。如下是 J 形状的这一模除的示例，该形状有 4 种可能的旋转。

```
>>> 0 % 4
0
>>> 1 % 4
1
>>> 2 % 4
2
>>> 3 % 4
3
>>> 5 % 4
1
>>> 6 % 4
2
>>> 7 % 4
3
>>> 8 % 4
0
>>>
```

```
240.                    if not isValidPosition(board, fallingPiece):
241.                        fallingPiece['rotation'] =
(fallingPiece['rotation'] - 1) % len(SHAPES[fallingPiece['shape']])
```

如果由于新的旋转位置与游戏板上已有的一些方块重叠而导致新的旋转位置无效，那么，我们想要通过从 fallingPiece['rotation']减去 1 而切换回最初的旋转。我们也可以使用 len(SHAPES[fallingPiece['shape']])来模除它，以便如果新的值是-1，模除将其改为列表中的最后一次旋转。如下是模除一个负数的示例。

```
>>> -1 % 4
3
```

```
242.                elif (event.key == K_q): # rotate the other direction
243.                    fallingPiece['rotation'] = (fallingPiece['rotation'] -
1) % len(SHAPES[fallingPiece['shape']])
244.                    if not isValidPosition(board, fallingPiece):
```

```
245.                        fallingPiece['rotation'] =
(fallingPiece['rotation'] + 1) % len(SHAPES[fallingPiece['shape']])
```

第 242 行到第 245 行所做的事情和第 238 行到第 241 行相同，只不过它们所处理的是玩家已经按下了 Q 键的情况，这个键将按照相反的方向旋转砖块。在这种情况下，我们从 fallingPiece['rotation']减去 1（这在第 243 行进行），而不是加上 1。

```
247.            # making the block fall faster with the down key
248.            elif (event.key == K_DOWN or event.key == K_s):
249.                movingDown = True
250.                if isValidPosition(board, fallingPiece, adjY=1):
251.                    fallingPiece['y'] += 1
252.                lastMoveDownTime = time.time()
```

如果按下了向下箭头键或者 S 键，说明玩家想要砖块下落得比正常速度更快一些。第 251 行将砖块在游戏板上向下移动一个空格（但是，只有在这是一个有效的空格的时候，才会这么做）。movingDown 变量设置为 True，并且 lastMoveDownTime 重新设置为当前时间。随后将检查这些变量，以确保只要按下向下箭头键或 S 键的时候，砖块就会以较快的速率下落。

7.15 找到底部

```
254.            # move the current block all the way down
255.            elif event.key == K_SPACE:
256.                movingDown = False
257.                movingLeft = False
258.                movingRight = False
259.                for i in range(1, BOARDHEIGHT):
260.                    if not isValidPosition(board, fallingPiece,
adjY=i):
261.                        break
262.                fallingPiece['y'] += i - 1
```

当玩家按下空格键的时候，砖块会立即向下坠落，直到碰到游戏板或着陆。程序首先需要计算砖块能够移动多少个空格才会着陆。第 256 行到第 258 行将会把所有移动变量设置为 False（这会使得程序后面部分的代码认为用户已经松开了曾经按下的一个箭头键）。之所以这么做，是因为代码将会把砖块移动到一个绝对的底部并且开始下落下一个砖块，并且我们不想出现这样的情况，即只不过因为玩家曾经在敲击空格键的时候按下了某一个箭头键，新的下落砖块马上就开始移动。

要搞清楚砖块最远能够落到哪里，我们应该先调用 isValidPosition() 并为 adjY 参数传入整数 1。如果 isValidPosition() 返回 False，我们知道，砖块没办法落得更远并且已经到达了底部。如果 isValidPosition() 返回 True，那我们知道它还可以下落一个空格。

在这种情况下，我们应该将 adjY 设置为 2 来调用 isValidPosition()。如果它再次返回 True，我们将 adjY 设置为 3 来调用 isValidPosition()，依次类推。这就是为什么第 259 行的 for 循环是这样做的原因，即以不断增加的整数值传递给 adjY 来调用 isValidPosition()，直到该函数调用返回 false 为止。此时，我们知道，i 中的值是超过底部 1 个空格的位置。这就是为什么第 262 行将 fallingPiece['y'] 增加 i-1 而不是 i 的原因。

还要注意，第 259 行的 for 循环的 range() 的第二个参数设置为 BOARDHEIGHT，因为这是砖块在必然碰到游戏板底部之前所能够下落的最大的量。

7.16 通过按下按键来移动

```
264.        # handle moving the block because of user input
265.        if (movingLeft or movingRight) and time.time() - lastMoveSidewaysTime > MOVESIDEWAYSFREQ:
266.            if movingLeft and isValidPosition(board, fallingPiece, adjX=-1):
267.                fallingPiece['x'] -= 1
268.            elif movingRight and isValidPosition(board, fallingPiece, adjX=1):
269.                fallingPiece['x'] += 1
270.            lastMoveSidewaysTime = time.time()
```

还记得吧，在第 227 行，如果玩家按下了向左箭头键，将 movingLeft 变量设置为 True（第 233 行也是相同的，如果玩家按下了向右箭头键，将 movingRight 设置为 True）。如果用户松开了这些键，还将该移动变量设置回 False（参见第 217 行和第 219 行）。

当玩家按下向左箭头或向右箭头键的时候，还发生了一件事情，就是将 lastMoveSidewaysTime 变量设置为当前时间（也就是 time.time() 的返回值）。如果玩家继续按住了箭头键而没有松开，那么 movingLeft 或 movingRight 变量还将继续设置为 True。

如果玩家按住按键超过了 0.15 秒（存储在 MOVESIDEWAYSFREQ 中的值是浮点数 0.15），那么，表达式 time.time() - lastMoveSidewaysTime > MOVESIDEWAYSFREQ 将为 True。如果玩家按住了箭头键并且超过了 0.15 秒，第 265 行的条件为 True，并且在这种情况下，我们应该将下落的砖块向左或向右移动，即便用户并没有再次按下箭头键。

这是非常有用的，因为让玩家重复地按箭头键以不断地让下落的砖块在游戏板上移动多个空格，这可能会变得令人厌烦。相反，他们可以只是按住一个箭头键，砖块将持续移动直到他们松开按键。当发生这种松开按键的情况的时候，第 216 行到第 221 行将会把移动变量设置为 False，从而使得第 265 行的条件为 False。这就可以防止下落的砖块移动得太频繁。

为了展示为什么当 MOVESIDEWAYSFREQ 中的秒数经过之后，time.time() - lastMoveSidewaysTime > MOVESIDEWAYSFREQ 会返回 True，可以运行如下这个简短的程序。

```
import time

WAITTIME = 4
begin = time.time()

while True:
    now = time.time()
    message = '%s, %s, %s' % (begin, now, (now - begin))
    if now - begin > WAITTIME:
        print(message + ' PASSED WAIT TIME!')
    else:
        print(message + ' Not yet...')
    time.sleep(0.2)
```

该程序拥有一个无限循环，要终止该循环，按下 Ctrl-C 键。程序的输出将如下所示。

```
1322106392.2, 1322106392.2, 0.0 Not yet...
1322106392.2, 1322106392.42, 0.219000101089 Not yet...
1322106392.2, 1322106392.65, 0.449000120163 Not yet...
1322106392.2, 1322106392.88, 0.680999994278 Not yet...
1322106392.2, 1322106393.11, 0.910000085831 Not yet...
1322106392.2, 1322106393.34, 1.1400001049 Not yet...
1322106392.2, 1322106393.57, 1.3710000515 Not yet...
1322106392.2, 1322106393.83, 1.6360001564 Not yet...
1322106392.2, 1322106394.05, 1.85199999809 Not yet...
1322106392.2, 1322106394.28, 2.08000016212 Not yet...
1322106392.2, 1322106394.51, 2.30900001526 Not yet...
1322106392.2, 1322106394.74, 2.54100012779 Not yet...
1322106392.2, 1322106394.97, 2.76999998093 Not yet...
1322106392.2, 1322106395.2, 2.99800014496 Not yet...
1322106392.2, 1322106395.42, 3.22699999809 Not yet...
1322106392.2, 1322106395.65, 3.45600008965 Not yet...
1322106392.2, 1322106395.89, 3.69200015068 Not yet...
1322106392.2, 1322106396.12, 3.92100000381 Not yet...
1322106392.2, 1322106396.35, 4.14899992943 PASSED WAIT TIME!
1322106392.2, 1322106396.58, 4.3789999485 PASSED WAIT TIME!
```

第 7 章 Tetromino

```
1322106392.2, 1322106396.81, 4.60700011253 PASSED WAIT TIME!
1322106392.2, 1322106397.04, 4.83700013161 PASSED WAIT TIME!
1322106392.2, 1322106397.26, 5.06500005722 PASSED WAIT TIME!
Traceback (most recent call last):
  File "C:\timetest.py", line 13, in <module>
    time.sleep(0.2)
KeyboardInterrupt
```

每一行输出的第一个数字是当程序初次启动时 time.time() 的返回值（这个值总是不变的）。第二个数字是 time.time() 最近的返回值（这个值会随着循环的每一次迭代而不断更新）。第三个数字是当前的时间减去起始的时间，即从 begin = time.time() 这行代码执行后所经过的秒数。

如果这个数字大于 4，代码将会开始打印"PASSED WAIT TIME"，而不是"Not yet..."。游戏程序就是通过这样而获知从一行代码运行之后是否经过了一定的时间。

在 Tetromino 程序中，time.time() – lastMoveSidewaysTime 表达式将会求得从上一次 lastMoveSidewaysTime 设置为当前时间之后所经过的秒数。如果这个值比 MOVESIDEWAYSFREQ 中的值要大，我们知道，是时间将下落的砖块多移动一个空格了。

不要忘了将 lastMoveSidewaysTime 再次更新为当前时间。这正是第 270 行代码所做的事情。

```
272.            if movingDown and time.time() - lastMoveDownTime > MOVEDOWNFREQ and isValidPosition(board, fallingPiece, adjY=1):
273.                fallingPiece['y'] += 1
274.                lastMoveDownTime = time.time()
```

第 272 行到第 274 行所做的事情几乎与第 265 行到第 270 行代码所做的事情相同，只不过是将移动下落的砖块向下移动。有一个单独的移动变量（movingDown）和"最后时间"变量（lastMoveDownTime），以及一个不同的"移动频率"变量（MOVEDOWNFREQ）。

7.17 让砖块"自然"落下

```
276.            # let the piece fall if it is time to fall
277.            if time.time() - lastFallTime > fallFreq:
278.                # see if the piece has landed
279.                if not isValidPosition(board, fallingPiece, adjY=1):
280.                    # falling piece has landed, set it on the board
281.                    addToBoard(board, fallingPiece)
282.                    score += removeCompleteLines(board)
283.                    level, fallFreq = calculateLevelAndFallFreq(score)
284.                    fallingPiece = None
```

```
285.        else:
286.            # piece did not land, just move the block down
287.            fallingPiece['y'] += 1
288.            lastFallTime = time.time()
```

砖块自然地向下移动（落下）的速率由 lastFallTime 变量记录。从下落的砖块上一次下落一格，如果经过了足够的时间，第 279 行到第 288 行将会让砖块下落一格。

如果第 279 行的条件是 True，那么，砖块已经着陆了。调用 addToBoard() 将会使得在游戏板数据结构中产生砖块的部分（以便将来的砖块能够在其上着陆），并且 removeCompleteLines() 调用将负责删除掉游戏板上任何已经填充完整的行，并且将方块向下推动。removeCompleteLines() 函数还会返回一个整数值，表明消除了多少行，以便我们将这个数字加到得分上。

由于分数已经修改了，我们调用 calculateLevelAndFallFreq() 函数来更新当前的关卡以及砖块下落的频度。最后，我们将 fallingPiece 变量设置为 None，以表示下一个砖块应该变为新的下落砖块，并且应该生成一个随机的新砖块作为下一个砖块（这在游戏循环开始的第 195 行到第 199 行完成）。

如果砖块没有着陆，我们直接将其 Y 位置向下设置一个空格（在第 287 行），并且将 lastFallTime 重置为当前时间（第 288 行）。

7.18 将所有内容绘制到屏幕上

```
290.    # drawing everything on the screen
291.    DISPLAYSURF.fill(BGCOLOR)
292.    drawBoard(board)
293.    drawStatus(score, level)
294.    drawNextPiece(nextPiece)
295.    if fallingPiece != None:
296.        drawPiece(fallingPiece)
297.
298.    pygame.display.update()
299.    FPSCLOCK.tick(FPS)
```

既然游戏循环已经处理了所有的事件并且更新了游戏状态，游戏循环只需要将游戏状态绘制到屏幕上。大多数的绘制由其他的函数处理，因此，游戏循环代码只需要调用这些函数就行了。然后，调用 pygame.display.update() 以使得显示 Surface 真正出现在计算机屏幕上，并且调用 tick() 方法给游戏添加一个短暂的暂停以使得游戏不会运行得太快。

7.19 制作文本的快捷函数 makeTextObjs()

```
302. def makeTextObjs(text, font, color):
303.     surf = font.render(text, True, color)
304.     return surf, surf.get_rect()
```

makeTextObjs()函数只不过为我们提供了一条捷径。给定文本、Font 对象和一个 Color 对象，该函数调用 render()并且返回用于该文本的 Surface 和 Rect 对象。这使得我们不必在每次需要 Surface 和 Rect 对象的时候输入代码来创建它们。

7.20 相同的旧的 terminate()函数

```
307. def terminate():
308.     pygame.quit()
309.     sys.exit()
```

terminate()函数和在前面的游戏程序中的作用相同。

7.21 使用 checkForKeyPress()函数等待按键事件

```
312. def checkForKeyPress():
313.     # Go through event queue looking for a KEYUP event.
314.     # Grab KEYDOWN events to remove them from the event queue.
315.     checkForQuit()
316.
317.     for event in pygame.event.get([KEYDOWN, KEYUP]):
318.         if event.type == KEYDOWN:
319.             continue
320.         return event.key
321.     return None
```

checkForKeyPress()函数和它在 Wormy 游戏中所做的事情相同。首先，它调用 checkForQuit()来处理任何的 QUIT 事件（或者是专门针对 Esc 键的 KEYUP 事件），如果有任何这样的事件，就会终止程序。然后，它从事件队列中提取出所有的 KEYUP 和 KEYDOWN 事件。它会忽略掉任何的 KEYDOWN 事件（针对 pygame.event.get()指定了 KEYDOWN，从而从事件队列中清除掉该类事件）。如果事件队列中没有 KEYUP 事件，那么，该函数返回 None。

7.22 通用文本屏幕函数 showTextScreen()

```
324. def showTextScreen(text):
325.     # This function displays large text in the
326.     # center of the screen until a key is pressed.
327.     # Draw the text drop shadow
328.     titleSurf, titleRect = makeTextObjs(text, BIGFONT, TEXTSHADOWCOLOR)
329.     titleRect.center = (int(WINDOWWIDTH / 2), int(WINDOWHEIGHT / 2))
330.     DISPLAYSURF.blit(titleSurf, titleRect)
331.
332.     # Draw the text
333.     titleSurf, titleRect = makeTextObjs(text, BIGFONT, TEXTCOLOR)
334.     titleRect.center = (int(WINDOWWIDTH / 2) - 3, int(WINDOWHEIGHT / 2) - 3)
335.     DISPLAYSURF.blit(titleSurf, titleRect)
336.
337.     # Draw the additional "Press a key to play." text.
338.     pressKeySurf, pressKeyRect = makeTextObjs('Press a key to play.', BASICFONT, TEXTCOLOR)
339.     pressKeyRect.center = (int(WINDOWWIDTH / 2), int(WINDOWHEIGHT / 2) + 100)
340.     DISPLAYSURF.blit(pressKeySurf, pressKeyRect)
```

我们将创建一个名为 showTextScreen() 的通用函数，而不是分别为开始屏幕和游戏结束屏幕创建函数。showTextScreen() 函数将会把我们传递给 text 参数的任何文本进行绘制，此外，还会额外地显示文本"Press a key to play"。

注意，第328行到第330行先以暗灰色绘制文本，随后第333行到第335行再次绘制相同的文本，只不过向左和向上偏移3个像素。这会创建一种"下拉阴影"的效果，使得文本看上去更漂亮。你可以通过注释掉第328行到第330行，看看没有下拉阴影的文本，以比较一下差异。

showTextScreen() 将用于开始屏幕、游戏结束屏幕，并且也用于暂停屏幕（本章后面将会介绍暂停屏幕）。

```
342.     while checkForKeyPress() == None:
343.         pygame.display.update()
344.         FPSCLOCK.tick()
```

我们想要文本停留在屏幕上直到用户按下一个键。这个小的循环将会持续调用 pygame.display.update() 和 FPSCLOCK.tick()，直到 checkForKeyPress() 返回一个不是 None 的值。当用户按下一个键的时候，会发生这种情况。

7.23　checkForQuit()函数

```
347. def checkForQuit():
348.     for event in pygame.event.get(QUIT): # get all the QUIT events
349.         terminate() # terminate if any QUIT events are present
350.     for event in pygame.event.get(KEYUP): # get all the KEYUP events
351.         if event.key == K_ESCAPE:
352.             terminate() # terminate if the KEYUP event was for the Esc key
353.         pygame.event.post(event) # put the other KEYUP event objects back
```

　　checkForQuit()函数可能会调用来处理将会导致游戏终止的任何事件。如果事件队列中有任何的 QUI 事件的话（这在第 348 行和第 349 行处理），或者如果有针对 Esc 键的一个 KEYUP 事件的话，就会发生这种情况。玩家应该能够在任何时候按下 Esc 键来退出程序。

　　由于第 350 行的 pygame.event.get()调用提取了所有的 KEYUP 事件（包括针对 Esc 键以外的键的事件），如果该事件不是针对 Esc 键的，我们想要通过调用 pygame.event. post()函数将其放回到事件队列中。

7.24　calculateLevelAndFallFreq()函数

```
356. def calculateLevelAndFallFreq(score):
357.     # Based on the score, return the level the player is on and
358.     # how many seconds pass until a falling piece falls one space.
359.     level = int(score / 10) + 1
360.     fallFreq = 0.27 - (level * 0.02)
361.     return level, fallFreq
```

　　每次玩家填满一行，其分数都将增加 1 分。每增加 10 分，游戏就进入下一关卡，砖块下落得会更快。关卡和下落的频度都是通过传递给该函数的分数来计算的。

　　要计算关卡，我们使用 int()来舍入除以 10 以后的分数。因此，如果分数是 0 到 9 之间的任何数字，int()调用会将其舍入到 0。代码这里的+1 部分，是因为我们想要第一个关卡作为第 1 关，而不是第 0 关。当分数达到 10 分的时候，int(10 / 10)将会计算为 1，并且+1 将会使得关卡变为 2。图 7-6 展示了分数从 1 到 34 时关卡值的变化。

　　为了计算下落的频度，我们首先有一个基准时间 0.27（这意味着每 0.27 秒，砖块会自然地下落一次）。然后，我们将关卡值乘以 0.02，并且从基准时间 0.27 中减去它。因此，对于关卡 1，我们从 0.27 中减去 0.02 * 1 (0.02) 得到 0.25。在关卡 2 中，我们减去 0.02 * 2 (0.04) 得到 0.23。我们可以这样来看待算式的 level * 0.02 部分，对于每一个关卡来说，砖块下落的速度都会比之前的关卡快上 0.02 秒。

7.24 calculateLevelAndFallFreq()函数

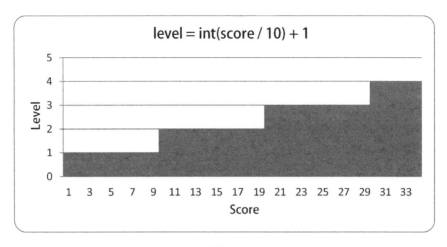

图 7-6

我们也可以通过图 7-7 来表示在游戏的每一个关卡中砖块下落得有多快。

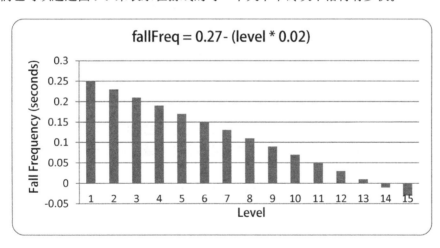

图 7-7

你可以看到，在关卡 14，下落的频度已经小于 0。这不会引起代码中的任何 Bug，因为第 277 行检查从下落的砖块上一次下降一个空格开始所经过的时间，是否比所计算的下落频度要大。因此，如果下落的频度为负值，那么，第 227 行中的条件将总是为 True，并且砖块将在游戏循环的每一次迭代的时候落下。对于第 14 关及其以上的关卡，砖块将无法下落得更快。

如果 FPS 设置为 25，这意味着在达到第 14 关的时候，下落的砖块将每秒钟下降 25 个空格。考虑一下，游戏板只有 20 个空格那么高，这意味着，玩家将会只有少于 1 秒的时间来设置每一个砖块。

如果想要砖块以一个较慢的速率开始（如果你能明白我的意思的话）下落，可以修改 calculateLevelAndFallFreq()所使用的算式。例如，假设第 360 行如下所示。

```
360.        fallFreq = 0.27 - (level * 0.01)
```

在上面的例子中，砖块将会在每一关下落得比前一关快 0.01 秒，而不是 0.02 秒。图形看上去如图 7-8 所示（图形中最开始的一行总是灰色的）。

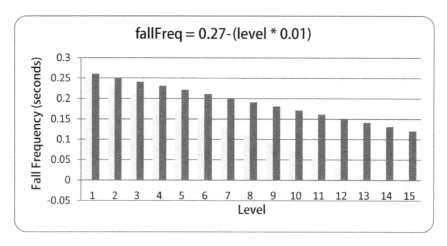

图 7-8

正如你所看到的，用这个新的算式，第 14 关只是像最初的第 7 关那么难。可以通过修改 calculateLevelAndFallFreq()中的算式，将游戏修改得更难或者更容易。

7.25　用函数 getNewPiece()产生新的砖块

```
363. def getNewPiece():
364.     # return a random new piece in a random rotation and color
365.     shape = random.choice(list(SHAPES.keys()))
366.     newPiece = {'shape': shape,
367.                 'rotation': random.randint(0, len(SHAPES[shape]) - 1),
368.                 'x': int(BOARDWIDTH / 2) - int(TEMPLATEWIDTH / 2),
369.                 'y': -2, # start it above the board (i.e. less than 0)
370.                 'color': random.randint(0, len(COLORS)-1)}
371.     return newPiece
```

getNewPiece()函数产生一个随机的砖块，放置于游戏板的顶部。首先，为了随机地选择砖块的形状，我们通过在第 365 行调用 list(SHAPES.keys())来创建所有可能的形状的一个列表。keys()字典方法返回"dict_keys"数据类型的一个值，必须先使用 list()函数将其转换为一个列表值，然后才能将其传递给 random.choice()。这是因为 random.choice()函数只接受列表值作为其参数。然后，random.choice()函数随机地返回列表中的一项的值。砖块数据结构只不过是带有键'shape'、'rotation'、'x'、'y'和'color'的一个字典值。

'rotation'键的值是 0 和比该形状拥有的可能的旋转数目少 1 之间的一个随机整数。可以通过表达式 len(SHAPES[shape])得到一个形状所拥有的旋转的数目。

注意，我们没有把字符串值的列表（就像存储在 S_SHAPE_TEMPLATE 常量中的那个一样）存储到每一个砖块数据结构中，以表示每个砖块的方块。相反，我们只是存储了形状和旋转的一个索引，并将其命名为 PIECES 常量。

'x'键的值总是要设置为游戏板的中间（也考虑到砖块自身的宽度，这可以从TEMPLATEWIDTH 常量获取）。'y'键的值总是要设置为-2，以便将其放置到游戏板上面一点点（游戏板的首行是第 0 行）。

由于 COLORS 常量是不同颜色的一个元组，选取从 0 到 COLORS 的长度（减去 1）之间的一个随机数，将会给我们一个随机的索引值以用于砖块的颜色。一旦 newPiece 字典中所有的值都设置了，getNewPiece()函数将返回 newPiece。

7.26 给游戏板数据结构添加砖块

```
374. def addToBoard(board, piece):
375.     # fill in the board based on piece's location, shape, and rotation
376.     for x in range(TEMPLATEWIDTH):
377.         for y in range(TEMPLATEHEIGHT):
378.             if SHAPES[piece['shape']][piece['rotation']][y][x] != BLANK:
379.                 board[x + piece['x']][y + piece['y']] = piece['color']
```

游戏板数据结构是用来记录之前着陆的砖块占据了哪些矩形空格的一种数据表示。当前下落的砖块在游戏板数据结构中并没有标记。addToBoard()函数所做的事情是接受一个砖块数据结构，并且将其方块添加到游戏板数据结构中。这在一个砖块着陆之后进行。

第 376 行到第 377 行的嵌套 for 循环，遍历了砖块数据结构中的每一个空格，并且如果找到空格中的一个方块（第 378 行），它将其添加到游戏板中（第 379 行）。

7.27 创建一个新的游戏板数据结构

```
382. def getBlankBoard():
383.     # create and return a new blank board data structure
```

第 7 章　Tetromino

```
384.    board = []
385.    for i in range(BOARDWIDTH):
386.        board.append([BLANK] * BOARDHEIGHT)
387.    return board
```

用于游戏板的数据结构相当简单，它是值的列表的一个列表。如果值和 BLANK 中的值相同，那么，这是一个空的空格。如果这个值是一个整数，那么，它表示一个方块，其颜色由 COLORS 常量列表中的整数索引确定，即 0 表示蓝色，1 表示绿色，2 表示红色，3 表示黄色。

要创建一个空白的游戏板，使用列表复制来创建 BLANK 值的一个列表，以表示一个列。这在第 386 行完成。针对游戏板中的每一列，都创建这样的一个列表（这就是 5 行的 for 循环所做的事情）。

7.28　isOnBoard()和 isValidPosition()函数

```
390. def isOnBoard(x, y):
391.    return x >= 0 and x < BOARDWIDTH and y < BOARDHEIGHT
```

isOnBoard()是一个简单的函数，它要检查所传递的 XY 坐标所表示的有效值是否存在于游戏板之上。只要 XY 坐标都不小于 0，并且不大于或等于 BOARDWIDTH 和 BOARDHEIGHT 常量，该函数就返回 True。

```
394. def isValidPosition(board, piece, adjX=0, adjY=0):
395.    # Return True if the piece is within the board and not colliding
396.    for x in range(TEMPLATEWIDTH):
397.        for y in range(TEMPLATEHEIGHT):
398.            isAboveBoard = y + piece['y'] + adjY < 0
399.            if isAboveBoard or SHAPES[piece['shape']][piece['rotation']][y][x] == BLANK:
400.                continue
```

给 isValidPosition()函数一个游戏板数据结构和一个砖块数据结构，如果砖块中所有的方块都在游戏板之上，并且没有和游戏板上的任何方块重叠，那么，该函数就返回 True。这通过接受砖块的 XY 坐标（实际上是 5×5 的砖块的左上角方块的坐标），并且将该坐标添加到砖块数据结构中而做到。图 7-9 有助于说明这一点。

在图 7-9 中左边的游戏板上，下落的砖块，即下落的砖块的左上角的 XY 坐标是(2, 3)，位于游戏板之上。但是，下落的砖块的坐标系统中的方块，拥有它们自己的坐标系统。要找到这些砖块的"游戏板"坐标系统，我们只需要将下落的砖块的左上角的"游戏板"坐标和

7.28 isOnBoard()和 isValidPosition()函数

方块的"砖块"坐标相加。

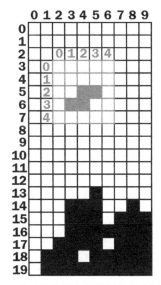 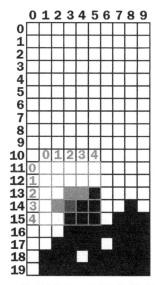

下落砖块处于有效位置的游戏板　　　　下落砖块处于无效位置的游戏板

图 7-9

在左边的游戏板中，下落的砖块的方块位于如下的"砖块"坐标中。

(2, 2) (3, 2) (1, 3) (2, 3)

当我们给这些坐标加上(2, 3)坐标（砖块在游戏板上的坐标）的时候，如下所示。

(2 + 2, 2 + 3) (3 + 2, 2 + 3) (1 + 2, 3 + 3) (2 + 2, 3 + 3)

在把(2, 3)坐标相加之后，这些方块位于如下的"游戏板"坐标中。

(4, 5) (5, 5) (3, 6) (4, 6)

现在，我们可以算出下落的砖块的方块的游戏板坐标了，我们可以看到它们是否与游戏板上已经着陆的方块重叠。第 396 行和第 397 行的嵌套的 for 循环遍历了下落的砖块上的每一个可能的坐标。

我们想要检查下落的砖块上的一个方块是否跑到游戏板之外了，或者是否与游戏板上的一个方块重叠了（尽管有一种例外的情况是方块在游戏板之外的上方位置。当下落的砖块刚开始下落的时候，可能会有这种情况）。第 398 行创建了一个名为 isAboveBoard 的变量，如果下落的砖块上的方块的坐标的 x 和 y 位置在游戏板之上的话，将其设置为 True。否则的话，将其设置为 False。

第 399 行的 if 语句检查砖块所在的空格是否在游戏板之上或者是空白。如果任何一项为 True，那么，代码将执行 continue 语句，并且进入下一次迭代（注意，第 399 行末尾是[y][x]，而不是[x][y]。这是因为 PIECES 数据结构的坐标是相反的。参见 7.6 节）。

第 7 章 Tetromino

```
401.            if not isOnBoard(x + piece['x'] + adjX, y + piece['y'] + adjY):
402.                return False
403.            if board[x + piece['x'] + adjX][y + piece['y'] + adjY] != BLANK:
404.                return False
405.    return True
```

第 401 行的 if 语句检查砖块的方块是否不在游戏板之上。第 403 行的 if 语句检查到砖块的方块所在的游戏板空格是否不为空白。如果这些条件中的任何一个为 True，isValidPosition()函数将返回 False。注意，这些 if 语句还根据传递给该函数的 adjX 和 adjY 参数调整了坐标。

如果代码执行了嵌套的 for 循环并且没有找到返回 False 的理由，那么，砖块的位置必须是有效的，因此，函数在第 405 行返回 True。

7.29 检查、删除和填满一行

```
407. def isCompleteLine(board, y):
408.    # Return True if the line filled with boxes with no gaps.
409.    for x in range(BOARDWIDTH):
410.        if board[x][y] == BLANK:
411.            return False
412.    return True
```

isCompleteLine 对参数 y 所指定的行执行一次简单的检查。当游戏板上的一行中的每一个空格都被一个方块填满的时候，我们认为这一行"填满"了。第 409 行的 for 循环遍历该行的每一个空格。如果一个空格是空白的（判断标准是它拥有和 BLANK 常量相同的值），那么，该函数返回 False。

```
415. def removeCompleteLines(board):
416.    # Remove any completed lines on the board, move everything above them down, and return the number of complete lines.
417.    numLinesRemoved = 0
418.    y = BOARDHEIGHT - 1 # start y at the bottom of the board
419.    while y >= 0:
```

removeCompleteLines()函数将会找出所传递的游戏板数据结构中的任何填满的行，删除这些行，然后，将其游戏板上该行之上的所有方块都向下移动一行。该函数将返回所删除的行的数目（这通过 numLinesRemoved 变量来记录），以便可以将这个数据添加到得分之中。

7.29 检查、删除和填满一行

该函数的工作方式是使用从最低的行开始的 y 变量，即 BOARDHEIGHT - 1，运行从第 419 行开始的一个循环。无论何时，当 y 所指定的行没有填满的时候，y 将递减到下一个最高的行。当 y 变为-1 的时候，循环最终停止。

```
420.            if isCompleteLine(board, y):
421.                # Remove the line and pull boxes down by one line.
422.                for pullDownY in range(y, 0, -1):
423.                    for x in range(BOARDWIDTH):
424.                        board[x][pullDownY] = board[x][pullDownY-1]
425.                # Set very top line to blank.
426.                for x in range(BOARDWIDTH):
427.                    board[x][0] = BLANK
428.                numLinesRemoved += 1
429.                # Note on the next iteration of the loop, y is the same.
430.                # This is so that if the line that was pulled down is also
431.                # complete, it will be removed.
432.            else:
433.                y -= 1 # move on to check next row up
434.        return numLinesRemoved
```

如果 y 所指定的行填满了，isCompleteLine()将返回 True。在这种情况下，程序需要将所删除的行之上的每一行的值都复制到最近的、较低的一行上。这就是第 422 行的 for 循环所做的事情（这就是为什么从 y 开始，而不是从 0 开始调用 range()的原因。还要注意，这里使用了 range()函数的 3 个参数的形式，因此，它返回的列表从 y 开始，到 0 结束，并且在每一次迭代只有都"增加"-1）。

让我们看一下如下的示例。为了节省篇幅，这里只显示了游戏板中最上面的 5 行。第 3 行是一个填满的行，这意味着在其之上的所有行（第 2 行、第 1 行和第 0 行）都必须"向下拉"。首先，将第 2 行向下复制到第 3 行。在图 7-10 中，右边的游戏板展示了完成该操作之后的样子。

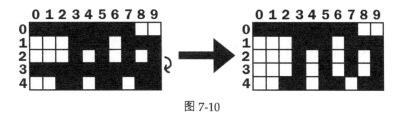

图 7-10

第 424 行代码的这种"下拉"实际上只是将较高行的值复制到其下面的一行。在第 2 行复制到第 3 行之后，第 1 行复制到第 2 行，然后是第 0 行复制到第 1 行，如图 7-11 所示。

第 0 行（最顶部的那一行）之上并没有需要向下复制的一行。但是，并不需要有一行复制到第 0 行之上，只需要将第 0 行的所有空格都设置为 BLANK。这是第 426 行和第 427 行代码所做的事情。在此之后，游戏板从图 7-12 左边所示的样子，变为图 7-12 右边所示的样子。

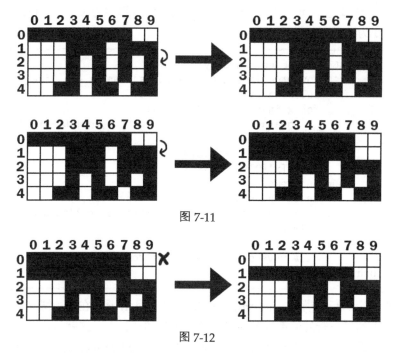

图 7-11

图 7-12

在删除了填满的行之后，代码执行到了第 419 行所开始的 while 循环的底部，因此，执行跳回到了循环的开始处。注意，删除一行并且向下拉动一行的时候，y 变量根本没有机会发生变化。因此，在下一次迭代的时候，y 变量指向和前面相同的一行。

这是需要的，因为如果有两个填满的行，那么，第二个填满的行也会被下拉，并且也会被删除。那么，代码随后将删除这个填满的行，然后进入到下一次迭代中。只有当这里没有一个填满的行的时候，y 变量才会在 433 行递减。一旦 y 变量一路递减到了 0，执行将会退出该 while 循环。

7.30 将游戏板坐标转换为像素坐标

```
437. def convertToPixelCoords(boxx, boxy):
438.     # Convert the given xy coordinates of the board to xy
439.     # coordinates of the location on the screen.
440.     return (XMARGIN + (boxx * BOXSIZE)), (TOPMARGIN + (boxy * BOXSIZE))
```

这个辅助函数将游戏板的方块坐标转换为像素坐标。该函数的工作方式与之前的游戏程序中所使用的其他的"转换坐标"函数相同。

7.31 在游戏板上或屏幕上的其他位置绘制方块

```
443. def drawBox(boxx, boxy, color, pixelx=None, pixely=None):
444.     # draw a single box (each tetromino piece has four boxes)
445.     # at xy coordinates on the board. Or, if pixelx & pixely
446.     # are specified, draw to the pixel coordinates stored in
447.     # pixelx & pixely (this is used for the "Next" piece).
448.     if color == BLANK:
449.         return
450.     if pixelx == None and pixely == None:
451.         pixelx, pixely = convertToPixelCoords(boxx, boxy)
452.     pygame.draw.rect(DISPLAYSURF, COLORS[color], (pixelx + 1, pixely + 1, BOXSIZE - 1, BOXSIZE - 1))
453.     pygame.draw.rect(DISPLAYSURF, LIGHTCOLORS[color], (pixelx + 1, pixely + 1, BOXSIZE - 4, BOXSIZE - 4))
```

drawBox()函数在屏幕上绘制一个单独的方块。该函数能够接受游戏板坐标的参数 boxx 和 boxy，表示应该绘制方块的位置。然而，如果指定了 pixelx 和 pixely 参数，那么这些像素坐标将覆盖 boxx 和 boxy 参数。pixelx 和 pixely 参数用来绘制"Next"砖块的方块，它并不在游戏板之上。

如果没有设置 pixelx 和 pixely 参数，那么，当函数第一次调用的时候，它们将默认地设置为 None。随后，第 450 行的 if 语句将会使用 convertToPixelCoords()的返回值来覆盖 None 值。这一调用会得到 boxx 和 boxy 所指定的游戏板坐标的像素坐标。

这段代码不会使用颜色填充整个方块的空格。为了使得砖块的方块之间有一条黑色的边框线，pygame.draw.rect()调用的 left 和 top 中都添加了一个+1，而 width 和 height 参数都加了一个 -1。为了绘制出突出显示的方块，首先在第 452 行使用较暗的颜色来绘制方块。然后，在第 453 行，在较暗的方块的顶部绘制一个稍微小一些的、明亮的方块。

7.32 将所有内容绘制到屏幕上

```
456. def drawBoard(board):
457.     # draw the border around the board
458.     pygame.draw.rect(DISPLAYSURF, BORDERCOLOR, (XMARGIN - 3, TOPMARGIN - 7, (BOARDWIDTH * BOXSIZE) + 8, (BOARDHEIGHT * BOXSIZE) + 8), 5)
459.
460.     # fill the background of the board
461.     pygame.draw.rect(DISPLAYSURF, BGCOLOR, (XMARGIN, TOPMARGIN, BOXSIZE * BOARDWIDTH, BOXSIZE * BOARDHEIGHT))
```

第 7 章 Tetromino

```
462.      # draw the individual boxes on the board
463.      for x in range(BOARDWIDTH):
464.          for y in range(BOARDHEIGHT):
465.              drawBox(x, y, board[x][y])
```

drawBoard()函数负责调用绘制函数，绘制游戏板边框和游戏板上的方块。首先，是要将游戏板于其上的 DISPLAYSURF，后面跟着游戏板的背景颜色。然后，针对游戏板上的每一个空格调用 drawBox()。drawBox()函数足够智能，如果 board[x][y]为 BLANK 的话，它会保留该方块为空格。

7.33　绘制得分和关卡文本

```
468. def drawStatus(score, level):
469.      # draw the score text
470.      scoreSurf = BASICFONT.render('Score: %s' % score, True, TEXTCOLOR)
471.      scoreRect = scoreSurf.get_rect()
472.      scoreRect.topleft = (WINDOWWIDTH - 150, 20)
473.      DISPLAYSURF.blit(scoreSurf, scoreRect)
474.
475.      # draw the level text
476.      levelSurf = BASICFONT.render('Level: %s' % level, True, TEXTCOLOR)
477.      levelRect = levelSurf.get_rect()
478.      levelRect.topleft = (WINDOWWIDTH - 150, 50)
479.      DISPLAYSURF.blit(levelSurf, levelRect)
```

drawStatus()函数负责绘制"Score:"和"Level:"信息的文本，它们会出现在屏幕的右上角。

7.34　在游戏板上或屏幕的其他位置绘制一个砖块

```
482. def drawPiece(piece, pixelx=None, pixely=None):
483.      shapeToDraw = SHAPES[piece['shape']][piece['rotation']]
484.      if pixelx == None and pixely == None:
485.          # if pixelx & pixely hasn't been specified, use the location
stored in the piece data structure
486.          pixelx, pixely = convertToPixelCoords(piece['x'], piece['y'])
487.
488.      # draw each of the blocks that make up the piece
489.      for x in range(TEMPLATEWIDTH):
490.          for y in range(TEMPLATEHEIGHT):
```

```
491.            if shapeToDraw[y][x] != BLANK:
492.                drawBox(None, None, piece['color'], pixelx + (x *
BOXSIZE), pixely + (y * BOXSIZE))
```

drawPiece()函数将会根据传递给它的砖块数据结构来绘制一个砖块的方块。这个函数用于绘制下落的砖块和"Next"砖块。由于砖块数据结构将包含所有的形状、位置、旋转和颜色信息，只需要将砖块数据结构传递给该函数就够了。

然而，"Next"砖块并不会绘制到游戏板上。在这种情况下，我们忽略砖块数据结构中包含的位置信息，而是让drawPiece()函数的调用者为pixelx和pixely参数传递实参，以指定应该将砖块确切地绘制到窗口上的什么位置。

如果没有传入pixelx和pixely参数，第484行和第486行将会使用convertToPixelCoords()调用的返回值来覆盖这些变量。

第489行和第490行的嵌套的for循环随后将针对需要绘制的砖块的每一个方块调用drawBox()。

7.35 绘制"Next"砖块

```
495. def drawNextPiece(piece):
496.     # draw the "next" text
497.     nextSurf = BASICFONT.render('Next:', True, TEXTCOLOR)
498.     nextRect = nextSurf.get_rect()
499.     nextRect.topleft = (WINDOWWIDTH - 120, 80)
500.     DISPLAYSURF.blit(nextSurf, nextRect)
501.     # draw the "next" piece
502.     drawPiece(piece, pixelx=WINDOWWIDTH-120, pixely=100)
503.
504.
505. if __name__ == '__main__':
506.     main()
```

drawNextPiece()负责绘制位于屏幕右上角的"Next"砖块。它通过调用drawPiece()函数，并且为drawPiece()的pixelx和pixely参数传递实参，做到这一点。

这是最后一个函数。第505行和第506行在所有的函数定义执行之后运行，并且调用main()函数开始程序的主要部分。

7.36 本章小结

Tetromino游戏（是较为流行的"Tetris"的一个翻版）的玩法很容易用语言向人们说清

第 7 章　Tetromino

楚"砖块从游戏板的顶部落下，玩家移动和旋转它们以便它们能够组成填满的行"。填满的行将会消失（让玩家得分），并且其上面的行将会向下移动。游戏持续进行，直到砖块填满了整个游戏板而玩家失败。

用简单易懂的语言说明它是一回事情，但是，当我们必须告诉计算机到底做些什么的时候，还有很多细节需要完善。最初的 Tetris 游戏，是由苏联的 Alex Pajitnov 一个人单枪匹马地设计和编程实现的。这个游戏很简单，有趣，并且令人着迷。这曾经是最流行的视频游戏之一，销售了 1 亿份，并且很多人创建了其翻版和变体。

所有这些都是由了解如何编程的一个人所创建的。

有了正确的思路和一些编程知识，你就可以创建有趣的、难以置信的游戏。经过一些实践，你就能够把自己的游戏思想变成像 Tetris 这样流行的、真正的程序。

要进行额外的编程练习，你可以从 http://invpy.com/buggy/tetromino 下载带 Bug 的 Tetromino 版本，并尝试搞清楚如何修正 Bug。

本书的 Web 站点上还有 Tetromino 游戏的一个变体 "Pentomino"，该游戏版本中的砖块由 5 个方块构成。还有一个叫作 "Tetromino for Idiots" 的简单版本，其中的砖块都是由一个方块构成的，如图 7-13 所示。

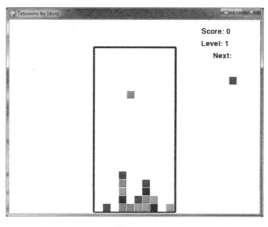

图 7-13

可以通过如下的链接下载这些版本。
- http://invpy.com/pentomino.py
- http://invpy.com/tetrominoforidiots.py

第 8 章 Squirrel Eat Squirrel

8.1 如何玩 Squirrel Eat Squirrel

Squirrel Eat Squirrel 有点模仿"Katamari Damacy"游戏。玩家控制着一个小松鼠，它必须在屏幕上跳来跳去地吃掉较小的松鼠而躲开较大的松鼠，如图 8-1 所示。每次玩家的松鼠吃掉一个比它更小的松鼠的时候，自己都会长大。如果玩家的松鼠碰撞到一个比自己更大的松鼠，它会丢掉一个生命值。当松鼠变成一种叫作 Omega Squirrel 的巨型松鼠的时候，玩家获胜。如果玩家 3 次碰到更大的松鼠，就会失败。我并不是真地很确定自己如何产生这种松鼠彼此吃掉的视频游戏的想法。有时候，我也觉得有点奇怪。

图 8-1

8.2 Squirrel Eat Squirrel 的设计

这款游戏中有 3 种数据结构，它们都表示为字典值。这些类型分别是玩家的松鼠对象、敌人的松鼠对象和草对象。游戏中每次只有一个玩家的松鼠对象。

注意：从技术上讲，"对象"有时候在面向对象编程中有特定的含义。Python 也具有 OOP 的特性，但是，本书并不介绍这些。从技术上讲，像"Rect 对象"或"Surface 对象"这样的 Pygame 对象也都是对象。但是在本书中，我们将要使用"对象"这个术语表示存在于游戏世界中的物体。实际上，玩家的松鼠、敌人的松鼠和草"对象"，它们都只是字典值。

所有的这些对象在它们的字典中都拥有如下的键：'x'、'y'和'rect'。'x'和'y'键的值给出了对象的左上角在游戏世界坐标系（game world coordinate）中的坐标。这些坐标和像素坐标不

第 8 章　Squirrel Eat Squirrel

同（像素坐标由'rect'键的值记录）。当学习相机的概念的时候，我们会介绍游戏世界和像素坐标之间的区别。

此外，玩家的松鼠、敌人的松鼠和草有它们自己的键，这些将在源代码开始处的大量注释中说明。

8.3　Squirrel Eat Squirrel 的源代码

可以从 http://invpy.com/ squirrel.py 下载 Squirrel Eat Squirrel 的源代码。如果得到任何的错误消息，查看一下错误消息中提到的代码行，并且检查代码是否有任何录入错误。可以将代码复制并粘贴到位于 http://invpy.com/diff/squirrel 的 Web 表单中，看看你的代码和本书中的代码是否有区别。

也可通过如下链接下载所需的图像文件。

- http://invpy.com/gameicon.png
- http://invpy.com/squirrel.png
- http://invpy.com/grass1.png
- http://invpy.com/grass2.png
- http://invpy.com/grass3.png
- http://invpy.com/grass4.png

```
1.  # Squirrel Eat Squirrel (a 2D Katamari Damacy clone)
2.  # By Al Sweigart al@inventwithpython.com
3.  # http://inventwithpython.com/pygame
4.  # Creative Commons BY-NC-SA 3.0 US
5.
6.  import random, sys, time, math, pygame
7.  from pygame.locals import *
8.
9.  FPS = 30 # frames per second to update the screen
10. WINWIDTH = 640 # width of the program's window, in pixels
11. WINHEIGHT = 480 # height in pixels
12. HALF_WINWIDTH = int(WINWIDTH / 2)
13. HALF_WINHEIGHT = int(WINHEIGHT / 2)
14.
15. GRASSCOLOR = (24, 255, 0)
16. WHITE = (255, 255, 255)
17. RED = (255, 0, 0)
18.
```

```
19. CAMERASLACK = 90      # how far from the center the squirrel moves before moving the camera
20. MOVERATE = 9          # how fast the player moves
21. BOUNCERATE = 6        # how fast the player bounces (large is slower)
22. BOUNCEHEIGHT = 30     # how high the player bounces
23. STARTSIZE = 25        # how big the player starts off
24. WINSIZE = 300         # how big the player needs to be to win
25. INVULNTIME = 2        # how long the player is invulnerable after being hit in seconds
26. GAMEOVERTIME = 4      # how long the "game over" text stays on the screen in seconds
27. MAXHEALTH = 3         # how much health the player starts with
28.
29. NUMGRASS = 80         # number of grass objects in the active area
30. NUMSQUIRRELS = 30     # number of squirrels in the active area
31. SQUIRRELMINSPEED = 3  # slowest squirrel speed
32. SQUIRRELMAXSPEED = 7  # fastest squirrel speed
33. DIRCHANGEFREQ = 2     # % chance of direction change per frame
34. LEFT = 'left'
35. RIGHT = 'right'
36.
37. """
38. This program has three data structures to represent the player, enemy squirrels, and grass background objects. The data structures are dictionaries with the following keys:
39.
40. Keys used by all three data structures:
41.     'x' - the left edge coordinate of the object in the game world (not a pixel coordinate on the screen)
42.     'y' - the top edge coordinate of the object in the game world (not a pixel coordinate on the screen)
43.     'rect' - the pygame.Rect object representing where on the screen the object is located.
44. Player data structure keys:
45.     'surface' - the pygame.Surface object that stores the image of the squirrel which will be drawn to the screen.
46.     'facing' - either set to LEFT or RIGHT, stores which direction the player is facing.
47.     'size' - the width and height of the player in pixels. (The width & height are always the same.)
48.     'bounce' - represents at what point in a bounce the player is in. 0 means standing (no bounce), up to BOUNCERATE (the completion of the bounce)
49.     'health' - an integer showing how many more times the player can be hit by a larger squirrel before dying.
50. Enemy Squirrel data structure keys:
```

第 8 章　Squirrel Eat Squirrel

```
51.        'surface' - the pygame.Surface object that stores the image of the
squirrel which will be drawn to the screen.
52.        'movex' - how many pixels per frame the squirrel moves horizontally. A
negative integer is moving to the left, a positive to the right.
53.        'movey' - how many pixels per frame the squirrel moves vertically. A
negative integer is moving up, a positive moving down.
54.        'width' - the width of the squirrel's image, in pixels
55.        'height' - the height of the squirrel's image, in pixels
56.        'bounce' - represents at what point in a bounce the player is in. 0
means standing (no bounce), up to BOUNCERATE (the completion of the bounce)
57.        'bouncerate' - how quickly the squirrel bounces. A lower number means
a quicker bounce.
58.        'bounceheight' - how high (in pixels) the squirrel bounces
59. Grass data structure keys:
60.        'grassImage' - an integer that refers to the index of the
pygame.Surface object in GRASSIMAGES used for this grass object
61. """
62.
63. def main():
64.     global FPSCLOCK, DISPLAYSURF, BASICFONT, L_SQUIR_IMG, R_SQUIR_IMG,
GRASSIMAGES
65.
66.     pygame.init()
67.     FPSCLOCK = pygame.time.Clock()
68.     pygame.display.set_icon(pygame.image.load('gameicon.png'))
69.     DISPLAYSURF = pygame.display.set_mode((WINWIDTH, WINHEIGHT))
70.     pygame.display.set_caption('Squirrel Eat Squirrel')
71.     BASICFONT = pygame.font.Font('freesansbold.ttf', 32)
72.
73.     # load the image files
74.     L_SQUIR_IMG = pygame.image.load('squirrel.png')
75.     R_SQUIR_IMG = pygame.transform.flip(L_SQUIR_IMG, True, False)
76.     GRASSIMAGES = []
77.     for i in range(1, 5):
78.         GRASSIMAGES.append(pygame.image.load('grass%s.png' % i))
79.
80.     while True:
81.         runGame()
82.
83.
84. def runGame():
85.     # set up variables for the start of a new game
86.     invulnerableMode = False  # if the player is invulnerable
87.     invulnerableStartTime = 0 # time the player became invulnerable
88.     gameOverMode = False      # if the player has lost
89.     gameOverStartTime = 0     # time the player lost
```

8.3 Squirrel Eat Squirrel 的源代码

```
90.     winMode = False          # if the player has won
91.
92.     # create the surfaces to hold game text
93.     gameOverSurf = BASICFONT.render('Game Over', True, WHITE)
94.     gameOverRect = gameOverSurf.get_rect()
95.     gameOverRect.center = (HALF_WINWIDTH, HALF_WINHEIGHT)
96.
97.     winSurf = BASICFONT.render('You have achieved OMEGA SQUIRREL!', True, WHITE)
98.     winRect = winSurf.get_rect()
99.     winRect.center = (HALF_WINWIDTH, HALF_WINHEIGHT)
100.
101.    winSurf2 = BASICFONT.render('(Press "r" to restart.)', True, WHITE)
102.    winRect2 = winSurf2.get_rect()
103.    winRect2.center = (HALF_WINWIDTH, HALF_WINHEIGHT + 30)
104.
105.    # camerax and cameray are where the middle of the camera view is
106.    camerax = 0
107.    cameray = 0
108.
109.    grassObjs = []    # stores all the grass objects in the game
110.    squirrelObjs = [] # stores all the non-player squirrel objects
111.    # stores the player object:
112.    playerObj = {'surface': pygame.transform.scale(L_SQUIR_IMG, (STARTSIZE, STARTSIZE)),
113.                 'facing': LEFT,
114.                 'size': STARTSIZE,
115.                 'x': HALF_WINWIDTH,
116.                 'y': HALF_WINHEIGHT,
117.                 'bounce':0,
118.                 'health': MAXHEALTH}
119.
120.    moveLeft  = False
121.    moveRight = False
122.    moveUp    = False
123.    moveDown  = False
124.
125.    # start off with some random grass images on the screen
126.    for i in range(10):
127.        grassObjs.append(makeNewGrass(camerax, cameray))
128.        grassObjs[i]['x'] = random.randint(0, WINWIDTH)
129.        grassObjs[i]['y'] = random.randint(0, WINHEIGHT)
130.
131.    while True: # main game loop
132.        # Check if we should turn off invulnerability
```

第 8 章　Squirrel Eat Squirrel

```
133.            if invulnerableMode and time.time() - invulnerableStartTime > INVULNTIME:
134.                invulnerableMode = False
135.
136.            # move all the squirrels
137.            for sObj in squirrelObjs:
138.                # move the squirrel, and adjust for their bounce
139.                sObj['x'] += sObj['movex']
140.                sObj['y'] += sObj['movey']
141.                sObj['bounce'] += 1
142.                if sObj['bounce'] > sObj['bouncerate']:
143.                    sObj['bounce'] = 0 # reset bounce amount
144.
145.                # random chance they change direction
146.                if random.randint(0, 99) < DIRCHANGEFREQ:
147.                    sObj['movex'] = getRandomVelocity()
148.                    sObj['movey'] = getRandomVelocity()
149.                    if sObj['movex'] > 0: # faces right
150.                        sObj['surface'] = pygame.transform.scale(R_SQUIR_IMG, (sObj['width'], sObj['height']))
151.                    else: # faces left
152.                        sObj['surface'] = pygame.transform.scale(L_SQUIR_IMG, (sObj['width'], sObj['height']))
153.
154.
155.            # go through all the objects and see if any need to be deleted.
156.            for i in range(len(grassObjs) - 1, -1, -1):
157.                if isOutsideActiveArea(camerax, cameray, grassObjs[i]):
158.                    del grassObjs[i]
159.            for i in range(len(squirrelObjs) - 1, -1, -1):
160.                if isOutsideActiveArea(camerax, cameray, squirrelObjs[i]):
161.                    del squirrelObjs[i]
162.
163.            # add more grass & squirrels if we don't have enough.
164.            while len(grassObjs) < NUMGRASS:
165.                grassObjs.append(makeNewGrass(camerax, cameray))
166.            while len(squirrelObjs) < NUMSQUIRRELS:
167.                squirrelObjs.append(makeNewSquirrel(camerax, cameray))
168.
169.            # adjust camerax and cameray if beyond the "camera slack"
170.            playerCenterx = playerObj['x'] + int(playerObj['size'] / 2)
171.            playerCentery = playerObj['y'] + int(playerObj['size'] / 2)
172.            if (camerax + HALF_WINWIDTH) - playerCenterx > CAMERASLACK:
173.                camerax = playerCenterx + CAMERASLACK - HALF_WINWIDTH
174.            elif playerCenterx - (camerax + HALF_WINWIDTH) > CAMERASLACK:
175.                camerax = playerCenterx - CAMERASLACK - HALF_WINWIDTH
```

```
176.            if (cameray + HALF_WINHEIGHT) - playerCentery > CAMERASLACK:
177.                cameray = playerCentery + CAMERASLACK - HALF_WINHEIGHT
178.            elif playerCentery - (cameray + HALF_WINHEIGHT) > CAMERASLACK:
179.                cameray = playerCentery - CAMERASLACK - HALF_WINHEIGHT
180.
181.            # draw the green background
182.            DISPLAYSURF.fill(GRASSCOLOR)
183.
184.            # draw all the grass objects on the screen
185.            for gObj in grassObjs:
186.                gRect = pygame.Rect( (gObj['x'] - camerax,
187.                                      gObj['y'] - cameray,
188.                                      gObj['width'],
189.                                      gObj['height']) )
190.                DISPLAYSURF.blit(GRASSIMAGES[gObj['grassImage']], gRect)
191.
192.
193.            # draw the other squirrels
194.            for sObj in squirrelObjs:
195.                sObj['rect'] = pygame.Rect( (sObj['x'] - camerax,
196.                                             sObj['y'] - cameray - 
getBounceAmount(sObj['bounce'], sObj['bouncerate'], sObj['bounceheight']),
197.                                             sObj['width'],
198.                                             sObj['height']) )
199.                DISPLAYSURF.blit(sObj['surface'], sObj['rect'])
200.
201.
202.            # draw the player squirrel
203.            flashIsOn = round(time.time(), 1) * 10 % 2 == 1
204.            if not gameOverMode and not (invulnerableMode and flashIsOn):
205.                playerObj['rect'] = pygame.Rect( (playerObj['x'] - camerax,
206.                                                  playerObj['y'] - cameray - 
getBounceAmount(playerObj['bounce'], BOUNCERATE, BOUNCEHEIGHT),
207.                                                  playerObj['size'],
208.                                                  playerObj['size']) )
209.                DISPLAYSURF.blit(playerObj['surface'], playerObj['rect'])
210.
211.
212.            # draw the health meter
213.            drawHealthMeter(playerObj['health'])
214.
215.            for event in pygame.event.get(): # event handling loop
216.                if event.type == QUIT:
217.                    terminate()
218.
219.                elif event.type == KEYDOWN:
```

第 8 章　Squirrel Eat Squirrel

```
220.                if event.key in (K_UP, K_w):
221.                    moveDown = False
222.                    moveUp = True
223.                elif event.key in (K_DOWN, K_s):
224.                    moveUp = False
225.                    moveDown = True
226.                elif event.key in (K_LEFT, K_a):
227.                    moveRight = False
228.                    moveLeft = True
229.                    if playerObj['facing'] == RIGHT: # change player image
230.                        playerObj['surface'] =
pygame.transform.scale(L_SQUIR_IMG, (playerObj['size'], playerObj['size']))
231.                    playerObj['facing'] = LEFT
232.                elif event.key in (K_RIGHT, K_d):
233.                    moveLeft = False
234.                    moveRight = True
235.                    if playerObj['facing'] == LEFT: # change player image
236.                        playerObj['surface'] =
pygame.transform.scale(R_SQUIR_IMG, (playerObj['size'], playerObj['size']))
237.                    playerObj['facing'] = RIGHT
238.                elif winMode and event.key == K_r:
239.                    return
240.
241.            elif event.type == KEYUP:
242.                # stop moving the player's squirrel
243.                if event.key in (K_LEFT, K_a):
244.                    moveLeft = False
245.                elif event.key in (K_RIGHT, K_d):
246.                    moveRight = False
247.                elif event.key in (K_UP, K_w):
248.                    moveUp = False
249.                elif event.key in (K_DOWN, K_s):
250.                    moveDown = False
251.
252.                elif event.key == K_ESCAPE:
253.                    terminate()
254.
255.        if not gameOverMode:
256.            # actually move the player
257.            if moveLeft:
258.                playerObj['x'] -= MOVERATE
259.            if moveRight:
260.                playerObj['x'] += MOVERATE
261.            if moveUp:
262.                playerObj['y'] -= MOVERATE
263.            if moveDown:
```

8.3 Squirrel Eat Squirrel 的源代码

```
264.                playerObj['y'] += MOVERATE
265.
266.            if (moveLeft or moveRight or moveUp or moveDown) or playerObj['bounce'] != 0:
267.                playerObj['bounce'] += 1
268.
269.                if playerObj['bounce'] > BOUNCERATE:
270.                    playerObj['bounce'] = 0 # reset bounce amount
271.
272.            # check if the player has collided with any squirrels
273.            for i in range(len(squirrelObjs)-1, -1, -1):
274.                sqObj = squirrelObjs[i]
275.                if 'rect' in sqObj and playerObj['rect'].colliderect(sqObj['rect']):
276.                    # a player/squirrel collision has occurred
277.
278.                    if sqObj['width'] * sqObj['height'] <= playerObj['size']**2:
279.                        # player is larger and eats the squirrel
280.                        playerObj['size'] += int( (sqObj['width'] * sqObj['height'])**0.2 ) + 1
281.                        del squirrelObjs[i]
282.
283.                        if playerObj['facing'] == LEFT:
284.                            playerObj['surface'] = pygame.transform.scale(L_SQUIR_IMG, (playerObj['size'], playerObj['size']))
285.                        if playerObj['facing'] == RIGHT:
286.                            playerObj['surface'] = pygame.transform.scale(R_SQUIR_IMG, (playerObj['size'], playerObj['size']))
287.
288.                        if playerObj['size'] > WINSIZE:
289.                            winMode = True # turn on "win mode"
290.
291.                    elif not invulnerableMode:
292.                        # player is smaller and takes damage
293.                        invulnerableMode = True
294.                        invulnerableStartTime = time.time()
295.                        playerObj['health'] -= 1
296.                        if playerObj['health'] == 0:
297.                            gameOverMode = True # turn on "game over mode"
298.                            gameOverStartTime = time.time()
299.        else:
300.            # game is over, show "game over" text
301.            DISPLAYSURF.blit(gameOverSurf, gameOverRect)
302.            if time.time() - gameOverStartTime > GAMEOVERTIME:
303.                return # end the current game
```

第 8 章　Squirrel Eat Squirrel

```
304.
305.            # check if the player has won.
306.            if winMode:
307.                DISPLAYSURF.blit(winSurf, winRect)
308.                DISPLAYSURF.blit(winSurf2, winRect2)
309.
310.            pygame.display.update()
311.            FPSCLOCK.tick(FPS)
312.
313.
314.
315.
316. def drawHealthMeter(currentHealth):
317.     for i in range(currentHealth): # draw red health bars
318.         pygame.draw.rect(DISPLAYSURF, RED,   (15, 5 + (10 * MAXHEALTH) - i
* 10, 20, 10))
319.     for i in range(MAXHEALTH): # draw the white outlines
320.         pygame.draw.rect(DISPLAYSURF, WHITE, (15, 5 + (10 * MAXHEALTH) - i
* 10, 20, 10), 1)
321.
322.
323. def terminate():
324.     pygame.quit()
325.     sys.exit()
326.
327.
328. def getBounceAmount(currentBounce, bounceRate, bounceHeight):
329.     # Returns the number of pixels to offset based on the bounce.
330.     # Larger bounceRate means a slower bounce.
331.     # Larger bounceHeight means a higher bounce.
332.     # currentBounce will always be less than bounceRate
333.     return int(math.sin( (math.pi / float(bounceRate)) * currentBounce ) *
bounceHeight)
334.
335. def getRandomVelocity():
336.     speed = random.randint(SQUIRRELMINSPEED, SQUIRRELMAXSPEED)
337.     if random.randint(0, 1) == 0:
338.         return speed
339.     else:
340.         return -speed
341.
342.
343. def getRandomOffCameraPos(camerax, cameray, objWidth, objHeight):
344.     # create a Rect of the camera view
345.     cameraRect = pygame.Rect(camerax, cameray, WINWIDTH, WINHEIGHT)
346.     while True:
```

8.3 Squirrel Eat Squirrel 的源代码

```
347.            x = random.randint(cameraX - WINWIDTH, cameraX + (2 * WINWIDTH))
348.            y = random.randint(cameraY - WINHEIGHT, cameraY + (2 * WINHEIGHT))
349.            # create a Rect object with the random coordinates and use colliderect()
350.            # to make sure the right edge isn't in the camera view.
351.            objRect = pygame.Rect(x, y, objWidth, objHeight)
352.            if not objRect.colliderect(cameraRect):
353.                return x, y
354.
355.
356.    def makeNewSquirrel(cameraX, cameraY):
357.        sq = {}
358.        generalSize = random.randint(5, 25)
359.        multiplier = random.randint(1, 3)
360.        sq['width']  = (generalSize + random.randint(0, 10)) * multiplier
361.        sq['height'] = (generalSize + random.randint(0, 10)) * multiplier
362.        sq['x'], sq['y'] = getRandomOffCameraPos(cameraX, cameraY, sq['width'], sq['height'])
363.        sq['movex'] = getRandomVelocity()
364.        sq['movey'] = getRandomVelocity()
365.        if sq['movex'] < 0: # squirrel is facing left
366.            sq['surface'] = pygame.transform.scale(L_SQUIR_IMG, (sq['width'], sq['height']))
367.        else: # squirrel is facing right
368.            sq['surface'] = pygame.transform.scale(R_SQUIR_IMG, (sq['width'], sq['height']))
369.        sq['bounce'] = 0
370.        sq['bouncerate'] = random.randint(10, 18)
371.        sq['bounceheight'] = random.randint(10, 50)
372.        return sq
373.
374.
375.    def makeNewGrass(cameraX, cameraY):
376.        gr = {}
377.        gr['grassImage'] = random.randint(0, len(GRASSIMAGES) - 1)
378.        gr['width']  = GRASSIMAGES[0].get_width()
379.        gr['height'] = GRASSIMAGES[0].get_height()
380.        gr['x'], gr['y'] = getRandomOffCameraPos(cameraX, cameraY, gr['width'], gr['height'])
381.        gr['rect'] = pygame.Rect( (gr['x'], gr['y'], gr['width'], gr['height']) )
382.        return gr
383.
384.
385.    def isOutsideActiveArea(cameraX, cameraY, obj):
386.        # Return False if cameraX and cameraY are more than
```

第 8 章　Squirrel Eat Squirrel

```
387.        # a half-window length beyond the edge of the window.
388.        boundsLeftEdge = camerax - WINWIDTH
389.        boundsTopEdge = cameray - WINHEIGHT
390.        boundsRect = pygame.Rect(boundsLeftEdge, boundsTopEdge, WINWIDTH * 3,
WINHEIGHT * 3)
391.        objRect = pygame.Rect(obj['x'], obj['y'], obj['width'], obj['height'])
392.        return not boundsRect.colliderect(objRect)
393.
394.
395.    if __name__ == '__main__':
396.        main()
```

8.4　常用设置代码

```
1.  # Squirrel Eat Squirrel (a 2D Katamari Damacy clone)
2.  # By Al Sweigart al@inventwithpython.com
3.  # http://inventwithpython.com/pygame
4.  # Creative Commons BY-NC-SA 3.0 US
5.
6.  import random, sys, time, math, pygame
7.  from pygame.locals import *
8.
9.  FPS = 30 # frames per second to update the screen
10. WINWIDTH = 640 # width of the program's window, in pixels
11. WINHEIGHT = 480 # height in pixels
12. HALF_WINWIDTH = int(WINWIDTH / 2)
13. HALF_WINHEIGHT = int(WINHEIGHT / 2)
14.
15. GRASSCOLOR = (24, 255, 0)
16. WHITE = (255, 255, 255)
17. RED = (255, 0, 0)
```

程序的开始处给几个常量变量赋值。这个程序频繁地使用窗口的宽度和高度的一半，以至于要专门用 HALF_WINWIDTH 和 HALF_WINHEIGHT 变量来存储这些数字。

```
19. CAMERASLACK = 90      # how far from the center the squirrel moves before
moving the camera
```

相机延迟（camera slack）稍后介绍。基本上，它的意思是当玩家松鼠移动到距离窗口中心 90 像素那么远的时候，相机开始跟着移动。

```
20. MOVERATE = 9          # how fast the player moves
21. BOUNCERATE = 6        # how fast the player bounces (large is slower)
```

```
22. BOUNCEHEIGHT = 30       # how high the player bounces
23. STARTSIZE = 25          # how big the player starts off
24. WINSIZE = 300           # how big the player needs to be to win
25. INVULNTIME = 2          # how long the player is invulnerable after being hit in seconds
26. GAMEOVERTIME = 4        # how long the "game over" text stays on the screen in seconds
27. MAXHEALTH = 3           # how much health the player starts with
28.
29. NUMGRASS = 80           # number of grass objects in the active area
30. NUMSQUIRRELS = 30       # number of squirrels in the active area
31. SQUIRRELMINSPEED = 3    # slowest squirrel speed
32. SQUIRRELMAXSPEED = 7    # fastest squirrel speed
33. DIRCHANGEFREQ = 2       # % chance of direction change per frame
34. LEFT = 'left'
35. RIGHT = 'right'
```

这些常量后面的注释说明了它们的用途。

8.5 描述数据结构

```
37. """
38. This program has three data structures to represent the player, enemy squirrels, and grass background objects. The data structures are dictionaries with the following keys:
39.
40. Keys used by all three data structures:
41.     'x' - the left edge coordinate of the object in the game world (not a pixel coordinate on the screen)
42.     'y' - the top edge coordinate of the object in the game world (not a pixel coordinate on the screen)
43.     'rect' - the pygame.Rect object representing where on the screen the object is located.
44. Player data structure keys:
45.     'surface' - the pygame.Surface object that stores the image of the squirrel which will be drawn to the screen.
46.     'facing' - either set to LEFT or RIGHT, stores which direction the player is facing.
47.     'size' - the width and height of the player in pixels. (The width & height are always the same.)
48.     'bounce' - represents at what point in a bounce the player is in. 0 means standing (no bounce), up to BOUNCERATE (the completion of the bounce)
49.     'health' - an integer showing how many more times the player can be hit by a larger squirrel before dying.
50. Enemy Squirrel data structure keys:
```

第 8 章 Squirrel Eat Squirrel

```
51.     'surface' - the pygame.Surface object that stores the image of the
squirrel which will be drawn to the screen.
52.     'movex' - how many pixels per frame the squirrel moves horizontally. A
negative integer is moving to the left, a positive to the right.
53.     'movey' - how many pixels per frame the squirrel moves vertically. A
negative integer is moving up, a positive moving down.
54.     'width' - the width of the squirrel's image, in pixels
55.     'height' - the height of the squirrel's image, in pixels
56.     'bounce' - represents at what point in a bounce the player is in. 0
means standing (no bounce), up to BOUNCERATE (the completion of the bounce)
57.     'bouncerate' - how quickly the squirrel bounces. A lower number means
a quicker bounce.
58.     'bounceheight' - how high (in pixels) the squirrel bounces
59. Grass data structure keys:
60.     'grassImage' - an integer that refers to the index of the
pygame.Surface object in GRASSIMAGES used for this grass object
61. """
```

第 37 行到第 61 行的注释是一个很长的、跨多行的字符串。它们描述了玩家松鼠、敌人松鼠和草对象中的键。在 Python 中，多行字符串值本身是当作一个多行注释来使用的。

8.6 main()函数

```
63. def main():
64.     global FPSCLOCK, DISPLAYSURF, BASICFONT, L_SQUIR_IMG, R_SQUIR_IMG,
GRASSIMAGES
65.
66.     pygame.init()
67.     FPSCLOCK = pygame.time.Clock()
68.     pygame.display.set_icon(pygame.image.load('gameicon.png'))
69.     DISPLAYSURF = pygame.display.set_mode((WINWIDTH, WINHEIGHT))
70.     pygame.display.set_caption('Squirrel Eat Squirrel')
71.     BASICFONT = pygame.font.Font('freesansbold.ttf', 32)
```

main()函数的前几行和我们在之前的游戏程序中所见到过的设置代码是相同的。pygame.display.set_icon()是一个 Pygame 函数，它负责设置窗口的标题栏的图标（就像 pygame.display.set_caption() 负责设置标题栏的标题文本一样）。pygame.display.set_icon()只有一个参数，是一个小图像的 Surface 对象。理想的图像大小是 32 像素×32 像素，尽管你也可以使用其他大小的图像。该图像将压缩为一个较小的大小，一般用作窗口的图标。

8.7 pygame.transform.flip()函数

```
73.     # load the image files
74.     L_SQUIR_IMG = pygame.image.load('squirrel.png')
75.     R_SQUIR_IMG = pygame.transform.flip(L_SQUIR_IMG, True, False)
76.     GRASSIMAGES = []
77.     for i in range(1, 5):
78.         GRASSIMAGES.append(pygame.image.load('grass%s.png' % i))
```

玩家的松鼠的图像和敌人的松鼠的图像都在第 74 行从 *squirrel.png* 加载。确保这个 PNG 文件和 *squirrel.py* 位于相同的目录中，否则的话，你将会得到一条错误消息：pygame.error: Couldn't open squirrel.png。

squirrel.png（可以从 http://invpy.com/squirrel.png 下载）中的图像是一只面朝左边的松鼠。我们还需要包含了面朝右边的松鼠的图片的一个 Surface 对象。我们可以调用 pygame.transform.flip()函数，而不是再创建一个 PNG 图像文件。该函数有 3 个参数：带有翻转的图像的 Surface 对象，表示进行水平翻转的一个 Boolean 值，以及表示进行垂直翻转的一个 Boolean 值。通过给第 2 个参数传递 True，而给第 3 个参数传递 False，函数所返回的 Surface 对象所拥有的图像，就会是面朝右边的松鼠。我们传入到 L_SQUIR_IMG 中的最初的 Surface 对象并没有改变。

水平翻转和垂直翻转后的图像如图 8-2 所示。

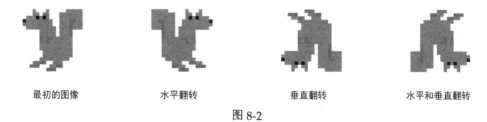

最初的图像　　　水平翻转　　　垂直翻转　　　水平和垂直翻转

图 8-2

```
80.     while True:
81.         runGame()
```

在 main()中的初始设置完成之后，通过调用 runGame()开始游戏。

8.8 更为详细的游戏状态

```
84. def runGame():
85.     # set up variables for the start of a new game
```

第 8 章 Squirrel Eat Squirrel

```
86.     invulnerableMode = False       # if the player is invulnerable
87.     invulnerableStartTime = 0      # time the player became invulnerable
88.     gameOverMode = False           # if the player has lost
89.     gameOverStartTime = 0          # time the player lost
90.     winMode = False                # if the player has won
```

Squirrel Eat Squirrel 游戏还有一些变量记录了游戏状态。我们将在后面的代码中用到这些变量的时候加以详细说明。

8.9 常用的文本创建代码

```
92.     # create the surfaces to hold game text
93.     gameOverSurf = BASICFONT.render('Game Over', True, WHITE)
94.     gameOverRect = gameOverSurf.get_rect()
95.     gameOverRect.center = (HALF_WINWIDTH, HALF_WINHEIGHT)
96.
97.     winSurf = BASICFONT.render('You have achieved OMEGA SQUIRREL!', True, WHITE)
98.     winRect = winSurf.get_rect()
99.     winRect.center = (HALF_WINWIDTH, HALF_WINHEIGHT)
100.
101.    winSurf2 = BASICFONT.render('(Press "r" to restart.)', True, WHITE)
102.    winRect2 = winSurf2.get_rect()
103.    winRect2.center = (HALF_WINWIDTH, HALF_WINHEIGHT + 30)
```

这些变量包含了带有"Game Over"、"You have achieved OMEGA SQUIRREL"和"(Press "r" to restart.)"等文本的 Surface 对象，当游戏结束之后，这些文本会出现在屏幕之上。

8.10 相机

```
105.    # camerax and cameray are where the middle of the camera view is
106.    camerax = 0
107.    cameray = 0
```

camerax 和 cameray 变量记录了相机的游戏坐标。把游戏世界假想为一个无限的 2D 空间。当然，这可能无法放入任何的屏幕之中。我们只能够在屏幕上绘制这个无限 2D 空间的一部分。我们将这部分区域称为一个相机（camera），因为我们的屏幕就好像是通过一个相机所看到的游戏世界的某个区域一样。图 8-3 所示是游戏世界的一幅图片（一个无限的绿色的领域）以及相机所能够看到的区域。

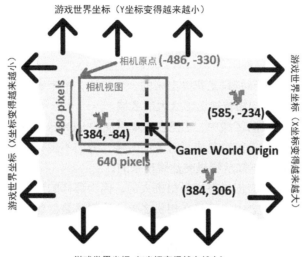

图 8-3

正如你所看到的，游戏世界的 XY 坐标永远地、持续地变得更大和更小。游戏世界的原点就是游戏世界坐标(0, 0)所在的位置。你可以看到，3 个松鼠位于（游戏世界坐标系中的）(–384, –84)、(384, 306)和(585, –234)。但是，我们只能够在屏幕上显示 640 像素 × 480 像素的区域（尽管如果我们给 pygame.display.set_mode()函数传递不同的数字，是可以修改这一区域的），因此，我们需要记录相机的原点位于游戏世界坐标中的何处。在图 8-3 中，相机位于游戏世界坐标中的(–486, –330)。

图 8-4 展示了相同的区域和松鼠，只不过所有内容都以相机坐标给出。

相机所能看到的区域（叫作相机视图，camera view），其中心位于（也就是原点，origin）游戏世界坐标的(–486, –330)。由于相机看到的内容显示于玩家的屏幕之上，"相机"坐标和"像素"坐标是相同的。要得到松鼠的像素坐标，即松鼠出现在屏幕上的什么位置，接受松鼠的游戏坐标，并且用其减去相机的原点的游戏坐标。

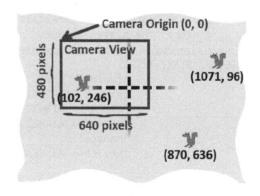

图 8-4

因此，左边的松鼠的游戏坐标为(–384, –84)，但是出现在屏幕上像素坐标(102, 246)处（对于 X 坐标，–384 –（–486）= 102；对于 Y 坐标，–84 –（–330）= 246）。

第 8 章　Squirrel Eat Squirrel

当做相同的计算求得另外两只松鼠的像素坐标的时候，我们发现它们位于屏幕范围之外。这就是为什么它们没有出现在相机视图之中的原因。

8.11　"活动区域"

"活动区域"（active area）只是我提出来的一个名称，用来表示游戏世界的这样一个区域，即相机视图加上其四周的相机视图那么大的区域。

计算某个物体是否在活动区域中，这将在本章后面的 isOutsideActiveArea() 函数部分介绍。当我们创建了新的敌人松鼠对象或草对象的时候，我们不想将其创建在相机视图之中，从而使得它们看上去好像是不知从哪里突然跳出来的。

但是，我们也不想将它们创建得离开相机太远，因为那样的话，它们可能根本不会出现在相机视图中。在活动区域之内，但是在相机之外，这似乎是创建松鼠对象和草对象的安全的区域。

此外，当松鼠对象和草对象超出了活动区域之外的时候，它们已经太远了，可以删除它们以不再占用太多内存。太远的对象是不再需要的，因为它们再回到相机视图的可能性很小。

如果你曾经在 Super Nintendo 上玩过 Super Mario World，有一个很好的视频说明了 Super Mario World 的相机系统的工作方式。可以通过 http://invpy.com/mariocamera 找到该视频。

8.12　记录游戏世界中的物体的位置

```
109.    grassObjs = []    # stores all the grass objects in the game
110.    squirrelObjs = [] # stores all the non-player squirrel objects
111.    # stores the player object:
112.    playerObj = {'surface': pygame.transform.scale(L_SQUIR_IMG, (STARTSIZE, STARTSIZE)),
113.                 'facing': LEFT,
114.                 'size': STARTSIZE,
115.                 'x': HALF_WINWIDTH,
116.                 'y': HALF_WINHEIGHT,
117.                 'bounce':0,
118.                 'health': MAXHEALTH}
119.
120.    moveLeft  = False
121.    moveRight = False
122.    moveUp    = False
123.    moveDown  = False
```

grassObjs 变量保存了游戏中的所有的草对象的一个列表。当创建一个新的草对象的时候，会将它们添加到这个列表中。当删除草对象的时候，会从列表中删除它们。squirrelObjs 变量以及敌人松鼠对象也是如此。

playerObj 变量不是一个列表，其自身只是一个字典值。第 120 行到第 123 行的移动变量，记录了按下了哪一个箭头方向键（或 WASD 键），这就像在前面几个游戏程序中一样。

8.13 从一些草开始

```
125.    # start off with some random grass images on the screen
126.    for i in range(10):
127.        grassObjs.append(makeNewGrass(camerax, cameray))
128.        grassObjs[i]['x'] = random.randint(0, WINWIDTH)
129.        grassObjs[i]['y'] = random.randint(0, WINHEIGHT)
```

一开始的时候，活动区域上应该有一些草对象在屏幕上可见。makeNewGrass()函数将创建并返回一个草对象，它随机地位于活动区域中的某处，但却在相机视图之外。这就是通常调用 makeNewGrass() 的时候想要的效果，但是，由于想要确保有一些草对象在屏幕之上，X 和 Y 坐标将会被覆盖。

8.14 游戏循环

```
131.    while True: # main game loop
```

游戏循环就像前面几个游戏程序中的一样，将会负责事件处理、更新游戏状态，并且将所有内容绘制到屏幕上。

8.15 检查去掉保护状态

```
132.        # Check if we should turn off invulnerability
133.        if invulnerableMode and time.time() - invulnerableStartTime > INVULNTIME:
134.            invulnerableMode = False
```

当玩家被一个敌人松鼠碰到并且没有死掉的时候，我们让玩家在数秒钟内变为受保护的状态（因为 INVULNTIME 常量设置为 2）。在此过程中，玩家松鼠会闪烁，并且不会受到任何其他松鼠的伤害。如果"受保护模式"的时间到了，第 134 行将 invulnerableMode 设置为 False。

第 8 章　Squirrel Eat Squirrel

8.16　移动敌人松鼠

```
136.        # move all the squirrels
137.        for sObj in squirrelObjs:
138.            # move the squirrel, and adjust for their bounce
139.            sObj['x'] += sObj['movex']
140.            sObj['y'] += sObj['movey']
```

敌人松鼠根据它们的'movex'和'movey'键中的值来移动。如果这些值是正的，松鼠朝着右边或下边移动。如果这些值是负的，它们向左或向上移动。值越大，它们在游戏循环的每一次迭代中移动的越远（这意味着它们移动得更快）。

第 137 行的 for 循环将对 squirrelObjs 列表中的每一个松鼠对象应用这一移动代码。首先，第 139 行和第 140 行将调整它们的'x'和'y'键值。

```
141.            sObj['bounce'] += 1
142.            if sObj['bounce'] > sObj['bouncerate']:
143.                sObj['bounce'] = 0 # reset bounce amount
```

sObj['bounce']中的值在游戏循环的每一次迭代中针对每一个松鼠都会递增。当这个值为 0 的时候，松鼠刚开始跳动。当这个值等于 sObj['bouncerate']之中的值的时候，它已经增加到了尽头（这就是为什么一个较小的 sObj['bouncerate']值可以产生更快的跳动的原因）。如果 sObj['bouncerate']等于 3，那么，只需要在游戏循环中进行 3 次迭代就可以让松鼠达到一次全速的跳动。如果 sObj['bouncerate']等于 10，则需要 10 次迭代。

当 sObj['bounce']变得比 sObj['bouncerate']大的时候，随后需要将其重置为 0。这就是第 142 行和第 143 行所做的事情。

```
145.            # random chance they change direction
146.            if random.randint(0, 99) < DIRCHANGEFREQ:
147.                sObj['movex'] = getRandomVelocity()
148.                sObj['movey'] = getRandomVelocity()
149.                if sObj['movex'] > 0: # faces right
150.                    sObj['surface'] = pygame.transform.scale(R_SQUIR_IMG, (sObj['width'], sObj['height']))
151.                else: # faces left
152.                    sObj['surface'] = pygame.transform.scale(L_SQUIR_IMG, (sObj['width'], sObj['height']))
```

在游戏循环的每一次迭代中，都有 2%的机会让松鼠随机地改变速度和方向。在第 146 行，random.randint(0, 99)调用随机地选取了 100 以内的一个整数。如果这个整数小于

DIRCHANGEFREQ（在第 33 行设置为 2），那么，将为 sObj['movex'] 和 sObj['movey'] 设置一个新的值。由于这意味着松鼠可能会改变方向，sObj['surface'] 中的 Surface 对象应该会被刚好朝左或朝右并且缩放为该松鼠那么大的一个新的对象所替代。这是第 149 行和第 152 行所确定的事情。注意，第 150 行获取了从 R_SQUIR_IMG 缩放得到的一个 Surface 对象，第 152 行获取了从 L_SQUIR_IMG 缩放得到的一个对象。

8.17 删除较远的草对象和松鼠对象

```
155.        # go through all the objects and see if any need to be deleted.
156.        for i in range(len(grassObjs) - 1, -1, -1):
157.            if isOutsideActiveArea(camerax, cameray, grassObjs[i]):
158.                del grassObjs[i]
159.        for i in range(len(squirrelObjs) - 1, -1, -1):
160.            if isOutsideActiveArea(camerax, cameray, squirrelObjs[i]):
161.                del squirrelObjs[i]
```

在游戏循环的每一次迭代中，代码都将检查所有的草对象和敌人松鼠对象，看看它们是否在活动区域之外。isOutsideActiveArea() 函数接受相机的当前坐标（这存储在 camerax 和 cameray 之中），以及草对象、敌人松鼠对象，并且如果该对象没有位于活动区域中，函数返回 True。如果是这种情况，在第 158 行（针对草对象）或第 161 行（针对松鼠对象）删除该对象。这就是为什么当玩家移动到距离松鼠对象和草对象足够远的时候（或者当敌人松鼠对象移动到距离玩家足够远的时候），会删除掉松鼠对象和草对象。这确保了在玩家附近总是有很多的松鼠对象和草对象。

8.18 当从列表中删除项的时候，反向遍历列表

删除松鼠对象和草对象是使用 del 操作符来完成的。然而，注意，第 156 行和第 159 行的 for 循环传递了参数给 range() 函数，以便从最后一项的索引开始计数并且随后递减 1，而不是按照通常所做的那样自增 1，直到到达数字 −1。和常规的方式相比，我们向后遍历类表的索引。之所以这么做，是因为我们所遍历的列表正是要从中删除项的列表。

要看看为什么需要逆向遍历，假设有如下的列表值。

```
animals = ['cat', 'mouse', 'dog', 'horse']
```

因此，我们想要编写代码来从这个列表中删除字符串 'dog' 的任何实例。我们可能会想到编写如下所示的代码。

第 8 章　Squirrel Eat Squirrel

```
for i in range(len(animals)):
    if animals[i] == 'dog':
        del animals[i]
```

但是，如果运行这段代码，我们将会得到如下所示的一个 IndexError 错误。

```
Traceback (most recent call last):
  File "<stdin>", line 2, in <module>
IndexError: list index out of range
```

要搞清楚为什么会发生这一错误，我们先来浏览一下代码。首先，animals 列表应该设置为['cat', 'mouse', 'dog', 'horse']，并且 len(animals)应该返回 4。这意味着，调用 range(4)将会导致 for 循环使用值 0、1、2 和 3 进行遍历。

当 for 循环将 i 设置为 2 进行遍历时，if 语句的条件将会是 True，并且 del animals[i]语句将会删除 animals[2]。这意味着，后续的动物列表将会是['cat', 'mouse', 'horse']。'dog'之后的所有的项的索引都减少了 1，因为'dog'值已经删除了。

但是，在 for 循环的下一次迭代中，i 设置为 3。但 animals[3]已经越界了，因此，animals 列表的有效索引不再是 0 到 3，而是 0 到 2。最初对 range()的调用是针对其中有 4 项的一个列表。该列表的长度变了，但是 for 循环还是按照最初的长度而设置的。

然而，如果从列表的最后一个索引遍历直到 0 的话，我们不会遇到这个问题。如下的程序从 animals 列表删除了'dog'字符串，而不会引发 IndexError 错误。

```
animals = ['cat', 'mouse', 'dog', 'horse']
for i in range(len(animals) - 1, -1, -1):
    if animals[i] == 'dog':
        del animals[i]
```

这段代码之所以不会引发错误，是因为这个 for 循环按照 3、2、1 和 0 的顺序遍历。在第一次迭代的时候，代码检查 animals[3]是否等于'dog'。它不是（animals[3]是'horse'），因此继续移动到下一次迭代。检查 animals[2]是否等于'dog'。它是的，因此删除 animals[2]。

删除 animals[2]之后, animals 列表设置为['cat', 'mouse', 'horse']。在下一次迭代的时候，i 设置为 1。animals[1]有一个值（'mouse'值），因此，不会导致错误。列表中'dog'之后的项的索引都向下移动了 1 位，这无关紧要，因为我们从列表的末尾开始并且向前移动，所有的那些项都已经检查过了。

类似的，我们从 grassObjs 和 squirrelObjs 删除草对象和松鼠对象也不会有错误，因为第 156 行到第 159 行的 for 循环也是逆向遍历的。

8.19 添加新的草对象和松鼠对象

```
163.        # add more grass & squirrels if we don't have enough.
164.        while len(grassObjs) < NUMGRASS:
165.            grassObjs.append(makeNewGrass(camerax, cameray))
166.        while len(squirrelObjs) < NUMSQUIRRELS:
167.            squirrelObjs.append(makeNewSquirrel(camerax, cameray))
```

还记得吧，在程序的开始处，NUMGRASS 常量设置为 80，而 NUMSQUIRRELS 常量设置为 30。设置这些变量，以便我们能够确保任何时候在活动区域都有足够多的草对象和松鼠对象。如果 grassObjs 和 squirrelObjs 的长度分别小于 NUMGRASS 或 NUMSQUIRRELS，将创建新的草对象和松鼠对象。创建这些对象的 makeNewGrass() 和 makeNewSquirrel() 函数，将在本章后面介绍。

8.20 相机延迟以及移动相机视图

```
169.        # adjust camerax and cameray if beyond the "camera slack"
170.        playerCenterx = playerObj['x'] + int(playerObj['size'] / 2)
171.        playerCentery = playerObj['y'] + int(playerObj['size'] / 2)
172.        if (camerax + HALF_WINWIDTH) - playerCenterx > CAMERASLACK:
173.            camerax = playerCenterx + CAMERASLACK - HALF_WINWIDTH
174.        elif playerCenterx - (camerax + HALF_WINWIDTH) > CAMERASLACK:
175.            camerax = playerCenterx - CAMERASLACK - HALF_WINWIDTH
176.        if (cameray + HALF_WINHEIGHT) - playerCentery > CAMERASLACK:
177.            cameray = playerCentery + CAMERASLACK - HALF_WINHEIGHT
178.        elif playerCentery - (cameray + HALF_WINHEIGHT) > CAMERASLACK:
179.            cameray = playerCentery - CAMERASLACK - HALF_WINHEIGHT
```

当玩家移动的时候，相机的位置（作为整数存储在 camerax 和 cameray 变量中）需要更新。我把相机更新之前玩家所能够移动的像素数称为"相机延迟（camera slack）"。第 19 行将 CAMERASLACK 设置为 90，由此，程序会认为在相机开始更新并跟着松鼠移动之前，玩家松鼠可以移动 90 个像素。

要理解第 172、第 174、第 176 和第 178 行中的 if 语句，你应该注意(camerax + HALF_WINWIDTH)和(cameray+HALF_WINHEIGHT)都是当前的屏幕中心的 XY 游戏世界坐标。playerCenterx 和 playerCentery 设置为玩家松鼠位置的中心，这也是游戏世界坐标。

对于第 172 行，如果屏幕中心的 X 坐标减去玩家的中心的 X 坐标比 CAMERASLACK 值大，这意味着玩家在相机中心右侧的距离已经超过了相机延迟，相机应该跟着移动。

第 8 章　Squirrel Eat Squirrel

Camerax 值需要更新，以便玩家松鼠刚好位于相机延迟的边缘。这就是为什么第 173 行将 camerax 设置为 playerCenterx + CAMERASLACK – HALF_WINWIDTH 的原因。注意，camerax 变量改变了，而不是 playerObj['x']的值改变了。我们想要移动相机而不是玩家。其他 3 个 if 条件按照相同的逻辑，分别针对左边、上边和下边。

8.21　绘制背景、草、松鼠和生命值指示

```
181.        # draw the green background
182.        DISPLAYSURF.fill(GRASSCOLOR)
```

第 182 行的代码开始绘制显示 Surface 对象的内容。首先，第 182 行绘制了绿色的背景。这将会覆盖 Surface 上之前的内容，从而使得我们可以从头开始绘制帧。

```
184.        # draw all the grass objects on the screen
185.        for gObj in grassObjs:
186.            gRect = pygame.Rect( (gObj['x'] - camerax,
187.                                  gObj['y'] - cameray,
188.                                  gObj['width'],
189.                                  gObj['height']) )
190.            DISPLAYSURF.blit(GRASSIMAGES[gObj['grassImage']], gRect)
```

第 185 行的 for 循环遍历了 grassObjs 列表中的所有草对象，并且创建了一个 Rect 对象，其中存储了 x、y、width 和 height 信息。该 Rect 对象存储在名为 gRect 的一个变量中。在第 190 行， blit()方法调用使用 gRect，以将草的图像绘制到显示 Surface 上。注意，gObj['grassImage']只包含一个整数，它是 GRASSIMAGES 的一个索引。GRASSIMAGES 是包含了所有草的图像的 Surface 对象的列表。Surface 对象占用了比单个整数更多的内存，并且所有的草对象都具有形式类似于 gObj['grassImage']的值。因此，只是将每一幅草图像在 GRASSIMAGES 中存储一次，并且直接存储草对象中的存储整数，这么做是有意义的。

```
193.        # draw the other squirrels
194.        for sObj in squirrelObjs:
195.            sObj['rect'] = pygame.Rect( (sObj['x'] - camerax,
196.                                         sObj['y'] - cameray -
getBounceAmount(sObj['bounce'], sObj['bouncerate'], sObj['bounceheight']),
197.                                         sObj['width'],
198.                                         sObj['height']) )
199.            DISPLAYSURF.blit(sObj['surface'], sObj['rect'])
```

绘制所有敌人松鼠游戏对象的这个 for 循环，和前面的 for 循环类似，只不过它所创建的 Rect 对象存储在了松鼠字典的'rect'键的值中。代码这么做的原因是我们稍后将使用这个

8.21 绘制背景、草、松鼠和生命值指示

Rect 对象来检查敌人松鼠是否与玩家的松鼠碰撞。

注意，Rect 构造函数的上方的参数不是 sObj['y'] – cameray，而是 sObj['y'] - cameray - getBounceAmount(sObj['bounce'],sObj['bouncerate'], sObj ['bounceheight'])。getBounceAmount()函数将返回上方值应该提升的像素数。

此外，没有松鼠图像的 Surface 对象的通用列表，这和草游戏对象及 GRASSIMAGES 不同。每一个敌人松鼠游戏对象，都有其自己的 Surface 对象，存储在'surface'键中。这是因为松鼠图像可能会缩放为不同的大小。

```
202.            # draw the player squirrel
203.            flashIsOn = round(time.time(), 1) * 10 % 2 == 1
```

在绘制了草和敌人松鼠之后，这段代码将绘制玩家松鼠。然而，也有一种情况我们可能会跳过玩家松鼠的绘制。当玩家的松鼠与一只较大的敌人松鼠碰撞，玩家受到伤害并闪烁一会儿，以表示玩家临时处于受保护的模式。这个闪烁效果是这样完成的，即在游戏循环中的某一次迭代中绘制玩家松鼠，而在另一些迭代中则不绘制。

玩家松鼠会在游戏循环迭代中绘制十分之一秒，随后在游戏循环迭代中又不绘制十分之一秒。只要玩家是受保护的（这在代码中意味着 invulnerableMode 变量设置为 True），就这样不断地重复。我们的代码会让闪烁持续 2 秒，因为在第 25 行的 INVULNTIME 常量变量中存储的是 2。

为了判断是否闪烁，第 202 行从 time.time()抓取了当前时间。让我们以该函数返回 1323926893.622 为例。这个值传递给 round()，它将其舍入到小数点后面一位（因此 1 作为第 2 个参数传递给了 round()）。这意味着 round()将返回值 1323926893.6。

随后，这个值乘以 10，变成了 13239268936。一旦有了这个整数，可以使用最初在第 3 章 Memory Puzzle 游戏中讨论过的"模除 2"的技巧，看看它是一个偶数，还是奇数。13239268936 % 2 等于 0，这意味着 flashIsOn 将设置为 False，因为 0 == 1 为 False。

实际上，time.time()将保持该返回值，并且最终将 False 放入到 flashIsOn，直到 1323926893.700，也就是下一个十分之一秒。这就是为什么 flashIsOn 变量将持续为 False 达到十分之一秒的时间，随后，在下一个十分之秒中 flashIsOn 为 True（不管在该十分之一秒之中发生了多少次迭代）的原因。

```
204.            if not gameOverMode and not (invulnerableMode and flashIsOn):
205.                playerObj['rect'] = pygame.Rect( (playerObj['x'] - camerax,
206.                                                  playerObj['y'] - cameray -
getBounceAmount(playerObj['bounce'], BOUNCERATE, BOUNCEHEIGHT),
207.                                                  playerObj['size'],
```

第 8 章　Squirrel Eat Squirrel

```
208.                                              playerObj['size']) )
209.            DISPLAYSURF.blit(playerObj['surface'], playerObj['rect'])
```

在我们绘制玩家松鼠之前，有 3 件事情必须为 True。游戏当前必须在进行中（这意味着 gameOverMode 为 False），并且玩家不处于受保护状态，并且没有闪烁（这意味着 invulnerableMode 和 flashIsOn 为 False）。

绘制玩家松鼠的代码，几乎与绘制敌人松鼠的代码相同。

```
212.            # draw the health meter
213.            drawHealthMeter(playerObj['health'])
```

drawHealthMeter()函数在屏幕的左上角绘制一个指示器，告诉玩家他的松鼠再被碰到几次就会死掉。本章后面将会介绍这个函数。

8.22　事件处理循环

```
215.        for event in pygame.event.get(): # event handling loop
216.            if event.type == QUIT:
217.                terminate()
```

在事件处理循环中，首先检查是否生成了 QUIT 事件。如果是的，那么，程序将会终止。

```
219.            elif event.type == KEYDOWN:
220.                if event.key in (K_UP, K_w):
221.                    moveDown = False
222.                    moveUp = True
223.                elif event.key in (K_DOWN, K_s):
224.                    moveUp = False
225.                    moveDown = True
```

如果按下向上箭头键或向下箭头键（或者其对等的 WASD 键），那么，该方向的移动变量（moveRight、moveDown 等）应该设置为 True，并且相反方向的移动变量应该设置为 False。

```
226.                elif event.key in (K_LEFT, K_a):
227.                    moveRight = False
228.                    moveLeft = True
229.                    if playerObj['facing'] == RIGHT: # change player image
230.                        playerObj['surface'] = pygame.transform.scale(L_SQUIR_IMG, (playerObj['size'], playerObj['size']))
231.                    playerObj['facing'] = LEFT
```

```
232.            elif event.key in (K_RIGHT, K_d):
233.                moveLeft = False
234.                moveRight = True
235.                if playerObj['facing'] == LEFT: # change player image
236.                    playerObj['surface'] =
pygame.transform.scale(R_SQUIR_IMG, (playerObj['size'], playerObj['size']))
237.                    playerObj['facing'] = RIGHT
```

当按下向左箭头键或向右箭头键的时候，也应该设置 moveLeft 和 moveRight 变量。此外，playerObj['facing']中的值也应该更新为 LEFT 或 RIGHT。如果玩家松鼠现在面朝一个新的方向，playerObj['surface']的值应该用一个面朝新的方向、正确缩放的松鼠图像来替代。如果按下了向左箭头键，运行第 229 行的代码，并且检查玩家的松鼠是否是面朝右边的。如果是这样的话，将玩家松鼠图像的一个新的、缩放的 Surface 对象存储到 playerObj['surface']中。第 23 行的 elif 语句的代码处理相反的情况。

```
238.            elif winMode and event.key == K_r:
239.                return
```

如果玩家通过长到足够大而赢得了游戏（在这种情况下，winMode 将设置为 True）并且按下了 R 键，那么 runGame()应该返回。这将会导致当前的游戏结束，并且下次调用 runGame()的时候将开始一次新的游戏。

```
241.        elif event.type == KEYUP:
242.            # stop moving the player's squirrel
243.            if event.key in (K_LEFT, K_a):
244.                moveLeft = False
245.            elif event.key in (K_RIGHT, K_d):
246.                moveRight = False
247.            elif event.key in (K_UP, K_w):
248.                moveUp = False
249.            elif event.key in (K_DOWN, K_s):
250.                moveDown = False
```

如果玩家松开了任何的箭头方向键或者 WASD 键，代码应该将该方向的移动变量设置为 False。这将会停止松鼠在该方向上的任何移动。

```
252.            elif event.key == K_ESCAPE:
253.                terminate()
```

如果按下的键是 Esc 键，则终止程序。

第 8 章 Squirrel Eat Squirrel

8.23 移动玩家并考虑跳动

```
255.        if not gameOverMode:
256.            # actually move the player
257.            if moveLeft:
258.                playerObj['x'] -= MOVERATE
259.            if moveRight:
260.                playerObj['x'] += MOVERATE
261.            if moveUp:
262.                playerObj['y'] -= MOVERATE
263.            if moveDown:
264.                playerObj['y'] += MOVERATE
```

只有在游戏没有结束的时候，第 255 行的 if 语句中的代码将来回移动玩家的松鼠（这就是为什么在玩家的松鼠死掉之后，再按下箭头方向键不会有任何效果的原因）。根据哪一个移动变量设置为 True，playerObj 字典应该通过 MOVERATE 修改其 playerObj['x']和 playerObj['y']值（这就是为什么一个较大的 MOVERATE 值会让松鼠移动得更快的原因）。

```
266.            if (moveLeft or moveRight or moveUp or moveDown) or playerObj['bounce'] != 0:
267.                playerObj['bounce'] += 1
268.
269.            if playerObj['bounce'] > BOUNCERATE:
270.                playerObj['bounce'] = 0 # reset bounce amount
```

playerObj['bounce']记录了玩家松鼠位于跳动的哪个位置。这个变量存储着从 0 到 BOUNCERATE 的一个整数值。就像敌人松鼠的跳动值一样，playerObj['bounce']值为 0 意味着玩家松鼠位于一次跳动的起点，而其值为 BOUNCERATE 意味着玩家松鼠位于跳动的终点。

只要玩家在移动，或者玩家已经停止了移动，但是松鼠还没有完成其当前的跳动，那么玩家的松鼠就要跳动。第 266 行的 if 语句中捕获了这一条件。如果任何的移动变量设置为 True 或者当前的 playerObj['bounce']不为 0（这意味着玩家当前在跳动之中），那么，应该在第 267 行递增该变量。

由于 playerObj['bounce']变量只能够在 0 到 BOUNCERATE 的范围，如果递增它使得其 BOUNCERATE 大，应该将其重置为 0。

8.24 碰撞检测：吃或被吃

```
272.            # check if the player has collided with any squirrels
273.            for i in range(len(squirrelObjs)-1, -1, -1):
274.                sqObj = squirrelObjs[i]
```

第 273 行的 for 循环将针对 squirrelObjs 中的每一个敌人松鼠游戏对象运行代码。注意，第 273 行中 range() 的参数从 squirrelObjs 的最后一个索引开始并递减。这是因为，for 循环中的代码最终可能删除这些敌人松鼠对象（如果玩家的松鼠最终吃掉了它们的话），因此，从列表的末尾向前来遍历这一点很重要。这么做的原因在前面的 8.18 节中介绍过。

```
275.                if 'rect' in sqObj and playerObj['rect'].colliderect(sqObj['rect']):
276.                    # a player/squirrel collision has occurred
277.
278.                    if sqObj['width'] * sqObj['height'] <= playerObj['size']**2:
279.                        # player is larger and eats the squirrel
280.                        playerObj['size'] += int( (sqObj['width'] * sqObj['height'])**0.2 ) + 1
281.                        del squirrelObjs[i]
```

如果玩家的松鼠等于或大于与它碰撞的敌人松鼠的大小，那么，玩家松鼠将吃掉敌人松鼠并长大。在第 280 行，根据敌人松鼠的大小计算后得到的数字，将会加到玩家松鼠对象的 'size' 键中，即玩家松鼠长大了。图 8-5 展示了不同大小的松鼠所带来的增长。注意，被吃掉的松鼠越大，导致的增长也越大。

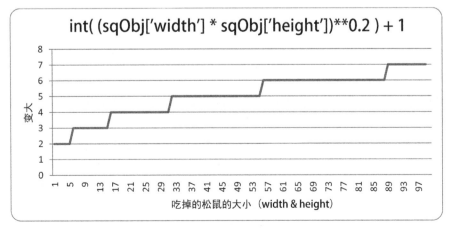

图 8-5

第 8 章　Squirrel Eat Squirrel

因此，根据该图，吃掉一个宽度和高度为 45 的松鼠（这是在 1600 像素的游戏区域中），将会导致玩家的宽度和高度都增加 5 个像素。

第 281 行从 squirrelObjs 列表中删除了被吃掉的松鼠对象，以使其从屏幕上消失或者更新其位置。

```
283.                    if playerObj['facing'] == LEFT:
284.                        playerObj['surface'] =
pygame.transform.scale(L_SQUIR_IMG, (playerObj['size'], playerObj['size']))
285.                    if playerObj['facing'] == RIGHT:
286.                        playerObj['surface'] =
pygame.transform.scale(R_SQUIR_IMG, (playerObj['size'], playerObj['size']))
```

既然松鼠变大了，那么玩家松鼠图像需要更新。这通过将 L_SQUIR_IMG 或 R_SQUIR_IMG 中的最初的松鼠图像传递给 pygame.transform.scale()函数来完成，该函数将返回图像的一个增大的版本。根据 playerObj['facing']等于 LEFT，还是 RIGHT，来确定我们将哪一个初始松鼠图像传递给该函数。

```
288.                    if playerObj['size'] > WINSIZE:
289.                        winMode = True # turn on "win mode"
```

玩家取得游戏的胜利的方式是让松鼠变得比 WINSIZE 常量变量中存储的整数值还要大。如果真是这样，将 winMode 变量设置为 True。该函数的其他部分的代码将负责显示祝贺消息，并且检查玩家是否按下了 R 键以重新开始游戏。

```
291.                elif not invulnerableMode:
292.                    # player is smaller and takes damage
293.                    invulnerableMode = True
294.                    invulnerableStartTime = time.time()
295.                    playerObj['health'] -= 1
296.                    if playerObj['health'] == 0:
297.                        gameOverMode = True # turn on "game over mode"
298.                        gameOverStartTime = time.time()
```

如果玩家的区域并不等于或大于敌人松鼠的区域，并且 invulnerableMode 没有设置为 True，那么玩家和这只较大的松鼠碰撞的时候会受到伤害。

为了防止玩家被同一只松鼠片刻之间多次伤害，我们在第 293 行将 invulnerable Mode 设置为 True，让玩家在简短的时间内处于受保护的状态，即不会受到敌人松鼠的进一步攻击。第 294 行将 invulnerableStartTime 设置为当前时间（这通过 time.time()返回），以便第 133 行和第 134 行知道什么时候将 invulnerableMode 设置为 False。

第 295 行将玩家的生命值减去 1。由于玩家的生命值有可能会为 0，第 296 行检查这一点，将 gameOverMode 设置为 True，并且将 gameOverStartTime 设置为当前时间。

8.25 游戏结束屏幕

```
299.        else:
300.            # game is over, show "game over" text
301.            DISPLAYSURF.blit(gameOverSurf, gameOverRect)
302.            if time.time() - gameOverStartTime > GAMEOVERTIME:
303.                return # end the current game
```

当玩家死掉的时候，"Game Over"文本（它位于 gameOverSurf 变量中的 Surface 对象之上）将会显示在屏幕上数秒钟的时间，这个时间存储在 GAMEOVERTIME 常量中。一旦达到了这个时间，runGame()函数将返回。

在玩家死掉之后和新的游戏开始之前的数秒钟，将允许敌人松鼠继续开始动画并移动。在新游戏开始之前，Squirrel Eat Squirrel 中的游戏结束屏幕并不会等待玩家按下一个键。

8.26 获胜

```
305.        # check if the player has won.
306.        if winMode:
307.            DISPLAYSURF.blit(winSurf, winRect)
308.            DISPLAYSURF.blit(winSurf2, winRect2)
309.
310.        pygame.display.update()
311.        FPSCLOCK.tick(FPS)
```

如果玩家达到了一定的大小（这在 WINSIZE 常量中指定），在第 289 行，winMode 设置为 True。当玩家获胜的时候，这些事情都会发生："You have achieved OMEGA SQUIRREL!‖"文本（位于 winSurf 变量中所存储的 Surface 对象之上）和"(Press ―r‖ to restart.)"文本（位于 winSurf2 变量中所存储的 Surface 对象之上）将会出现在屏幕上。游戏继续，直到用户按下 R 键，此时，程序执行将从 runGame()返回。R 键的事件处理代码在第 238 行到第 239 行完成。

8.27 绘制图形化的生命值指标

```
316. def drawHealthMeter(currentHealth):
317.     for i in range(currentHealth): # draw red health bars
```

第 8 章 Squirrel Eat Squirrel

```
318.            pygame.draw.rect(DISPLAYSURF, RED,   (15, 5 + (10 * MAXHEALTH) - i * 10, 20, 10))
319.        for i in range(MAXHEALTH): # draw the white outlines
320.            pygame.draw.rect(DISPLAYSURF, WHITE, (15, 5 + (10 * MAXHEALTH) - i * 10, 20, 10), 1)
```

为了绘制生命值指标，第 317 行的第一个 for 循环绘制了一个实心的红色矩形，表示玩家的生命值的数量。第 319 行的 for 循环绘制了一个空心的白色矩形，表示玩家可能拥有的所有生命值（这是 MAXHEALTH 常量中所存储的一个整数值）。注意，drawHealthMeter() 中并没有调用 pygame.display.update()函数。

8.28 相同的旧的 terminate()函数

```
323. def terminate():
324.     pygame.quit()
325.     sys.exit()
```

terminate()函数的工作方式和之前游戏中的相同。

8.29 正弦函数

```
328. def getBounceAmount(currentBounce, bounceRate, bounceHeight):
329.     # Returns the number of pixels to offset based on the bounce.
330.     # Larger bounceRate means a slower bounce.
331.     # Larger bounceHeight means a higher bounce.
332.     # currentBounce will always be less than bounceRate
333.     return int(math.sin( (math.pi / float(bounceRate)) * currentBounce ) * bounceHeight)
334.
```

有一个叫作正弦函数的数学函数（它和编程中的函数类似，也是根据其参数"返回"或"求得"一个数字）。你可能已经在 math 类中了解过，但是如果还没有的话，这里将给出介绍。Python 将这个数学函数作为 math 模块中的一个 Python 函数使用。你可以传入一个整数或浮点数值给 math.sin()，它将返回一个叫作"正弦值"的浮点数值。

在交互式 shell 中，我们可以看看 math.sin()针对一些值的返回值。

```
>>> import math
>>> math.sin(1)
0.8414709848078965
>>> math.sin(2)
0.90929742682568171
>>> math.sin(3)
0.14112000805986721
>>> math.sin(4)
-0.7568024953079282
>>> math.sin(5)
-0.95892427466313845
```

似乎很难根据我们所传递的值来预测 math.sin() 将会返回什么(这使得你怀疑 math.sin() 的用途所在)。但是,如果用图形表示出从 1 到 10 的整数的正弦值,将会得到图 8-6 所示内容。

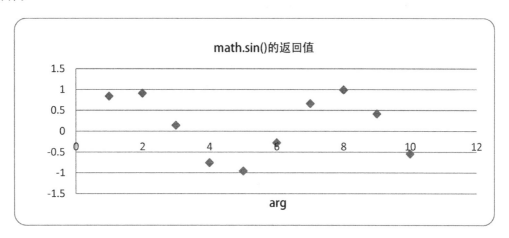

图 8-6

你可能会看到 math.sin() 返回了一个波浪线的样子。如果针对整数以外的数值（例如,1.5 和 2.5 等）计算出正弦值,并用线条将这些点连接起来,会更容易看到如图 8-7 所示的波浪线的样子。

实际上,如果继续给这幅图添加更多的数据点,将会看到如图 8-8 所示的正弦曲线。

注意,math.sin(0)返回 0,然后逐渐增加,直到 math.sin(3.14 / 2)返回 1,然后它开始减小,直到 math.sin(3.14) 返回 0。数值 3.14 在数学上是一个特殊的值,称为 pi。这个值也存储在 math 模块中的常量变量 pi 中（这就是为什么第 333 行使用了该变量 math.pi),从技术上讲,这就是浮点数 3.1415926535897931。由于想要对松鼠使用一个波浪线式的跳动,我们只要留意 math.sin()对于参数 0 到 3.14 的返回值就可以了,如图 8-9 所示。

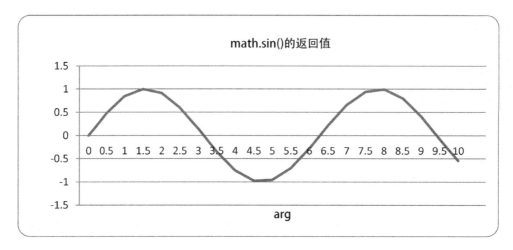

图 8-7

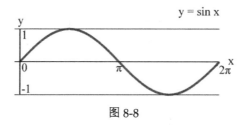

图 8-8

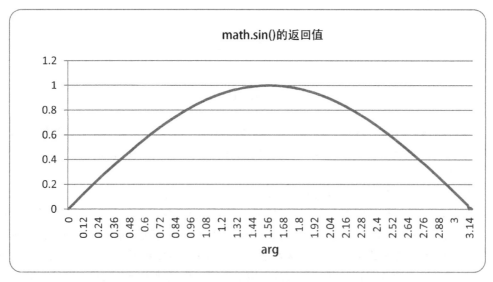

图 8-9

来看一下 getBounceAmount() 的返回值，并搞清楚它到底做些什么。

```
333.        return int(math.sin( (math.pi / float(bounceRate)) * currentBounce ) *
bounceHeight)
```

还记得吧,在第 21 行,我们将 BOUNCERATE 常量设置为 6。这意味着,我们的代码将只会将 playerObj['bounce']从 0 增加到 6,并且我们想要把从 0 到 3.14 的浮点值划分为 6 个部分,只需要做一次简单的除法就可以了,即 3.14 / 6 = 0.5235。图形之中的 3.14 的长度划分为用于"正弦波浪跳动"的 6 个部分的话,每一个部分都是 0.5235。

你可以看到,当 playerObj['bounce']等于 3 的时候(处于 0 到 6 的中间),传递给 math.sin()调用的值是 math.pi / 6 * 3,即 1.5707(0 到 3.1415 的一半)。然后,math.sin(1.5707)将返回 1.0,而这是正弦波浪线中最高的部分(正弦波浪线中最高的部分位于波浪的一半处)。

随着 playerObj['bounce']的值的增加,getBounceAmount()函数的返回值,将具有从 0 到 3.14 的正弦波浪线相同的跳动形状。如果想要让松鼠跳动得更高些,可以增加 BOUNCEHEIGHT 常量。如果想要让松鼠跳动低一些,可以增加 BOUNCERATE 常量。

正弦函数是三角数学中的一个概念。如果想要了解有关正弦波浪线的更多内容,可以通过 Wiki 页面 http://en.wikipedia.org/wiki/Sine 了解更多信息。

8.30 对 Python 2 的向后兼容

我们调用 float()来把 bounceRate 转换为一个浮点数,只是因为这样一来,程序在 Python 2 中也能工作。在 Python 3 中,除法操作符将得到一个浮点数值,即便操作数都是整数,如下所示。

```
>>> # Python version 3
...
>>> 10 / 5
2.0
>>> 10 / 4
2.5
>>>
```

然而,在 Python 2 中,只有当操作数是浮点数值的时候,除法操作符/才会求得一个浮点数值。如果两个操作数都是整数,那么,Python 2 中的除法操作符将求得一个整数(必要的时候会舍入),如下所示。

```
>>> # Python version 2
...
>>> 10 / 5
2
```

第 8 章　Squirrel Eat Squirrel

```
>>> 10 / 4
2
>>> 10 / 4.0
2.5
>>> 10.0 / 4
2.5
>>> 10.0 / 4.0
2.5
```

但是，如果总是使用 float() 函数将一个值转换为一个浮点值，那么，不管使用何种 Python 版本运行这段源代码，除法操作符都将求得一个浮点值。做出这些修改，以便我们的代码能够在较早的软件版本中使用，这种做法叫作向后兼容（backwards compatibility）。保持向后兼容很重要，因为并不是每个人都总是运行最新版本的软件，而你总是想要确保所编写的代码对于尽可能多的计算机都能够使用。

你并不总是能够使得 Python 3 的代码向后兼容 Python 2，但是，如果可能的话，应该这么做。否则的话，当使用 Python 2 的人试图运行你的游戏却得到错误消息的时候，他们会认为你的程序有 Bug。

通过 http://inventwithpython.com/appendixa.html，可以找到 Python 2 和 Python 3 之间的一些差异的一个列表。

8.31　getRandomVelocity() 函数

```
335. def getRandomVelocity():
336.     speed = random.randint(SQUIRRELMINSPEED, SQUIRRELMAXSPEED)
337.     if random.randint(0, 1) == 0:
338.         return speed
339.     else:
340.         return -speed
```

getRandomVelocity() 函数用来随机地确定敌人松鼠能够移动得多快。这个速率的范围在 SQUIRRELMINSPEED 和 SQUIRRELMAXSPEED 常量中设置，但是，最终，这个速度要么是负值（表示松鼠到达了最左边或最上边），要么是正数（表示松鼠到达了最右边或最下边）。这个随机速度总是有一半的机会为正或为负。

8.32　找到一个地方添加新的松鼠和草

```
343. def getRandomOffCameraPos(camerax, cameray, objWidth, objHeight):
344.     # create a Rect of the camera view
```

8.32 找到一个地方添加新的松鼠和草

```
345.      cameraRect = pygame.Rect(camerax, cameray, WINWIDTH, WINHEIGHT)
346.      while True:
347.          x = random.randint(camerax - WINWIDTH, camerax + (2 * WINWIDTH))
348.          y = random.randint(cameray - WINHEIGHT, cameray + (2 * WINHEIGHT))
349.          # create a Rect object with the random coordinates and use colliderect()
350.          # to make sure the right edge isn't in the camera view.
351.          objRect = pygame.Rect(x, y, objWidth, objHeight)
352.          if not objRect.colliderect(cameraRect):
353.              return x, y
```

当在游戏世界中创建新的松鼠对象和草对象的时候，我们想要让其位于活动区域中（以便它靠近玩家松鼠），但它们不是在相机视图中（以使得它们不会只是突然在屏幕中弹出）。为了做到这一点，我们创建了一个 Rect 对象来表示相机区域（使用 camerax、cameray、WINWIDTH 和 WINHEIGHT 常量）。

接下来，随机地生成位于活动区域之中的 XY 坐标的数字。活动区域的左边界和上边界，分别位于 camerax 和 cameray 的左边和上边 WINWIDTH 和 WINHEIGHT 像素的位置。因此，活动区域的左上角位于 camerax – WINWIDTH 和 cameray – WINHEIGHT。活动区域的宽度和高度，也是 WINWIDTH 和 WINHEIGHT 的大小的 3 倍，如图 8-10 所示（其中 WINWIDTH 设置为 640 像素，而 WINHEIGHT 设置为 480 像素）。

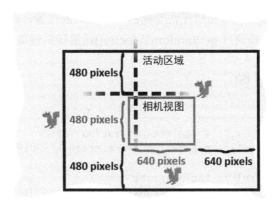

图 8-10

这意味着右边和下边将会是 camerax + (2 * WINWIDTH) 和 cameray + (2 * WINHEIGHT)。第 352 行将会检查，随机 XY 坐标是否与相机视图的 Rect 对象冲突。如果不冲突，那么，这些坐标将会返回。如果冲突，那么第 346 行的 while 循环将持续生成新的坐标，直到它找到可接受的一个。

8.33 创建敌人松鼠数据结构

```
356. def makeNewSquirrel(camerax, cameray):
357.     sq = {}
358.     generalSize = random.randint(5, 25)
359.     multiplier = random.randint(1, 3)
360.     sq['width']  = (generalSize + random.randint(0, 10)) * multiplier
361.     sq['height'] = (generalSize + random.randint(0, 10)) * multiplier
362.     sq['x'], sq['y'] = getRandomOffCameraPos(camerax, cameray, sq['width'], sq['height'])
363.     sq['movex'] = getRandomVelocity()
364.     sq['movey'] = getRandomVelocity()
```

创建敌人松鼠游戏对象，类似于生成草游戏对象。每个敌人松鼠的数据，也都存储在一个字典之中。宽度和高度在第 360 行到第 361 行设置为随机的大小。还使用了 generalSize 变量，以使得每个松鼠的宽度和高度不会彼此相差太大。否则的话，为宽度和高度使用完全随机的数字，可能会导致很高很瘦的松鼠和很矮很胖的松鼠。松鼠的宽度和高度，是用一个通用的大小加上 0 到 10 之间的一个随机数（一般有一些细微的变化），然后再乘以 multiplier 变量。

松鼠的最初的 XY 坐标将是相机无法看到的一个随机的位置，以防止松鼠"突然"出现在屏幕上。速度和方向也是通过 getRandomVelocity() 函数随机选取的。

8.34 翻转松鼠图像

```
365.     if sq['movex'] < 0: # squirrel is facing left
366.         sq['surface'] = pygame.transform.scale(L_SQUIR_IMG, (sq['width'], sq['height']))
367.     else: # squirrel is facing right
368.         sq['surface'] = pygame.transform.scale(R_SQUIR_IMG, (sq['width'], sq['height']))
369.     sq['bounce'] = 0
370.     sq['bouncerate'] = random.randint(10, 18)
371.     sq['bounceheight'] = random.randint(10, 50)
372.     return sq
```

L_SQUIR_IMG 和 R_SQUIR_IMG 常量所包含的 Surface 对象，面朝左边和面朝右边的松鼠图像分别位于其上。新的 Surface 对象将使用 pygame.transform.scale() 函数来产生，以匹配松鼠的宽度和高度（分别存储在 sq['width']和 sq['height']之中）。

此后，随机地生成 3 个和跳动相关的值（除了 sq['bounce']，它是 0，因为松鼠总是从跳动的起始位置开始），并且在 372 行返回字典。

8.35 创建草数据结构

```
375. def makeNewGrass(camerax, cameray):
376.     gr = {}
377.     gr['grassImage'] = random.randint(0, len(GRASSIMAGES) - 1)
378.     gr['width']  = GRASSIMAGES[0].get_width()
379.     gr['height'] = GRASSIMAGES[0].get_height()
380.     gr['x'], gr['y'] = getRandomOffCameraPos(camerax, cameray, gr['width'], gr['height'])
381.     gr['rect'] = pygame.Rect( (gr['x'], gr['y'], gr['width'], gr['height']) )
382.     return gr
```

草游戏对象是字典，它带有常用的'x'、'y'、'width'、'height'和'rect'键，还有'grassImage'键，后者是 0 到 GRASSIMAGES 列表的长度减去 1 之间的一个数字。这个数字将决定草游戏对象中拥有什么图像。例如，如果草对象的'grassImage'键是 3，那么，它将使用 GRASSIMAGES[3]所存储的 Surface 对象作为其图像。

8.36 检查是否在活动区域之外

```
385. def isOutsideActiveArea(camerax, cameray, obj):
386.     # Return False if camerax and cameray are more than
387.     # a half-window length beyond the edge of the window.
388.     boundsLeftEdge = camerax - WINWIDTH
389.     boundsTopEdge = cameray - WINHEIGHT
390.     boundsRect = pygame.Rect(boundsLeftEdge, boundsTopEdge, WINWIDTH * 3, WINHEIGHT * 3)
391.     objRect = pygame.Rect(obj['x'], obj['y'], obj['width'], obj['height'])
392.     return not boundsRect.colliderect(objRect)
```

如果你传递给 isOutsideActiveArea()的对象位于 camerax 和 cameray 参数所指定的"活动区域"之外，那么，该函数返回 True。记住，活动区域是围绕在相机视图之外的相机视图大小的一圈区域（相机视图的宽度和高度由 WINWIDTH 和 WINHEIGHT 设置），如图 8-11 所示。

我们可以创建一个 Rect 对象来表示一个活动区域，通过传递 camerax – WINWIDTH 作为左边界值，而 cameray – WINHEIGHT 作为上边界值，然后，WINWIDTH * 3 和

WINHEIGHT * 3 分别作为宽度和高度。一旦有了表示活动区域的一个 Rect 对象，我们可以使用 colliderect()方法来判断 obj 参数中的对象是否与活动区域 Rect 对象碰撞，即是否位于其中。

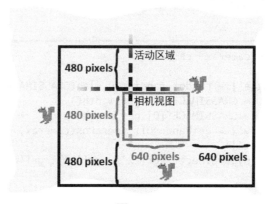

图 8-11

由于玩家松鼠、敌人松鼠和草对象都拥有 'x'、'y'、'width' 和 'height' 键，isOutsideActiveArea()代码可以用于这些游戏对象中的任何一种。

```
395. if __name__ == '__main__':
396.     main()
```

最后，在所有的函数定义之后，程序将运行 main()函数并开始游戏。

8.37 本章小结

Squirrel Eat Squirrel 是我的第一款一次拥有多个敌人在游戏板上移动的游戏。拥有多个敌人的关键是使用一个字典值，它带有用于每个敌人松鼠的相同的键，从而在游戏循环的一次迭代之中，相同的代码可以在每一个敌人松鼠上运行。

本章还介绍了相机的概念。我们前面的游戏并不需要相机，因为整个游戏世界刚好就是一个屏幕。然而，当你的游戏涉及让玩家在一个较大的游戏世界中移动的时候，需要编写代码来负责在游戏世界坐标系统和屏幕像素坐标系统之间的转换。

最后，还引入了数学正弦函数来处理真实的松鼠跳动（不管每次跳的有多高或多远）。要进行编程，你不需要了解太多的数学知识。在大多数情况下，只要知道加法、乘法和负数就很好了。然而，如果你学习了数学，往往会发现有几次机会来使用数学知识使得游戏更酷。

要进行更多的编程练习，可以从 http://invpy.com/buggy/squirrel 下载 Squirrel Eat Squirrel 的带有 Bug 的版本，并尝试搞清楚如何修正这些 Bug。

第 9 章　Star Pusher

9.1　如何玩 Star Pusher

Star Pusher 是 Sokoban 或 Box Pusher 的翻版。玩家在一个带有几个星星的屋子之中。在房间中的地面上的某些贴片精灵之上，有星星的标记。玩家必须搞清楚如何将星星推到带有星星标记的贴片之上。如果一颗星星的后面有一堵墙或者有另一个星星的话，玩家是无法推动它的。玩家不能拉动星星，因此，如果将一个星星推到了一个角落里，玩家就必须重新开始这一关。当所有的星星都已经推动到了带有星星标记的地板贴片之上时，这一关就完成了，并且开始下一关，如图 9-1 所示。

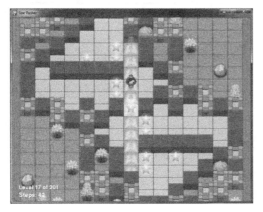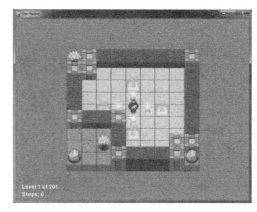

图 9-1

每个关卡都由贴片图像的一个 2D 栅格组成。贴片精灵（Tile sprite）是具有相同大小的图像，它们可以彼此相邻地放置，以构成较为复杂的图像。使用一些地板和墙壁的贴片，我们可以创建带有众多有趣的形状、大小不同的关卡。

关卡文件并没有包含在源代码中。相反，你可以自己创建关卡文件或者下载它们。可以从 http://invpy.com/starPusherLevels.tx 下载带有 201 个关卡的一个关卡文件。当你运行 Star Pusher 程序的时候，确保关卡文件位于和文件 *starpusher.py* 相同的文件夹中。否则，将会得到一条错误消息 AssertionError: Cannot find the level file: starPusherLevels.txt。这个关卡最初是由 David W. Skinner 设计的。你可以从他的 Web 站点 http://users.bentonrea.com/~sasquatch/sokoban/下载到更多的迷宫。

9.2 Star Pusher 的源代码

可以从 http://invpy.com/starpusher.py 下载 Star Pusher 的源代码。如果得到任何的错误消息，查看一下错误消息中提到的代码行，并且检查代码是否有任何录入错误。可以将代码复制并粘贴到位于 http://invpy.com/diff/starpusher 的 Web 表单中，看看你的代码和本书中的代码是否有区别。

可以从 http://invpy.com/starPusherLevels.txt 下载关卡文件。可以从 http://invpy.com/starPusherImages.zip 下载贴片。

此外，就像 Squirrel Eat Squirrel 游戏中的松鼠对象、草对象和敌人松鼠对象一样，当我在本章中说"地图对象"、"游戏状态对象"或"关卡对象"的时候，并不是表示面向对象编程意义中的对象。这些"对象"实际上只是字典值，但是，将其称为对象会更容易理解，因为它们表示游戏世界中的事物。

```
1.  # Star Pusher (a Sokoban clone)
2.  # By Al Sweigart al@inventwithpython.com
3.  # http://inventwithpython.com/pygame
4.  # Creative Commons BY-NC-SA 3.0 US
5.
6.  import random, sys, copy, os, pygame
7.  from pygame.locals import *
8.
9.  FPS = 30 # frames per second to update the screen
10. WINWIDTH = 800 # width of the program's window, in pixels
11. WINHEIGHT = 600 # height in pixels
12. HALF_WINWIDTH = int(WINWIDTH / 2)
13. HALF_WINHEIGHT = int(WINHEIGHT / 2)
14.
15. # The total width and height of each tile in pixels.
16. TILEWIDTH = 50
17. TILEHEIGHT = 85
18. TILEFLOORHEIGHT = 45
19.
20. CAM_MOVE_SPEED = 5 # how many pixels per frame the camera moves
21.
22. # The percentage of outdoor tiles that have additional
23. # decoration on them, such as a tree or rock.
24. OUTSIDE_DECORATION_PCT = 20
25.
26. BRIGHTBLUE     = (  0, 170, 255)
27. WHITE          = (255, 255, 255)
28. BGCOLOR = BRIGHTBLUE
29. TEXTCOLOR = WHITE
```

```
30.
31. UP = 'up'
32. DOWN = 'down'
33. LEFT = 'left'
34. RIGHT = 'right'
35.
36.
37. def main():
38.     global FPSCLOCK, DISPLAYSURF, IMAGESDICT, TILEMAPPING,
OUTSIDEDECOMAPPING, BASICFONT, PLAYERIMAGES, currentImage
39.
40.     # Pygame initialization and basic set up of the global variables.
41.     pygame.init()
42.     FPSCLOCK = pygame.time.Clock()
43.
44.     # Because the Surface object stored in DISPLAYSURF was returned
45.     # from the pygame.display.set_mode() function, this is the
46.     # Surface object that is drawn to the actual computer screen
47.     # when pygame.display.update() is called.
48.     DISPLAYSURF = pygame.display.set_mode((WINWIDTH, WINHEIGHT))
49.
50.     pygame.display.set_caption('Star Pusher')
51.     BASICFONT = pygame.font.Font('freesansbold.ttf', 18)
52.
53.     # A global dict value that will contain all the Pygame
54.     # Surface objects returned by pygame.image.load().
55.     IMAGESDICT = {'uncovered goal': pygame.image.load('RedSelector.png'),
56.                   'covered goal': pygame.image.load('Selector.png'),
57.                   'star': pygame.image.load('Star.png'),
58.                   'corner': pygame.image.load('Wall Block Tall.png'),
59.                   'wall': pygame.image.load('Wood Block Tall.png'),
60.                   'inside floor': pygame.image.load('Plain Block.png'),
61.                   'outside floor': pygame.image.load('Grass Block.png'),
62.                   'title': pygame.image.load('star_title.png'),
63.                   'solved': pygame.image.load('star_solved.png'),
64.                   'princess': pygame.image.load('princess.png'),
65.                   'boy': pygame.image.load('boy.png'),
66.                   'catgirl': pygame.image.load('catgirl.png'),
67.                   'horngirl': pygame.image.load('horngirl.png'),
68.                   'pinkgirl': pygame.image.load('pinkgirl.png'),
69.                   'rock': pygame.image.load('Rock.png'),
70.                   'short tree': pygame.image.load('Tree_Short.png'),
71.                   'tall tree': pygame.image.load('Tree_Tall.png'),
72.                   'ugly tree': pygame.image.load('Tree_Ugly.png')}
73.
74.     # These dict values are global, and map the character that appears
75.     # in the level file to the Surface object it represents.
```

第 9 章 Star Pusher

```
76.      TILEMAPPING = {'x': IMAGESDICT['corner'],
77.                     '#': IMAGESDICT['wall'],
78.                     'o': IMAGESDICT['inside floor'],
79.                     ' ': IMAGESDICT['outside floor']}
80.      OUTSIDEDECOMAPPING = {'1': IMAGESDICT['rock'],
81.                            '2': IMAGESDICT['short tree'],
82.                            '3': IMAGESDICT['tall tree'],
83.                            '4': IMAGESDICT['ugly tree']}
84.
85.      # PLAYERIMAGES is a list of all possible characters the player can be.
86.      # currentImage is the index of the player's current player image.
87.      currentImage = 0
88.      PLAYERIMAGES = [IMAGESDICT['princess'],
89.                      IMAGESDICT['boy'],
90.                      IMAGESDICT['catgirl'],
91.                      IMAGESDICT['horngirl'],
92.                      IMAGESDICT['pinkgirl']]
93.
94.      startScreen() # show the title screen until the user presses a key
95.
96.      # Read in the levels from the text file. See the readLevelsFile() for
97.      # details on the format of this file and how to make your own levels.
98.      levels = readLevelsFile('starPusherLevels.txt')
99.      currentLevelIndex = 0
100.
101.     # The main game loop. This loop runs a single level, when the user
102.     # finishes that level, the next/previous level is loaded.
103.     while True: # main game loop
104.         # Run the level to actually start playing the game:
105.         result = runLevel(levels, currentLevelIndex)
106.
107.         if result in ('solved', 'next'):
108.             # Go to the next level.
109.             currentLevelIndex += 1
110.             if currentLevelIndex >= len(levels):
111.                 # If there are no more levels, go back to the first one.
112.                 currentLevelIndex = 0
113.         elif result == 'back':
114.             # Go to the previous level.
115.             currentLevelIndex -= 1
116.             if currentLevelIndex < 0:
117.                 # If there are no previous levels, go to the last one.
118.                 currentLevelIndex = len(levels)-1
119.         elif result == 'reset':
120.             pass # Do nothing. Loop re-calls runLevel() to reset the level
121.
```

```
122.
123. def runLevel(levels, levelNum):
124.     global currentImage
125.     levelObj = levels[levelnum]
126.     mapObj = decorateMap(levelObj['mapObj'], levelObj['startState']['player'])
127.     gameStateObj = copy.deepcopy(levelObj['startState'])
128.     mapNeedsRedraw = True # set to True to call drawMap()
129.     levelSurf = BASICFONT.render('Level %s of %s' % (levelObj['levelNum'] + 1, totalNumOfLevels), 1, TEXTCOLOR)
130.     levelRect = levelSurf.get_rect()
131.     levelRect.bottomleft = (20, WINHEIGHT - 35)
132.     mapWidth = len(mapObj) * TILEWIDTH
133.     mapHeight = (len(mapObj[0]) - 1) * (TILEHEIGHT - TILEFLOORHEIGHT) + TILEHEIGHT
134.     MAX_CAM_X_PAN = abs(HALF_WINHEIGHT - int(mapHeight / 2)) + TILEWIDTH
135.     MAX_CAM_Y_PAN = abs(HALF_WINWIDTH - int(mapWidth / 2)) + TILEHEIGHT
136.
137.     levelIsComplete = False
138.     # Track how much the camera has moved:
139.     cameraOffsetX = 0
140.     cameraOffsetY = 0
141.     # Track if the keys to move the camera are being held down:
142.     cameraUp = False
143.     cameraDown = False
144.     cameraLeft = False
145.     cameraRight = False
146.
147.     while True: # main game loop
148.         # Reset these variables:
149.         playerMoveTo = None
150.         keyPressed = False
151.
152.         for event in pygame.event.get(): # event handling loop
153.             if event.type == QUIT:
154.                 # Player clicked the "X" at the corner of the window.
155.                 terminate()
156.
157.             elif event.type == KEYDOWN:
158.                 # Handle key presses
159.                 keyPressed = True
160.                 if event.key == K_LEFT:
161.                     playerMoveTo = LEFT
162.                 elif event.key == K_RIGHT:
163.                     playerMoveTo = RIGHT
164.                 elif event.key == K_UP:
```

第 9 章　Star Pusher

```
165.                    playerMoveTo = UP
166.                elif event.key == K_DOWN:
167.                    playerMoveTo = DOWN
168.
169.                # Set the camera move mode.
170.                elif event.key == K_a:
171.                    cameraLeft = True
172.                elif event.key == K_d:
173.                    cameraRight = True
174.                elif event.key == K_w:
175.                    cameraUp = True
176.                elif event.key == K_s:
177.                    cameraDown = True
178.
179.                elif event.key == K_n:
180.                    return 'next'
181.                elif event.key == K_b:
182.                    return 'back'
183.
184.                elif event.key == K_ESCAPE:
185.                    terminate() # Esc key quits.
186.                elif event.key == K_BACKSPACE:
187.                    return 'reset' # Reset the level.
188.                elif event.key == K_p:
189.                    # Change the player image to the next one.
190.                    currentImage += 1
191.                    if currentImage >= len(PLAYERIMAGES):
192.                        # After the last player image, use the first one.
193.                        currentImage = 0
194.                    mapNeedsRedraw = True
195.
196.            elif event.type == KEYUP:
197.                # Unset the camera move mode.
198.                if event.key == K_a:
199.                    cameraLeft = False
200.                elif event.key == K_d:
201.                    cameraRight = False
202.                elif event.key == K_w:
203.                    cameraUp = False
204.                elif event.key == K_s:
205.                    cameraDown = False
206.
207.        if playerMoveTo != None and not levelIsComplete:
208.            # If the player pushed a key to move, make the move
209.            # (if possible) and push any stars that are pushable.
210.            moved = makeMove(mapObj, gameStateObj, playerMoveTo)
```

```
211.
212.            if moved:
213.                # increment the step counter.
214.                gameStateObj['stepCounter'] += 1
215.                mapNeedsRedraw = True
216.
217.                if isLevelFinished(levelObj, gameStateObj):
218.                    # level is solved, we should show the "Solved!" image.
219.                    levelIsComplete = True
220.                    keyPressed = False
221.
222.        DISPLAYSURF.fill(BGCOLOR)
223.
224.        if mapNeedsRedraw:
225.            mapSurf = drawMap(mapObj, gameStateObj, levelObj['goals'])
226.            mapNeedsRedraw = False
227.
228.        if cameraUp and cameraOffsetY < MAX_CAM_X_PAN:
229.            cameraOffsetY += CAM_MOVE_SPEED
230.        elif cameraDown and cameraOffsetY > -MAX_CAM_X_PAN:
231.            cameraOffsetY -= CAM_MOVE_SPEED
232.        if cameraLeft and cameraOffsetX < MAX_CAM_Y_PAN:
233.            cameraOffsetX += CAM_MOVE_SPEED
234.        elif cameraRight and cameraOffsetX > -MAX_CAM_Y_PAN:
235.            cameraOffsetX -= CAM_MOVE_SPEED
236.
237.        # Adjust mapSurf's Rect object based on the camera offset.
238.        mapSurfRect = mapSurf.get_rect()
239.        mapSurfRect.center = (HALF_WINWIDTH + cameraOffsetX, HALF_WINHEIGHT + cameraOffsetY)
240.
241.        # Draw mapSurf to the DISPLAYSURF Surface object.
242.        DISPLAYSURF.blit(mapSurf, mapSurfRect)
243.
244.        DISPLAYSURF.blit(levelSurf, levelRect)
245.        stepSurf = BASICFONT.render('Steps: %s' % (gameStateObj['stepCounter']), 1, TEXTCOLOR)
246.        stepRect = stepSurf.get_rect()
247.        stepRect.bottomleft = (20, WINHEIGHT - 10)
248.        DISPLAYSURF.blit(stepSurf, stepRect)
249.
250.        if levelIsComplete:
251.            # is solved, show the "Solved!" image until the player
252.            # has pressed a key.
253.            solvedRect = IMAGESDICT['solved'].get_rect()
254.            solvedRect.center = (HALF_WINWIDTH, HALF_WINHEIGHT)
```

第 9 章 Star Pusher

```
255.                DISPLAYSURF.blit(IMAGESDICT['solved'], solvedRect)
256.
257.            if keyPressed:
258.                return 'solved'
259.
260.        pygame.display.update() # draw DISPLAYSURF to the screen.
261.        FPSCLOCK.tick()
262.
263.
274. def decorateMap(mapObj, startxy):
275.     """Makes a copy of the given map object and modifies it.
276.     Here is what is done to it:
277.         * Walls that are corners are turned into corner pieces.
278.         * The outside/inside floor tile distinction is made.
279.         * Tree/rock decorations are randomly added to the outside tiles.
280.
281.     Returns the decorated map object."""
282.
283.     startx, starty = startxy # Syntactic sugar
284.
285.     # Copy the map object so we don't modify the original passed
286.     mapObjCopy = copy.deepcopy(mapObj)
287.
288.     # Remove the non-wall characters from the map data
289.     for x in range(len(mapObjCopy)):
290.         for y in range(len(mapObjCopy[0])):
291.             if mapObjCopy[x][y] in ('$', '.', '@', '+', '*'):
292.                 mapObjCopy[x][y] = ' '
293.
294.     # Flood fill to determine inside/outside floor tiles.
295.     floodFill(mapObjCopy, startx, starty, ' ', 'o')
296.
297.     # Convert the adjoined walls into corner tiles.
298.     for x in range(len(mapObjCopy)):
299.         for y in range(len(mapObjCopy[0])):
300.
301.             if mapObjCopy[x][y] == '#':
302.                 if (isWall(mapObjCopy, x, y-1) and isWall(mapObjCopy, x+1, y)) or \
303.                     (isWall(mapObjCopy, x+1, y) and isWall(mapObjCopy, x, y+1)) or \
304.                     (isWall(mapObjCopy, x, y+1) and isWall(mapObjCopy, x-1, y)) or \
305.                     (isWall(mapObjCopy, x-1, y) and isWall(mapObjCopy, x, y-1)):
```

```
306.                         mapObjCopy[x][y] = 'x'
307.
308.                 elif mapObjCopy[x][y] == ' ' and random.randint(0, 99) <
OUTSIDE_DECORATION_PCT:
309.                     mapObjCopy[x][y] =
random.choice(list(OUTSIDEDECOMAPPING.keys()))
310.
311.     return mapObjCopy
312.
313.
314. def isBlocked(mapObj, gameStateObj, x, y):
315.     """Returns True if the (x, y) position on the map is
316.     blocked by a wall or star, otherwise return False."""
317.
318.     if isWall(mapObj, x, y):
319.         return True
320.
321.     elif x < 0 or x >= len(mapObj) or y < 0 or y >= len(mapObj[x]):
322.         return True # x and y aren't actually on the map.
323.
324.     elif (x, y) in gameStateObj['stars']:
325.         return True # a star is blocking
326.
327.     return False
328.
329.
330. def makeMove(mapObj, gameStateObj, playerMoveTo):
331.     """Given a map and game state object, see if it is possible for the
332.     player to make the given move. If it is, then change the player's
333.     position (and the position of any pushed star). If not, do nothing.
334.
335.     Returns True if the player moved, otherwise False."""
336.
337.     # Make sure the player can move in the direction they want.
338.     playerx, playery = gameStateObj['player']
339.
340.     # This variable is "syntactic sugar". Typing "stars" is more
341.     # readable than typing "gameStateObj['stars']" in our code.
342.     stars = gameStateObj['stars']
343.
344.     # The code for handling each of the directions is so similar aside
345.     # from adding or subtracting 1 to the x/y coordinates. We can
346.     # simplify it by using the xOffset and yOffset variables.
347.     if playerMoveTo == UP:
348.         xOffset = 0
349.         yOffset = -1
```

第9章 Star Pusher

```
350.        elif playerMoveTo == RIGHT:
351.            xOffset = 1
352.            yOffset = 0
353.        elif playerMoveTo == DOWN:
354.            xOffset = 0
355.            yOffset = 1
356.        elif playerMoveTo == LEFT:
357.            xOffset = -1
358.            yOffset = 0
359.
360.        # See if the player can move in that direction.
361.        if isWall(mapObj, playerx + xOffset, playery + yOffset):
362.            return False
363.        else:
364.            if (playerx + xOffset, playery + yOffset) in stars:
365.                # There is a star in the way, see if the player can push it.
366.                if not isBlocked(mapObj, gameStateObj, playerx + (xOffset*2), playery + (yOffset*2)):
367.                    # Move the star.
368.                    ind = stars.index((playerx + xOffset, playery + yOffset))
369.                    stars[ind] = (stars[ind][0] + xOffset, stars[ind][1] + yOffset)
370.                else:
371.                    return False
372.            # Move the player upwards.
373.            gameStateObj['player'] = (playerx + xOffset, playery + yOffset)
374.            return True
375.
376.
377. def startScreen():
378.     """Display the start screen (which has the title and instructions)
379.     until the player presses a key. Returns None."""
380.
381.     # Position the title image.
382.     titleRect = IMAGESDICT['title'].get_rect()
383.     topCoord = 50 # topCoord tracks where to position the top of the text
384.     titleRect.top = topCoord
385.     titleRect.centerx = HALF_WINWIDTH
386.     topCoord += titleRect.height
387.
388.     # Unfortunately, Pygame's font & text system only shows one line at
389.     # a time, so we can't use strings with \n newline characters in them.
390.     # So we will use a list with each line in it.
391.     instructionText = ['Push the stars over the marks.',
392.                        'Arrow keys to move, WASD for camera control, P to change character.',
```

```
393.                        'Backspace to reset level, Esc to quit.',
394.                        'N for next level, B to go back a level.']
395.
396.     # Start with drawing a blank color to the entire window:
397.     DISPLAYSURF.fill(BGCOLOR)
398.
399.     # Draw the title image to the window:
400.     DISPLAYSURF.blit(IMAGESDICT['title'], titleRect)
401.
402.     # Position and draw the text.
403.     for i in range(len(instructionText)):
404.         instSurf = BASICFONT.render(instructionText[i], 1, TEXTCOLOR)
405.         instRect = instSurf.get_rect()
406.         topCoord += 10 # 10 pixels will go in between each line of text.
407.         instRect.top = topCoord
408.         instRect.centerx = HALF_WINWIDTH
409.         topCoord += instRect.height # Adjust for the height of the line.
410.         DISPLAYSURF.blit(instSurf, instRect)
411.
412.     while True: # Main loop for the start screen.
413.         for event in pygame.event.get():
414.             if event.type == QUIT:
415.                 terminate()
416.             elif event.type == KEYDOWN:
417.                 if event.key == K_ESCAPE:
418.                     terminate()
419.                 return # user has pressed a key, so return.
420.
421.         # Display the DISPLAYSURF contents to the actual screen.
422.         pygame.display.update()
423.         FPSCLOCK.tick()
424.
425.
426. def readLevelsFile(filename):
427.     assert os.path.exists(filename), 'Cannot find the level file: %s' % (filename)
428.     mapFile = open(filename, 'r')
429.     # Each level must end with a blank line
430.     content = mapFile.readlines() + ['\r\n']
431.     mapFile.close()
432.
433.     levels = [] # Will contain a list of level objects.
434.     levelNum = 0
435.     mapTextLines = [] # contains the lines for a single level's map.
436.     mapObj = [] # the map object made from the data in mapTextLines
437.     for lineNum in range(len(content)):
```

```
438.            # Process each line that was in the level file.
439.            line = content[lineNum].rstrip('\r\n')
440.
441.            if ';' in line:
442.                # Ignore the ; lines, they're comments in the level file.
443.                line = line[:line.find(';')]
444.
445.            if line != '':
446.                # This line is part of the map.
447.                mapTextLines.append(line)
448.            elif line == '' and len(mapTextLines) > 0:
449.                # A blank line indicates the end of a level's map in the file.
450.                # Convert the text in mapTextLines into a level object.
451.
452.                # Find the longest row in the map.
453.                maxWidth = -1
454.                for i in range(len(mapTextLines)):
455.                    if len(mapTextLines[i]) > maxWidth:
456.                        maxWidth = len(mapTextLines[i])
457.                # Add spaces to the ends of the shorter rows. This
458.                # ensures the map will be rectangular.
459.                for i in range(len(mapTextLines)):
460.                    mapTextLines[i] += ' ' * (maxWidth - len(mapTextLines[i]))
461.
462.                # Convert mapTextLines to a map object.
463.                for x in range(len(mapTextLines[0])):
464.                    mapObj.append([])
465.                for y in range(len(mapTextLines)):
466.                    for x in range(maxWidth):
467.                        mapObj[x].append(mapTextLines[y][x])
468.
469.                # Loop through the spaces in the map and find the @, ., and $
470.                # characters for the starting game state.
471.                startx = None # The x and y for the player's starting position
472.                starty = None
473.                goals = [] # list of (x, y) tuples for each goal.
474.                stars = [] # list of (x, y) for each star's starting position.
475.                for x in range(maxWidth):
476.                    for y in range(len(mapObj[x])):
477.                        if mapObj[x][y] in ('@', '+'):
478.                            # '@' is player, '+' is player & goal
479.                            startx = x
480.                            starty = y
481.                        if mapObj[x][y] in ('.', '+', '*'):
482.                            # '.' is goal, '*' is star & goal
483.                            goals.append((x, y))
```

```
484.                    if mapObj[x][y] in ('$', '*'):
485.                        # '$' is star
486.                        stars.append((x, y))
487.
488.            # Basic level design sanity checks:
489.            assert startx != None and starty != None, 'Level %s (around
line %s) in %s is missing a "@" or "+" to mark the start point.' % (levelNum+1,
lineNum, filename)
490.            assert len(goals) > 0, 'Level %s (around line %s) in %s must
have at least one goal.' % (levelNum+1, lineNum, filename)
491.            assert len(stars) >= len(goals), 'Level %s (around line %s) in
%s is impossible to solve. It has %s goals but only %s stars.' % (levelNum+1,
lineNum, filename, len(goals), len(stars))
492.
493.            # Create level object and starting game state object.
494.            gameStateObj = {'player': (startx, starty),
495.                            'stepCounter': 0,
496.                            'stars': stars}
497.            levelObj = {'width': maxWidth,
498.                        'height': len(mapObj),
499.                        'mapObj': mapObj,
500.                        'goals': goals,
501.                        'startState': gameStateObj}
502.
503.            levels.append(levelObj)
504.
505.            # Reset the variables for reading the next map.
506.            mapTextLines = []
507.            mapObj = []
508.            gameStateObj = {}
509.            levelNum += 1
510.    return levels
511.
512.
513. def floodFill(mapObj, x, y, oldCharacter, newCharacter):
514.     """Changes any values matching oldCharacter on the map object to
515.     newCharacter at the (x, y) position, and does the same for the
516.     positions to the left, right, down, and up of (x, y), recursively."""
517.
518.     # In this game, the flood fill algorithm creates the inside/outside
519.     # floor distinction. This is a "recursive" function.
520.     # For more info on the Flood Fill algorithm, see:
521.     #   http://en.wikipedia.org/wiki/Flood_fill
522.     if mapObj[x][y] == oldCharacter:
523.         mapObj[x][y] = newCharacter
```

第9章 Star Pusher

```
524.
525.        if x < len(mapObj) - 1 and mapObj[x+1][y] == oldCharacter:
526.            floodFill(mapObj, x+1, y, oldCharacter, newCharacter) # call right
527.        if x > 0 and mapObj[x-1][y] == oldCharacter:
528.            floodFill(mapObj, x-1, y, oldCharacter, newCharacter) # call left
529.        if y < len(mapObj[x]) - 1 and mapObj[x][y+1] == oldCharacter:
530.            floodFill(mapObj, x, y+1, oldCharacter, newCharacter) # call down
531.        if y > 0 and mapObj[x][y-1] == oldCharacter:
532.            floodFill(mapObj, x, y-1, oldCharacter, newCharacter) # call up
533.
534.
535.    def drawMap(mapObj, gameStateObj, goals):
536.        """Draws the map to a Surface object, including the player and
537.        stars. This function does not call pygame.display.update(), nor
538.        does it draw the "Level" and "Steps" text in the corner."""
539.
540.        # mapSurf will be the single Surface object that the tiles are drawn
541.        # on, so that it is easy to position the entire map on the DISPLAYSURF
542.        # Surface object. First, the width and height must be calculated.
543.        mapSurfWidth = len(mapObj) * TILEWIDTH
544.        mapSurfHeight = (len(mapObj[0]) - 1) * (TILEHEIGHT - TILEFLOORHEIGHT) + TILEHEIGHT
545.        mapSurf = pygame.Surface((mapSurfWidth, mapSurfHeight))
546.        mapSurf.fill(BGCOLOR) # start with a blank color on the surface.
547.
548.        # Draw the tile sprites onto this surface.
549.        for x in range(len(mapObj)):
550.            for y in range(len(mapObj[x])):
551.                spaceRect = pygame.Rect((x * TILEWIDTH, y * (TILEHEIGHT - TILEFLOORHEIGHT), TILEWIDTH, TILEHEIGHT))
552.                if mapObj[x][y] in TILEMAPPING:
553.                    baseTile = TILEMAPPING[mapObj[x][y]]
554.                elif mapObj[x][y] in OUTSIDEDECOMAPPING:
555.                    baseTile = TILEMAPPING[' ']
556.
557.                # First draw the base ground/wall tile.
558.                mapSurf.blit(baseTile, spaceRect)
559.
560.                if mapObj[x][y] in OUTSIDEDECOMAPPING:
561.                    # Draw any tree/rock decorations that are on this tile.
562.                    mapSurf.blit(OUTSIDEDECOMAPPING[mapObj[x][y]], spaceRect)
563.                elif (x, y) in gameStateObj['stars']:
564.                    if (x, y) in goals:
565.                        # A goal AND star are on this space, draw goal first.
566.                        mapSurf.blit(IMAGESDICT['covered goal'], spaceRect)
567.                    # Then draw the star sprite.
```

```
568.                 mapSurf.blit(IMAGESDICT['star'], spaceRect)
569.             elif (x, y) in goals:
570.                 # Draw a goal without a star on it.
571.                 mapSurf.blit(IMAGESDICT['uncovered goal'], spaceRect)
572.
573.             # Last draw the player on the board.
574.             if (x, y) == gameStateObj['player']:
575.                 # Note: The value "currentImage" refers
576.                 # to a key in "PLAYERIMAGES" which has the
577.                 # specific player image we want to show.
578.                 mapSurf.blit(PLAYERIMAGES[currentImage], spaceRect)
579.
580.     return mapSurf
581.
582.
583. def isLevelFinished(levelObj, gameStateObj):
584.     """Returns True if all the goals have stars in them."""
585.     for goal in levelObj['goals']:
586.         if goal not in gameStateObj['stars']:
587.             # Found a space with a goal but no star on it.
588.             return False
589.     return True
590.
591.
592. def terminate():
593.     pygame.quit()
594.     sys.exit()
595.
596.
597. if __name__ == '__main__':
598.     main()
```

9.3 初始化设置

```
1. # Star Pusher (a Sokoban clone)
2. # By Al Sweigart al@inventwithpython.com
3. # http://inventwithpython.com/pygame
4. # Creative Commons BY-NC-SA 3.0 US
5.
6. import random, sys, copy, os, pygame
7. from pygame.locals import *
8.
9. FPS = 30 # frames per second to update the screen
```

第 9 章　Star Pusher

```
10. WINWIDTH = 800 # width of the program's window, in pixels
11. WINHEIGHT = 600 # height in pixels
12. HALF_WINWIDTH = int(WINWIDTH / 2)
13. HALF_WINHEIGHT = int(WINHEIGHT / 2)
14.
15. # The total width and height of each tile in pixels.
16. TILEWIDTH = 50
17. TILEHEIGHT = 85
18. TILEFLOORHEIGHT = 45
19.
20. CAM_MOVE_SPEED = 5 # how many pixels per frame the camera moves
21.
22. # The percentage of outdoor tiles that have additional
23. # decoration on them, such as a tree or rock.
24. OUTSIDE_DECORATION_PCT = 20
25.
26. BRIGHTBLUE = (  0, 170, 255)
27. WHITE      = (255, 255, 255)
28. BGCOLOR = BRIGHTBLUE
29. TEXTCOLOR = WHITE
30.
31. UP = 'up'
32. DOWN = 'down'
33. LEFT = 'left'
34. RIGHT = 'right'
```

这些常量将在程序的不同部分中用到。TILEWIDTH 和 TILEHEIGHT 变量表示每个贴片图像有 50 像素宽和 85 像素高。然而，当绘制到屏幕上的时候，这些贴片彼此重叠（本章稍后介绍这一点）。TILEFLOORHEIGHT 表示地板的贴片部分有 45 个像素的高度。普通地板图像的一个图形，如图 9-2 所示。

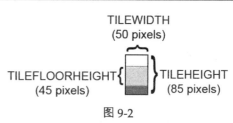

图 9-2

关卡的房间之外的草地贴片，有时候会有一些额外的装饰添加于其上（例如，树木或石头）。OUTSIDE_DECORATION_PCT 常量表示这些贴片中多大的百分比将随机地拥有这些装饰物。

```
37. def main():
38.     global FPSCLOCK, DISPLAYSURF, IMAGESDICT, TILEMAPPING,
OUTSIDEDECOMAPPING, BASICFONT, PLAYERIMAGES, currentImage
39.
40.     # Pygame initialization and basic set up of the global variables.
41.     pygame.init()
```

9.3 初始化设置

```
42.     FPSCLOCK = pygame.time.Clock()
43.
44.     # Because the Surface object stored in DISPLAYSURF was returned
45.     # from the pygame.display.set_mode() function, this is the
46.     # Surface object that is drawn to the actual computer screen
47.     # when pygame.display.update() is called.
48.     DISPLAYSURF = pygame.display.set_mode((WINWIDTH, WINHEIGHT))
49.
50.     pygame.display.set_caption('Star Pusher')
51.     BASICFONT = pygame.font.Font('freesansbold.ttf', 18)
```

这是程序开始处常用的 Pygame 设置。

```
53.     # A global dict value that will contain all the Pygame
54.     # Surface objects returned by pygame.image.load().
55.     IMAGESDICT = {'uncovered goal': pygame.image.load('RedSelector.png'),
56.                   'covered goal': pygame.image.load('Selector.png'),
57.                   'star': pygame.image.load('Star.png'),
58.                   'corner': pygame.image.load('Wall Block Tall.png'),
59.                   'wall': pygame.image.load('Wood Block Tall.png'),
60.                   'inside floor': pygame.image.load('Plain Block.png'),
61.                   'outside floor': pygame.image.load('Grass Block.png'),
62.                   'title': pygame.image.load('star_title.png'),
63.                   'solved': pygame.image.load('star_solved.png'),
64.                   'princess': pygame.image.load('princess.png'),
65.                   'boy': pygame.image.load('boy.png'),
66.                   'catgirl': pygame.image.load('catgirl.png'),
67.                   'horngirl': pygame.image.load('horngirl.png'),
68.                   'pinkgirl': pygame.image.load('pinkgirl.png'),
69.                   'rock': pygame.image.load('Rock.png'),
70.                   'short tree': pygame.image.load('Tree_Short.png'),
71.                   'tall tree': pygame.image.load('Tree_Tall.png'),
72.                   'ugly tree': pygame.image.load('Tree_Ugly.png')}
```

IMAGESDICT 是一个字典，其中存储了所有加载的图像。这使得它很容易在其他函数中使用，因为只有 IMAGESDICT 变量是需要成为全局的。如果我们将这些图像中的每一个都存储到一个单独的变量中，那么，所有的 18 个变量（因为游戏中用到了 18 个图像）就都需要成为全局的。将带有图像的所有 Surface 对象都包含到一个字典中，这会更容易处理。

```
74.     # These dict values are global, and map the character that appears
75.     # in the level file to the Surface object it represents.
76.     TILEMAPPING = {'x': IMAGESDICT['corner'],
77.                    '#': IMAGESDICT['wall'],
78.                    'o': IMAGESDICT['inside floor'],
79.                    ' ': IMAGESDICT['outside floor']}
```

第 9 章 Star Pusher

地图的数据结构只是单个字符串的一个 2D 列表。TILEMAPPING 字典把该地图数据结构中用到的字符串连接到了它们所表示的图像（在后面介绍 drawMap()函数的时候，这一点会更清楚）。

```
80.     OUTSIDEDECOMAPPING = {'1': IMAGESDICT['rock'],
81.                           '2': IMAGESDICT['short tree'],
82.                           '3': IMAGESDICT['tall tree'],
83.                           '4': IMAGESDICT['ugly tree']}
```

OUTSIDEDECOMAPPING 也是一个字典，它把地图数据结构中用到的字符串连接到所加载的图像。"外部装饰"图像绘制到门外的草地贴片之上。

```
85.     # PLAYERIMAGES is a list of all possible characters the player can be.
86.     # currentImage is the index of the player's current player image.
87.     currentImage = 0
88.     PLAYERIMAGES = [IMAGESDICT['princess'],
89.                     IMAGESDICT['boy'],
90.                     IMAGESDICT['catgirl'],
91.                     IMAGESDICT['horngirl'],
92.                     IMAGESDICT['pinkgirl']]
```

PLAYERIMAGES 列表存储了用于玩家的图像。currentImage 变量记录了当前选择的玩家图像的索引。例如，当 currentImage 设置为 0 的时候，就是 PLAYERIMAGES[0]，这是"公主"玩家图像，它会绘制到屏幕上。

```
94.     startScreen() # show the title screen until the user presses a key
95.
96.     # Read in the levels from the text file. See the readLevelsFile() for
97.     # details on the format of this file and how to make your own levels.
98.     levels = readLevelsFile('starPusherLevels.txt')
99.     currentLevelIndex = 0
```

startScreen()函数将持续显示最初的开始屏幕（这还包括针对游戏的说明），直到玩家按下一个键。当玩家按下一个键的时候，startScreen()函数返回，然后从关卡文件读取关卡。玩家从第一关开始，第一关是关卡列表中索引为 0 的关卡对象。

```
101.    # The main game loop. This loop runs a single level, when the user
102.    # finishes that level, the next/previous level is loaded.
103.    while True: # main game loop
104.        # Run the level to actually start playing the game:
105.        result = runLevel(levels, currentLevelIndex)
```

runLevel()函数处理游戏的所有动作。传递给它关卡对象的一个列表，以及该列表中要

9.3 初始化设置

玩的关卡的整数索引。当玩家完成了该关卡，runLevel()将返回如下的字符串之一：'solved'（因为玩家已经完成了将所有星星推入到目标位置）、'next'（因为玩家想要跳到下一关卡）、'back'（因为玩家想要回到前一关卡）和'reset'（因为玩家想要重新玩当前的关卡，他们可能把一个星星推到死角里了）。

```
107.        if result in ('solved', 'next'):
108.            # Go to the next level.
109.            currentLevelIndex += 1
110.            if currentLevelIndex >= len(levels):
111.                # If there are no more levels, go back to the first one.
112.                currentLevelIndex = 0
113.        elif result == 'back':
114.            # Go to the previous level.
115.            currentLevelIndex -= 1
116.            if currentLevelIndex < 0:
117.                # If there are no previous levels, go to the last one.
118.                currentLevelIndex = len(levels)-1
```

如果 runLevel()返回了字符串'solved'或'next'，那么我们需要将 levelNum 增加 1。如果增加 levelNum 以至于超出了已有的关卡数目，那么 levelNum 将重新设置为 0。

如果返回的是'back'，则相反，levelNum 减去 1。如果这使得 levelNum 变得小于 0，那么将其设置为最后一关，即 len(levels)-1。

```
119.        elif result == 'reset':
120.            pass # Do nothing. Loop re-calls runLevel() to reset the level
```

如果返回值是'reset'，那么，代码什么也不做。pass 语句什么也不做（就像一条注释一样），但是，还是需要它，因为 Python 解释器期待在一条 elif 的后面有一行缩进的代码。

我们也可以完全从源代码中删掉第 119 行和第 120 行，程序将会同样地工作。我们在这里包含它的原因是为了程序的可读性，以便后面在修改代码的时候，不会忘记 runLevel()也可能会返回'reset'字符串。

```
123. def runLevel(levels, levelNum):
124.     global currentImage
125.     levelObj = levels[levelNum]
126.     mapObj = decorateMap(levelObj['mapObj'], levelObj['startState']['player'])
127.     gameStateObj = copy.deepcopy(levelObj['startState'])
```

关卡列表包含了我们从关卡文件加载的所有关卡对象。当前关卡（这是 levelNum 所设置的值）的关卡对象存储在 levelObj 变量中。从 decorateMap()函数返回了一个地图对象（该对象在户内贴片和户外贴片之间做出一个区分，户外的贴片装饰有树木和石头）。为了记录玩家在玩这一关时的游戏状态，使用 copy.deepcopy()函数生成了 levelObj 中所存储的游戏状态对象的一个副本。

之所以生成该游戏状态副本，是因为存储在 levelObj['startState']中的游戏状态对象，表示了关卡刚一开始的时候的游戏状态，这时，我们还没有做任何的修改。否则的话，如果玩家重新开始该关卡，该关卡最初的游戏状态就丢失了。

使用 copy.deepcopy()函数，是因为游戏状态对象是拥有元组的一个字典。从技术上讲，这是包含了对元组的引用的字典（http://invpy.com/references 详细地介绍了引用）。使用一条赋值语句来生成字典的一个副本的话，将会产生引用的副本，而不是它们所引用的值的副本，因此，这个副本和最初的字典都指向相同的元组。

copy.deepcopy()函数通过生成字典中的实际的元组的副本，从而解决了这个问题。通过这种方式，我们可以保证修改一个字典的话，将不会影响到另一个字典。

```
128.        mapNeedsRedraw = True # set to True to call drawMap()
129.        levelSurf = BASICFONT.render('Level %s of %s' % (levelObj['levelNum']
+ 1, totalNumOfLevels), 1, TEXTCOLOR)
130.        levelRect = levelSurf.get_rect()
131.        levelRect.bottomleft = (20, WINHEIGHT - 35)
132.        mapWidth = len(mapObj) * TILEWIDTH
133.        mapHeight = (len(mapObj[0]) - 1) * (TILEHEIGHT - TILEFLOORHEIGHT) +
TILEHEIGHT
134.        MAX_CAM_X_PAN = abs(HALF_WINHEIGHT - int(mapHeight / 2)) + TILEWIDTH
135.        MAX_CAM_Y_PAN = abs(HALF_WINWIDTH - int(mapWidth / 2)) + TILEHEIGHT
136.
137.        levelIsComplete = False
138.        # Track how much the camera has moved:
139.        cameraOffsetX = 0
140.        cameraOffsetY = 0
141.        # Track if the keys to move the camera are being held down:
142.        cameraUp = False
143.        cameraDown = False
144.        cameraLeft = False
145.        cameraRight = False
```

在开始玩一个关卡的时候，会设置移动变量。mapWidth 和 mapHeight 变量是地图的像素大小。计算 mapHeight 的表达式有一点复杂，因为贴片是彼此重叠的。只有底部的一行贴片是完整的高度（这就是表达式的+ TILEHEIGHT 部分），所有其他的行的贴片（其数目是(len(mapObj[0]) - 1)）都是略微重叠的。这意味着它们实际上每个贴片都只有(TILEHEIGHT - TILEFLOORHEIGHT)个像素的高度。

Star Pusher 相机可以独立于玩家在地图上的移动而移动。这就是为什么相机需要其自己的一组"移动"变量：cameraUp、cameraDown、cameraLeft 和 cameraRight。cameraOffsetX 和 cameraOffsetY 变量记录了相机的位置。

9.3 初始化设置

```
147.    while True: # main game loop
148.        # Reset these variables:
149.        playerMoveTo = None
150.        keyPressed = False
151.
152.        for event in pygame.event.get(): # event handling loop
153.            if event.type == QUIT:
154.                # Player clicked the "X" at the corner of the window.
155.                terminate()
156.
```

playerMoveTo 变量将设置为玩家想要在地图上移动玩家角色的方向常量。keyPressed 变量记录了在游戏循环的这一次迭代中是否有任何的键按下。稍后当玩家打通了关卡的时候，将会检查该变量。

```
157.            elif event.type == KEYDOWN:
158.                # Handle key presses
159.                keyPressed = True
160.                if event.key == K_LEFT:
161.                    playerMoveTo = LEFT
162.                elif event.key == K_RIGHT:
163.                    playerMoveTo = RIGHT
164.                elif event.key == K_UP:
165.                    playerMoveTo = UP
166.                elif event.key == K_DOWN:
167.                    playerMoveTo = DOWN
168.
169.                # Set the camera move mode.
170.                elif event.key == K_a:
171.                    cameraLeft = True
172.                elif event.key == K_d:
173.                    cameraRight = True
174.                elif event.key == K_w:
175.                    cameraUp = True
176.                elif event.key == K_s:
177.                    cameraDown = True
178.
179.                elif event.key == K_n:
180.                    return 'next'
181.                elif event.key == K_b:
182.                    return 'back'
183.
184.                elif event.key == K_ESCAPE:
```

第 9 章 Star Pusher

```
185.                    terminate() # Esc key quits.
186.                elif event.key == K_BACKSPACE:
187.                    return 'reset' # Reset the level.
188.                elif event.key == K_p:
189.                    # Change the player image to the next one.
190.                    currentImage += 1
191.                    if currentImage >= len(PLAYERIMAGES):
192.                        # After the last player image, use the first one.
193.                        currentImage = 0
194.                    mapNeedsRedraw = True
195.
196.            elif event.type == KEYUP:
197.                # Unset the camera move mode.
198.                if event.key == K_a:
199.                    cameraLeft = False
200.                elif event.key == K_d:
201.                    cameraRight = False
202.                elif event.key == K_w:
203.                    cameraUp = False
204.                elif event.key == K_s:
205.                    cameraDown = False
```

这段代码负责处理当各种按键按下的时候做些什么。

```
207.        if playerMoveTo != None and not levelIsComplete:
208.            # If the player pushed a key to move, make the move
209.            # (if possible) and push any stars that are pushable.
210.            moved = makeMove(mapObj, gameStateObj, playerMoveTo)
211.
212.            if moved:
213.                # increment the step counter.
214.                gameStateObj['stepCounter'] += 1
215.                mapNeedsRedraw = True
216.
217.            if isLevelFinished(levelObj, gameStateObj):
218.                # level is solved, we should show the "Solved!" image.
219.                levelIsComplete = True
220.                keyPressed = False
```

　　如果 playerMoveTo 变量不再设置为 None，那么我们知道玩家想要移动。调用 makeMove()负责修改 gameStateObj 中的玩家位置的 XY 坐标，以及推动任何的星星。makeMove()的返回值存储在 moved 之中。如果这个值是 True,玩家角色将沿着该方向移动。如果这个值是 False，那么，玩家一定是试图向一面墙壁的贴片移动，或者玩家要推动的星星的背后有东西。在这种情况下，玩家不能移动，地图上也没有什么变化。

9.3 初始化设置

```
222.        DISPLAYSURF.fill(BGCOLOR)
223.
224.        if mapNeedsRedraw:
225.            mapSurf = drawMap(mapObj, gameStateObj, levelObj['goals'])
226.            mapNeedsRedraw = False
```

地图并不需要在游戏循环的每一次迭代中重新绘制。实际上，这个游戏程序很复杂，以至于如果这样做的话，会导致游戏略微有些（但不是很明显）滞后。真地只有在某些内容变化的时候（诸如玩家移动或者推动一个星星），才需要重新绘制地图。因此，只有当 mapNeedsRedraw 变量设置为 True 的时候，才需要通过调用一次 drawMap()函数来更新 mapSurf 变量中的 Surface 对象。

在第 225 行绘制了地图之后，mapNeedsRedraw 设置为 False。如果想要看看在游戏循环的每一帧上绘制地图会如何拖慢程序，可以注释掉第 226 行并回到程序。你将会注意到，相机的移动显著地变慢了。

```
228.        if cameraUp and cameraOffsetY < MAX_CAM_X_PAN:
229.            cameraOffsetY += CAM_MOVE_SPEED
230.        elif cameraDown and cameraOffsetY > -MAX_CAM_X_PAN:
231.            cameraOffsetY -= CAM_MOVE_SPEED
232.        if cameraLeft and cameraOffsetX < MAX_CAM_Y_PAN:
233.            cameraOffsetX += CAM_MOVE_SPEED
234.        elif cameraRight and cameraOffsetX > -MAX_CAM_Y_PAN:
235.            cameraOffsetX -= CAM_MOVE_SPEED
```

如果相机移动变量设置为 True，并且相机没有越过（例如，水平越过）MAX_CAM_X_PAN 和 MAX_CAM_Y_PAN 所设置的边界，那么，相机位置（存储在 cameraOffsetX 和 cameraOffsetY 中）应该移动 CAM_MOVE_SPEED 多个像素。

注意，在第 228 行和第 230 行有一条 if 和 elif 语句，用于将相机向上或向下移动，然后第 232 行和第 234 行上有另外的一条 if 和 elif 语句。通过这种方式，用户可以同时水平和垂直地移动相机。如果第 232 行是一条 elif 语句的话，这是不可能做到的。

```
237.        # Adjust mapSurf's Rect object based on the camera offset.
238.        mapSurfRect = mapSurf.get_rect()
239.        mapSurfRect.center = (HALF_WINWIDTH + cameraOffsetX,
HALF_WINHEIGHT + cameraOffsetY)
240.
241.        # Draw mapSurf to the DISPLAYSURF Surface object.
242.        DISPLAYSURF.blit(mapSurf, mapSurfRect)
243.
244.        DISPLAYSURF.blit(levelSurf, levelRect)
245.        stepSurf = BASICFONT.render('Steps: %s' %
```

```
(gameStateObj['stepCounter']), 1, TEXTCOLOR)
246.            stepRect = stepSurf.get_rect()
247.            stepRect.bottomleft = (20, WINHEIGHT - 10)
248.            DISPLAYSURF.blit(stepSurf, stepRect)
249.
250.            if levelIsComplete:
251.                # is solved, show the "Solved!" image until the player
252.                # has pressed a key.
253.                solvedRect = IMAGESDICT['solved'].get_rect()
254.                solvedRect.center = (HALF_WINWIDTH, HALF_WINHEIGHT)
255.                DISPLAYSURF.blit(IMAGESDICT['solved'], solvedRect)
256.
257.                if keyPressed:
258.                    return 'solved'
259.
260.            pygame.display.update() # draw DISPLAYSURF to the screen.
261.            FPSCLOCK.tick()
262.
263.
```

第237行到第261行定位相机，并且将地图和其他的图形绘制到DISPLAYSURF中的显示Surface对象上。如果关卡打通了，那么，还会在所有内容之上绘制获胜的图像。如果用户在此迭代中按下了一个键，keyPressed变量设置为True，并且runLevel()函数在此时返回。

```
264. def isWall(mapObj, x, y):
265.     """Returns True if the (x, y) position on
266.     the map is a wall, otherwise return False."""
267.     if x < 0 or x >= len(mapObj) or y < 0 or y >= len(mapObj[x]):
268.         return False # x and y aren't actually on the map.
269.     elif mapObj[x][y] in ('#', 'x'):
270.         return True # wall is blocking
271.     return False
```

如果在地图对象上，在传递给函数的XY坐标处有一面墙，isWall()函数返回True。墙对象在地图对象中表示为一个'x'或'#'字符串。

```
274. def decorateMap(mapObj, startxy):
275.     """Makes a copy of the given map object and modifies it.
276.     Here is what is done to it:
277.      * Walls that are corners are turned into corner pieces.
278.      * The outside/inside floor tile distinction is made.
279.      * Tree/rock decorations are randomly added to the outside tiles.
280.
281.     Returns the decorated map object."""
```

9.3 初始化设置

```
282.
283.        startx, starty = startxy # Syntactic sugar
284.
285.        # Copy the map object so we don't modify the original passed
286.        mapObjCopy = copy.deepcopy(mapObj)
```

decorateMap()函数修改数据结构 mapObj，以使其在地图文件中显得不一样。decorateMap()所修改的3处内容，在该函数最开始处的注释中加以说明。

```
288.        # Remove the non-wall characters from the map data
289.        for x in range(len(mapObjCopy)):
290.            for y in range(len(mapObjCopy[0])):
291.                if mapObjCopy[x][y] in ('$', '.', '@', '+', '*'):
292.                    mapObjCopy[x][y] = ' '
```

地图对象拥有表示玩家、目标和星星的字符。这些对于地图对象来说，都是必需的（在地图文件读取之后，这些都存储到其他的数据结构中），以免将其转换为空白的空格。

```
294.        # Flood fill to determine inside/outside floor tiles.
295.        floodFill(mapObjCopy, startx, starty, ' ', 'o')
```

floodFill()函数将把墙之中的所有贴片从' '字符修改为'o'字符。它利用过一种叫作递归（recursion）的编程概念来做到这一点，本章后面的9.6节将介绍这一概念。

```
297.        # Convert the adjoined walls into corner tiles.
298.        for x in range(len(mapObjCopy)):
299.            for y in range(len(mapObjCopy[0])):
300.
301.                if mapObjCopy[x][y] == '#':
302.                    if (isWall(mapObjCopy, x, y-1) and isWall(mapObjCopy, x+1, y)) or \
303.                       (isWall(mapObjCopy, x+1, y) and isWall(mapObjCopy, x, y+1)) or \
304.                       (isWall(mapObjCopy, x, y+1) and isWall(mapObjCopy, x-1, y)) or \
305.                       (isWall(mapObjCopy, x-1, y) and isWall(mapObjCopy, x, y-1)):
306.                        mapObjCopy[x][y] = 'x'
307.
308.                    elif mapObjCopy[x][y] == ' ' and random.randint(0, 99) < OUTSIDE_DECORATION_PCT:
309.                        mapObjCopy[x][y] = random.choice(list(OUTSIDEDECOMAPPING.keys()))
310.
311.        return mapObjCopy
```

第 9 章　Star Pusher

第 301 行的较大的、多行的 if 语句，检查当前的 XY 坐标的墙壁贴片是否是一个墙角贴片，通过检查和它相邻的墙壁贴面是否够成一个角落的形状来做到这一点。如果是墙角，将地图对象中表示常规墙壁的'#'字符串修改为表示墙角贴片的一个'x'字符串。

```
314. def isBlocked(mapObj, gameStateObj, x, y):
315.     """Returns True if the (x, y) position on the map is
316.     blocked by a wall or star, otherwise return False."""
317.
318.     if isWall(mapObj, x, y):
319.         return True
320.
321.     elif x < 0 or x >= len(mapObj) or y < 0 or y >= len(mapObj[x]):
322.         return True # x and y aren't actually on the map.
323.
324.     elif (x, y) in gameStateObj['stars']:
325.         return True # a star is blocking
326.
327.     return False
```

在 3 种情况下，地图上的一个空间将会被阻塞：如果那里有一个星星，那是一面墙，或者空间的坐标超过了地图的边界。isBlocked()检查这 3 种情况，并且如果 XY 坐标阻塞了，它返回 True，否则返回 False。

```
330. def makeMove(mapObj, gameStateObj, playerMoveTo):
331.     """Given a map and game state object, see if it is possible for the
332.     player to make the given move. If it is, then change the player's
333.     position (and the position of any pushed star). If not, do nothing.
334.
335.     Returns True if the player moved, otherwise False."""
336.
337.     # Make sure the player can move in the direction they want.
338.     playerx, playery = gameStateObj['player']
339.
340.     # This variable is "syntactic sugar". Typing "stars" is more
341.     # readable than typing "gameStateObj['stars']" in our code.
342.     stars = gameStateObj['stars']
343.
344.     # The code for handling each of the directions is so similar aside
345.     # from adding or subtracting 1 to the x/y coordinates. We can
346.     # simplify it by using the xOffset and yOffset variables.
347.     if playerMoveTo == UP:
348.         xOffset = 0
349.         yOffset = -1
350.     elif playerMoveTo == RIGHT:
```

```
351.            xOffset = 1
352.            yOffset = 0
353.        elif playerMoveTo == DOWN:
354.            xOffset = 0
355.            yOffset = 1
356.        elif playerMoveTo == LEFT:
357.            xOffset = -1
358.            yOffset = 0
359.
360.        # See if the player can move in that direction.
361.        if isWall(mapObj, playerx + xOffset, playery + yOffset):
362.            return False
363.        else:
364.            if (playerx + xOffset, playery + yOffset) in stars:
365.                # There is a star in the way, see if the player can push it.
366.                if not isBlocked(mapObj, gameStateObj, playerx + (xOffset*2), playery + (yOffset*2)):
367.                    # Move the star.
368.                    ind = stars.index((playerx + xOffset, playery + yOffset))
369.                    stars[ind] = (stars[ind][0] + xOffset, stars[ind][1] + yOffset)
370.                else:
371.                    return False
372.            # Move the player upwards.
373.            gameStateObj['player'] = (playerx + xOffset, playery + yOffset)
374.            return True
```

makeMove()函数进行检查,确保朝着一个特定的方向移动玩家是一次有效的移动。只要没有一面墙阻塞路径,或者不是星星的后面有一面墙或一个星星的情况,玩家就能够朝着该方向移动。gameStateObj 变量将会更新以反映这一点,并且会返回 True 值,以告诉函数的调用者,玩家已经移动了。

如果空格中有玩家想要移动的一个星星,该星星的位置也会改变,并且这一信息也会更新到 gameStateObj 变量中。"推动星星"就是这样实现的。

如果玩家朝着想要的方向移动被阻塞了,那么,gameStateObj 不会修改,并且该函数返回 False。

```
377. def startScreen():
378.     """Display the start screen (which has the title and instructions)
379.     until the player presses a key. Returns None."""
380.
381.     # Position the title image.
382.     titleRect = IMAGESDICT['title'].get_rect()
383.     topCoord = 50 # topCoord tracks where to position the top of the text
```

第 9 章　Star Pusher

```
384.        titleRect.top = topCoord
385.        titleRect.centerx = HALF_WINWIDTH
386.        topCoord += titleRect.height
387.
388.        # Unfortunately, Pygame's font & text system only shows one line at
389.        # a time, so we can't use strings with \n newline characters in them.
390.        # So we will use a list with each line in it.
391.        instructionText = ['Push the stars over the marks.',
392.                           'Arrow keys to move, WASD for camera control, P to change character.',
393.                           'Backspace to reset level, Esc to quit.',
394.                           'N for next level, B to go back a level.']
```

startScreen()函数需要在窗口的下方中央显示几条不同的文本消息。在 instructionText 列表中，我们将每一行文本存储为一个字符串。标题图像（作为一个 Surface 对象存储在 IMAGESDICT['title']中，最初是从 star_title.png 文件载入的）将放置在距离窗口顶部 50 个像素的地方。这就是为什么第 383 行在 topCoord 变量中存储了整数值 50。topCoord 变量将记录标题图像和说明文本的 Y 坐标。X 坐标总是要设置为让图像和文本居中放置，例如，在第 385 行为标题图像设置了 X 坐标。在第 386 行，给 topCoord 加上了图像的高度。通过这种方式，我们可以修改图像而不用修改开始屏幕代码。

```
396.        # Start with drawing a blank color to the entire window:
397.        DISPLAYSURF.fill(BGCOLOR)
398.
399.        # Draw the title image to the window:
400.        DISPLAYSURF.blit(IMAGESDICT['title'], titleRect)
401.
402.        # Position and draw the text.
403.        for i in range(len(instructionText)):
404.            instSurf = BASICFONT.render(instructionText[i], 1, TEXTCOLOR)
405.            instRect = instSurf.get_rect()
406.            topCoord += 10 # 10 pixels will go in between each line of text.
407.            instRect.top = topCoord
408.            instRect.centerx = HALF_WINWIDTH
409.            topCoord += instRect.height # Adjust for the height of the line.
410.            DISPLAYSURF.blit(instSurf, instRect)
```

第 400 行将标题图像复制到了显示 Surface 对象上。从第 403 行开始的 for 循环，将会渲染、定位并复制 instructionText 列表中的每一个说明字符串。topCoord 变量总是根据之前渲染的文本的大小（第 409 行）以及 10 个额外的像素（在第 406 行，以便每行文本之间有 10 个像素的间隔）来增加。

```
412.    while True: # Main loop for the start screen.
413.        for event in pygame.event.get():
414.            if event.type == QUIT:
415.                terminate()
416.            elif event.type == KEYDOWN:
417.                if event.key == K_ESCAPE:
418.                    terminate()
419.                return # user has pressed a key, so return.
420.
421.        # Display the DISPLAYSURF contents to the actual screen.
422.        pygame.display.update()
423.        FPSCLOCK.tick()
```

startScreen()中有一个游戏循环，从第 412 行开始，它处理表示程序是否应该终止或从 startScreen() 函数返回的事件。循环将持续调用 pygame.display.update() 和 FPSCLOCK.tick()，以保持开始屏幕显示在屏幕之上，直到玩家触发任何一个事件。

9.4 Star Pusher 中的数据结构

Star Pusher 拥有特定格式的关卡、地图和游戏状态数据结构。

9.4.1 游戏状态数据结构

游戏状态对象有一个字典，它带有 3 个键：'player'、'stepCounter'和'stars'。
- 'player'键中的值是两个整数的一个元组，表示玩家当前的 XY 位置。
- 'stepCounter'键中的值是一个整数，它记录了玩家在该关卡中移动了多少次（从而玩家可以尝试将来以更少的步骤来解决谜题）。
- 'stars'键的值是两个整数的元组的一个列表，这两个整数表示当前关卡的每一个星星的 XY 值。

9.4.2 地图数据结构

地图数据结构是一个简单的 2D 列表，其中的两个索引用来表示地图的 X 和 Y 坐标。这个列表的列表中的每个索引的值都是一个单字符的字符串，表示地图上位于每一个空格的贴片。
- '#'—— 木质的墙
- 'x'—— 墙角
- '@'—— 玩家在该关卡的起始空格

第 9 章　Star Pusher

- '.'—— 一个目标空格
- '$'—— 关卡开始的时候，带有一个星星的空格
- '+'—— 带有一个目标的空格，并且是玩家的起始空格
- '*'—— 在关卡开始的时候带有一个目标和一个星星的空格
- ' '—— 门外的草地空格
- 'o'—— 室内的地板空格（这是一个小写的字母 O，而不是一个零）
- '1'—— 草地上的一块岩石
- '2'—— 草地上的一颗矮树
- '3'—— 草地上的一颗高树
- '4'—— 草地上的一颗丑树

9.4.3　关卡数据结构

关卡对象包含了一个游戏状态对象（它将会是关卡刚开始时候使用的状态）、一个地图对象和一些其他的值。关卡对象自身是一个字典，带有如下的键。

- 键'width'的值是一个整数，表示整个地图有多少个贴片那么宽。
- 键'height'的值是一个整数，表示整个地图有多少个贴片那么高。
- 键'mapObj'的值是该关卡的地图对象。
- 键'goals'的值是两个整数的元组的一个列表，整数表示地图上的每一个目标空格的 XY 坐标。
- 键'startState'的值是一个游戏对象，用来表示关卡开始的时候星星和玩家的起始位置。

9.5　读取和写入文本文件

Python 拥有从玩家的硬盘读取文件的函数。对于用一个单独的文件保存每个关卡的所有数据来说，这非常有用。这也是一种很好的思路，因为要获取新的关卡，玩家不必修改游戏的源代码，而只需要能够下载新的关卡文件就行了。

9.5.1　文本文件和二进制文件

文本文件是包含了简单的文本数据的文件。文本文件可以通过 Windows 上的 Notepad 应用程序、Ubuntu 上的 Gedit 和 Mac OS X 上的 TextEdit 来创建。还有很多其他的称为文本编辑器的程序，也可以创建和修改文本文件。IDLE 自己的文件编辑器就是

一个文本编辑器。

文本编辑器和字处理程序（诸如 Microsoft Word、OpenOffice Writer 或 iWork Pages）之间的差异是文本编辑器只能打开文本。你不能设置文本的字体、大小或颜色（IDLE 根据 Python 代码的种类，自动设置文本的颜色，但你不能自行修改，因此，它仍然是一款文本编辑器）。文本文件和二进制文件的区别，对于这个游戏程序来说并不是很重要，你可以通过 http://invpy.com/textbinary 来加以了解。在本章中，你只需要知道，Star Pusher 程序只是处理文本文件。

9.5.2 写入文件

要创建一个文件，调用 open() 函数并且传递给它两个参数：表示文件名称的一个字符串，以及字符串'w'，后者告诉 open() 函数你想要以"write"模式打开该文件。open() 函数返回一个文件对象。

```
>>> textFile = open('hello.txt', 'w')
>>>
```

如果你在交互式 shell 中运行这段代码，将会在和 python.exe 程序位置相同的文件夹下（在 Windows 下，这个位置可能是 C:\Python32），创建该函数所要创建的 *hello.txt* 文件。如果从一个.py 程序中调用该 open() 函数，这个文件会创建于.py 文件所在的目录中。

"write"模式告诉 open()，如果该文件不存在的话，就创建它。如果确实存在，那么，open() 将删除该文件并创建一个新的、空的文件。这就像一条赋值语句一样，它可以创建一个新的变量，或者是覆盖已有的变量中的当前值。这可能会有一定的危险性。如果你意外地将一个重要的文件的文件名传递给了 open() 函数，并且以'w'作为第 2 个参数，那么，将会删除该文件。这可能会导致必须在计算机上重装操作系统，甚至可能导致启动核导弹发射。

文件对象有一个名为 write() 的方法，它可以用来把文本写入到文件中。只要给其传递一个字符串就可以了，就像给 print() 传递一个字符串一样。区别在于 write() 并不会自动地在字符串的末尾添加一个换行字符('\n')。如果想要添加一个换行，必须将其包含在一个字符串中。

```
>>> textFile = open('hello.txt', 'w')
>>> textFile.write('This will be the content of the file.\nHello world!\n')
>>>
```

要告诉 Python 你已经将内容写到了这个文件中，应该调用文件对象的 close() 方法（尽

第 9 章　Star Pusher

管在程序结束的时候，Python 会自动地关闭任何已经打开的文件对象）。

```
>>> textFile.close()
```

9.5.3　从文件读取

要从文件读取内容，给 open()函数传递字符串 'r'而不是'w'。然后，在文件对象上调用 readlines()方法以读取文件的内容。最后，调用 close()方法关闭文件。

```
>>> textFile = open('hello.txt', 'r')
>>> content = textFile.readlines()
>>> textFile.close()
```

readlines()方法返回字符串的一个列表：每个字符串针对文件中的文本的一行。

```
>>> content
['This will be the content of the file.\n', 'Hello world!\n']
>>>
```

如果想要再次从该文件读取内容，必须在文件对象上调用 close()，并且再次打开它。

作为 readlines()的替代，你也可以调用 read()方法，它将会把整个文件的内容都作为一个字符串值返回。

```
>>> textFile = open('hello.txt', 'r')
>>> content = textFile.read()
>>> content
'This will be the content of the file.\nHello world!\n'
```

要注意一点，如果漏掉了 open()函数的第二个参数，Python 会假设你想要以读取模式打开文件。因此，open('foobar.txt', 'r')和 open('foobar.txt')所做的事情完全相同。

9.5.4　Star Pusher 地图文件格式

我们需要特定格式的关卡文本文件。那么，什么字符分别表示墙、星星，或者玩家的开始位置呢？如果有多个关卡的地图，如何区别什么时候一个关卡的地图结束，而另一个关卡开始呢？

好在，我们所使用的地图文件格式已经定义好了。除了 Star Pusher 这款游戏，已经有很多的 Sokoban 游戏了（可以从 http://invpy.com/sokobanclones 找到更多游戏），并且，它

9.5 读取和写入文本文件

们都使用相同的地图格式。如果从 http://invpy.com/starPusherLevels.txt 下载了关卡文件，并且用文本编辑器打开它，应该会看到如下所示的内容。

```
; Star Pusher (Sokoban clone)
; http://inventwithpython.com/blog
; By Al Sweigart al@inventwithpython.com
;
; Everything after the ; is a comment and will be ignored by the game that
; reads in this file.
;
; The format is described at:
; http://sokobano.de/wiki/index.php?title=Level_format
;    @ - The starting position of the player.
;    $ - The starting position for a pushable star.
;    . - A goal where a star needs to be pushed.
;    + - Player & goal
;    * - Star & goal
;   (space) - an empty open space.
;    # - A wall.
;
; Level maps are separated by a blank line (I like to use a ; at the start
; of the line since it is more visible.)
;
; I tried to use the same format as other people use for their Sokoban games,
; so that loading new levels is easy. Just place the levels in a text file
; and name it "starPusherLevels.txt" (after renaming this file, of course).


; Starting demo level:
  ########
 ##      #
 #  .    #
 #   $   #
 # .$@$. #
 ####$   #
    #.   #
    #   ##
    #####
```

文件开始的注释说明了文件的格式。当你加载第一个关卡的时候，看上去如图 9-3 所示。

第 9 章　Star Pusher

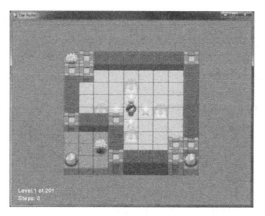

图 9-3

```
426. def readLevelsFile(filename):
427.     assert os.path.exists(filename), 'Cannot find the level file: %s' %
(filename)
```

如果传递给 os.path.exists()函数的字符串所指定的文件已经存在，该函数将返回 True。如果文件不存在，os.path.exists()将返回 False。

```
428.     mapFile = open(filename, 'r')
429.     # Each level must end with a blank line
430.     content = mapFile.readlines() + ['\r\n']
431.     mapFile.close()
432.
433.     levels = [] # Will contain a list of level objects.
434.     levelNum = 0
435.     mapTextLines = [] # contains the lines for a single level's map.
436.     mapObj = [] # the map object made from the data in mapTextLines
```

以读取方式打开的关卡文件的文件对象，存储在 mapFile 中。关卡文件中的所有的文本都存储在 content 变量中的一个字符串列表中，在其末尾添加了一个空行（这么做的原因稍后说明）。

在创建了关卡对象之后，它们将存储在 levels 列表中。levelNum 变量记录了在该关卡文件中可以找到多少个关卡。mapTextLines 列表是字符串列表，来自于针对单个地图的 content 列表（相对于存储关卡文件中所有地图的字符串的 content）。mapObj 变量将是一个 2D 列表。

```
437.     for lineNum in range(len(content)):
438.         # Process each line that was in the level file.
439.         line = content[lineNum].rstrip('\r\n')
```

9.5 读取和写入文本文件

第 437 行的 for 循环将遍历从关卡文件中读取的每一行，一次遍历一行。行号将存储在 lineNum 中，该行的文本字符串将存储在 line 中。字符串末尾的任何换行字符都会去掉。

```
441.        if ';' in line:
442.            # Ignore the ; lines, they're comments in the level file.
443.            line = line[:line.find(';')]
```

在地图文件中，在分号后面的任何文本，都会当作注释对待并且忽略掉。这就像是 Python 中表示注释的#符号一样。为了确保代码不会意外地把注释当作地图的一部分，修改了 line 变量以便它只能够由截止到分号字符（但不包括分号字符）的文本组成（注意，只是修改了 content 列表中的字符串，而不会修改硬盘上的关卡文件）。

```
445.        if line != '':
446.            # This line is part of the map.
447.            mapTextLines.append(line)
```

地图文件中还有针对多个关卡的地图。mapTextLines 列表将包含当前所加载的关卡的地图文件中的文本行。只要当前行不是空白行，该行就会添加到 mapTextLines 的末尾。

```
448.        elif line == '' and len(mapTextLines) > 0:
449.            # A blank line indicates the end of a level's map in the file.
450.            # Convert the text in mapTextLines into a level object.
```

当地图文件中有一个空白行的时候，这就表示当前关卡的地图结束了。此后的任何文本行都将是用于后续的关卡的。然而要注意，mapTextLines 中必须至少有一行，以便多个空白行在一起的时候不会被当作是多个关卡的开始和结束。

```
452.            # Find the longest row in the map.
453.            maxWidth = -1
454.            for i in range(len(mapTextLines)):
455.                if len(mapTextLines[i]) > maxWidth:
456.                    maxWidth = len(mapTextLines[i])
```

mapTextLines 中的所有字符串需要具有相同的长度（以便它们能够形成一个矩形），因此，它们应该补充一些额外的空白空格，直到其与最长的字符串具有相同的长度。For 循环遍历了 mapTextLines 中的每一个字符串，并且当它发现一个新的、最长的字符串的时候，就会更新 maxWidth。在该循环执行完之后，maxWidth 变量应该设置为 mapTextLines 中最长的字符串的长度。

第 9 章 Star Pusher

```
457.            # Add spaces to the ends of the shorter rows. This
458.            # ensures the map will be rectangular.
459.            for i in range(len(mapTextLines)):
460.                mapTextLines[i] += ' ' * (maxWidth - len(mapTextLines[i]))
```

第 459 行的 for 循环再次遍历了 mapTextLines 中的字符串，这一次它添加了足够的空格字符，将每一个字符串都补充得和 maxWidth 一样长。

```
462.            # Convert mapTextLines to a map object.
463.            for x in range(len(mapTextLines[0])):
464.                mapObj.append([])
465.            for y in range(len(mapTextLines)):
466.                for x in range(maxWidth):
467.                    mapObj[x].append(mapTextLines[y][x])
```

mapTextLines 变量只是存储了字符串的一个列表（列表中的每个字符串都表示一行，并且字符串中的每个字符都表示不同列的一个字符。这就是为什么 467 行将 Y 和 X 索引反向，就像是 Tetromino 游戏中的 SHAPES 数据结构一样）。但是，地图对象必须是单个字符串的列表的一个列表，如 mapObj[x][y] 引用位于 XY 坐标的贴片。第 463 行的 for 循环针对 mapTextLines 中的每一列，给 mapobj 添加了一个空的列表。

第 465 行和第 466 行的嵌套 for 循环，将会使用单个字符的字符串填充这些列表，以表示地图上的每一个贴片。这会创建出 Start Pusher 所使用的地图对象。

```
469.            # Loop through the spaces in the map and find the @, ., and $
470.            # characters for the starting game state.
471.            startx = None # The x and y for the player's starting position
472.            starty = None
473.            goals = [] # list of (x, y) tuples for each goal.
474.            stars = [] # list of (x, y) for each star's starting position.
475.            for x in range(maxWidth):
476.                for y in range(len(mapObj[x])):
477.                    if mapObj[x][y] in ('@', '+'):
478.                        # '@' is player, '+' is player & goal
479.                        startx = x
480.                        starty = y
481.                    if mapObj[x][y] in ('.', '+', '*'):
482.                        # '.' is goal, '*' is star & goal
483.                        goals.append((x, y))
484.                    if mapObj[x][y] in ('$', '*'):
485.                        # '$' is star
486.                        stars.append((x, y))
```

在创建了地图对象之后，第 475 行和第 476 行的嵌套的 for 循环将遍历每个空格，以找

9.5 读取和写入文本文件

到 3 个物体的 XY 坐标。

1. 玩家开始的位置。这将存储在 startx 和 starty 中,稍后将在第 494 行存储到游戏状态对象中。
2. 所有星星的开始位置。这将存储在 stars 列表中,稍后将在第 496 行存储到游戏状态对象中。
3. 所有目标的位置。这些将存储在 goals 列表中,随后将在第 500 行存储到关卡对象中。

记住,游戏状态对象包含了所有可能修改的内容。这就是为什么玩家的位置要存储在其中(因为玩家可能会移动),以及为什么星星要存储在其中(因为星星可能会被玩家推动)。但是,目标都存储在关卡对象中,因为它们不会移动。

```
488.            # Basic level design sanity checks:
489.            assert startx != None and starty != None, 'Level %s (around
line %s) in %s is missing a "@" or "+" to mark the start point.' % (levelNum+1,
lineNum, filename)
490.            assert len(goals) > 0, 'Level %s (around line %s) in %s must
have at least one goal.' % (levelNum+1, lineNum, filename)
491.            assert len(stars) >= len(goals), 'Level %s (around line %s) in
%s is impossible to solve. It has %s goals but only %s stars.' % (levelNum+1,
lineNum, filename, len(goals), len(stars))
```

此时,关卡已经读取并处理了。要确保该关卡能够正确地工作,还需要通过一些断言。如果这些断言中的任何条件为 False,那么,Python 将会产生一个错误(使用 assert 语句中的字符串),表明关卡文件中有错误。

第 489 行的第一个断言检查确保在地图上的某处有一个玩家开始起点。第 490 行的第二个断言检查确保在地图上的某处至少有一个(或多个)目标。第 491 行的第三个断言检查确保针对每个目标至少有一个星星(但是,星星的数目比目标的数目多也是允许的)。

```
493.            # Create level object and starting game state object.
494.            gameStateObj = {'player': (startx, starty),
495.                            'stepCounter': 0,
496.                            'stars': stars}
497.            levelObj = {'width': maxWidth,
498.                        'height': len(mapObj),
499.                        'mapObj': mapObj,
500.                        'goals': goals,
501.                        'startState': gameStateObj}
502.
503.            levels.append(levelObj)
```

最后,这些对象将存储到游戏状态对象中,游戏状态对象自身又存储到了关卡对象中。

第 9 章　Star Pusher

在第 503 行，关卡对象添加到了一个关卡对象的列表中。当所有的地图都已经处理之后，readLevelsFile()函数将返回这个 levels 列表。

```
505.            # Reset the variables for reading the next map.
506.            mapTextLines = []
507.            mapObj = []
508.            gameStateObj = {}
509.            levelNum += 1
510.    return levels
```

既然这个关卡已经处理了，变量 mapTextLines、mapObj 以及 gameStateObj，应该针对将要从关卡文件读取的下一个关卡重置为空白值。作为下一个关卡的关卡编号，levelNum 变量也应该增加 1。

9.6　递归函数

在了解 floodFill()函数是如何工作的之前，需要先了解递归。递归是一个简单的概念，一个递归函数就是（recursive function）调用其自身的函数，就像如下程序所示的函数（不要输入每一行开头的字母）。

```
A. def passFortyTwoWhenYouCallThisFunction(param):
B.     print('Start of function.')
C.     if param != 42:
D.         print('You did not pass 42 when you called this function.')
E.         print('Fine. I will do it myself.')
F.         passFortyTwoWhenYouCallThisFunction(42) # this is the recursive call
G.     if param == 42:
H.         print('Thank you for passing 42 when you called this function.')
I.     print('End of function.')
J.
K. passFortyTwoWhenYouCallThisFunction(41)
```

在你自己的程序中，不要使用像 passFortyTwoWhenYouCallThisFunction()这么长的函数名。我这么做有点傻。

当运行该程序的时候，在第 A 行执行 def 语句的时候，定义了该函数。将要执行的下一行代码是第 K 行，它调用 passFortyTwoWhenYouCallThisFunction()并且传入了 41。结果，该函数在第 F 行调用自身并且传入了 42。我们称这个调用为递归调用（recursive call.）。

程序的输出如下所示。

```
Start of function.
You did not pass 42 when you called this function.
Fine. I will do it myself.
Start of function.
Thank you for passing 42 when you called this function.
End of function.
End of function.
```

注意，"Start of function"和"End of function"文本出现了两次。让我们来搞清楚到底发生了什么，以及是按照什么顺序发生的。

在第 K 行，调用了该函数并且传入了 41 作为 param 参数。第 B 行打印出了"Start of function"。第 C 行的条件将会为 True（因为 41 != 42），因此第 C 行和第 D 行将打印出其消息。第 F 行将进行一次递归调用，并且传入 42 作为 param 参数。因此，执行再次从第 B 行开始并打印出"Start of function"。第 C 行的条件这一次为 False，因此，它跳转到第 G 行并且发现该条件为 True。这会导致第 H 行调用并在屏幕上显示"Thank you…"。然后，在函数的最后一行，也就是第 I 行，将会执行打印"End of function"，并且函数返回到调用它的那一行。

但是记住，调用函数的那行代码是第 F 行。并且，在最初的调用中，param 设置为 41 了。代码继续向下到第 G 行并检查该条件，该条件为 False（因为 41 == 42 为 False），因此它跳过了第 H 行的 print() 调用。相反，它在第 I 行运行了 print()，这会使得"End of function"第二次显示出来。

由于程序运行到达了函数的末尾，它返回到了调用该函数的那行代码，也就是第 K 行。在第 K 行之后，没有其他的代码行了，因此，程序终止了。注意，局部变量并不是相对于函数是局部的，对于一次特定的函数调用来说，也是的。

9.7 栈溢出

每次调用一个函数的时候，Python 解释器都记住了在哪一行进行调用。通过这种方式，当函数返回的时候，Python 知道从哪里开始继续执行。记录这一信息可能需要占用一些内存。通常来说，这不会很费事，看一下如下的代码。

```
def funky():
    funky()

funky()
```

如果你运行该程序，将会看到如下所示的大量输出。

第 9 章　Star Pusher

```
...
  File "C:\test67.py", line 2, in funky
    funky()
  File "C:\test67.py", line 2, in funky
    funky()
  File "C:\test67.py", line 2, in funky
    funky()
  File "C:\test67.py", line 2, in funky
    funky()
  File "C:\test67.py", line 2, in funky
    funky()
RuntimeError: maximum recursion depth exceeded
```

funky()函数什么也不做，只是调用自己。然后，在该调用中，函数再次调用自己。然后，它再次调用自己，然后又一次，然后再一次。每次它调用自己的时候，Python 都必须记住执行调用的代码行，以便当程序返回的时候能够从那里继续。但是，funky()函数不会返回，它只是不断地调用自己。这就像是一个无限循环 Bug，其中，程序不断运行并且不会停止。为了防止用尽内存，当你达到 1000 次调用的深度的时候，Python 将会产生一个错误，并且程序崩溃。这种类型的 Bug 叫作栈溢出（stack overflow）。

这段代码也会导致栈溢出，即便没有递归函数。

```
def spam():
    eggs()

def eggs():
    spam()

spam()
```

当你运行该程序的时候，它会导致如下的一个错误。

```
...
  File "C:\test67.py", line 2, in spam
    eggs()
  File "C:\test67.py", line 5, in eggs
    spam()
  File "C:\test67.py", line 2, in spam
    eggs()
  File "C:\test67.py", line 5, in eggs
    spam()
  File "C:\test67.py", line 2, in spam
    eggs()
RuntimeError: maximum recursion depth exceeded
```

9.8 使用基本条件防止栈溢出

为了防止栈溢出的 Bug，我们必须有一种基本条件（base case），其中函数将停止进行新的递归调用。如果没有基本条件，函数调用将不会停止，最终会导致栈溢出。如下是带有基本条件的一个递归函数的示例。基本条件就是 param 参数等于 2 的情况。

```
def fizz(param):
    print(param)
    if param == 2:
        return
    fizz(param - 1)

fizz(5)
```

当运行该程序的时候，输出如下所示。

```
5
4
3
2
```

该程序不会有栈溢出的错误，因为一旦 param 参数设置为 2，if 语句的条件就会为 True，并且该函数将会返回，然后，剩下的调用也将依次返回。

然而，如果你的代码不会满足基本条件，那么还是会导致栈溢出。如果我们将 fizz(5) 调用修改为 fizz(0)，那么，程序的输出将会如下所示。

```
  File "C:\rectest.py", line 5, in fizz
    fizz(param - 1)
  File "C:\rectest.py", line 5, in fizz
    fizz(param - 1)
  File "C:\rectest.py", line 5, in fizz
    fizz(param - 1)
  File "C:\rectest.py", line 2, in fizz
    print(param)
RuntimeError: maximum recursion depth exceeded
```

递归调用和基本条件用来执行漫水填充（flood fill）算法，下面将介绍该算法。

9.9 漫水填充算法

漫水填充算法在 Star Pusher 中用来修改关卡中的所有的墙内的地板贴片，将其从使用

第 9 章 Star Pusher

"室外地板"贴片图像（这是地图上所有贴片默认使用的）修改为使用"室内地板"贴片。第 295 行是最初的 floodFill() 调用。它将会把用' '字符串（这表示一个室外地板）表示的任何贴片，转换为一个'o'字符串（这表示一个室内地板）。

```
513. def floodFill(mapObj, x, y, oldCharacter, newCharacter):
514.     """Changes any values matching oldCharacter on the map object to
515.     newCharacter at the (x, y) position, and does the same for the
516.     positions to the left, right, down, and up of (x, y), recursively."""
517.
518.     # In this game, the flood fill algorithm creates the inside/outside
519.     # floor distinction. This is a "recursive" function.
520.     # For more info on the Flood Fill algorithm, see:
521.     #    http://en.wikipedia.org/wiki/Flood_fill
522.     if mapObj[x][y] == oldCharacter:
523.         mapObj[x][y] = newCharacter
```

第 522 行和第 523 行查看传递给 floodFill() 的 XY 坐标处的贴片，如果该贴片最初与 oldCharacter 字符串相同的话，就将其修改为 newCharacter。

```
525.     if x < len(mapObj) - 1 and mapObj[x+1][y] == oldCharacter:
526.         floodFill(mapObj, x+1, y, oldCharacter, newCharacter) # call right
527.     if x > 0 and mapObj[x-1][y] == oldCharacter:
528.         floodFill(mapObj, x-1, y, oldCharacter, newCharacter) # call left
529.     if y < len(mapObj[x]) - 1 and mapObj[x][y+1] == oldCharacter:
530.         floodFill(mapObj, x, y+1, oldCharacter, newCharacter) # call down
531.     if y > 0 and mapObj[x][y-1] == oldCharacter:
532.         floodFill(mapObj, x, y-1, oldCharacter, newCharacter) # call up
```

这 4 条 if 语句检查 XY 坐标的贴片的右边、左边、下面和上边是否与 oldCharacter 相同，如果相同的话，将会使用那些坐标对 floodFill() 进行递归调用。

为了更好地理解 floodFill() 函数是如何工作的，这里给出了一个版本，它不使用递归调用，而是使用 XY 坐标的一个列表来记录，应该检查地图上的哪一个空格并且可能需要将其修改为 newCharacter。

```
def floodFill(mapObj, x, y, oldCharacter, newCharacter):
    spacesToCheck = []
    if mapObj[x][y] == oldCharacter:
        spacesToCheck.append((x, y))
    while spacesToCheck != []:
        x, y = spacesToCheck.pop()
        mapObj[x][y] = newCharacter

        if x < len(mapObj) - 1 and mapObj[x+1][y] == oldCharacter:
            spacesToCheck.append((x+1, y)) # check right
```

```
            if x > 0 and mapObj[x-1][y] == oldCharacter:
                spacesToCheck.append((x-1, y)) # check left
            if y < len(mapObj[x]) - 1 and mapObj[x][y+1] == oldCharacter:
                spacesToCheck.append((x, y+1)) # check down
            if y > 0 and mapObj[x][y-1] == oldCharacter:
                spacesToCheck.append((x, y-1)) # check up
```

如果你想要阅读一个更为详细的使用猫和僵尸的递归示例教程，可以访问 http://invpy.com/recursivezombies。

9.10 绘制地图

```
535. def drawMap(mapObj, gameStateObj, goals):
536.     """Draws the map to a Surface object, including the player and
537.     stars. This function does not call pygame.display.update(), nor
538.     does it draw the "Level" and "Steps" text in the corner."""
539.
540.     # mapSurf will be the single Surface object that the tiles are drawn
541.     # on, so that it is easy to position the entire map on the DISPLAYSURF
542.     # Surface object. First, the width and height must be calculated.
543.     mapSurfWidth = len(mapObj) * TILEWIDTH
544.     mapSurfHeight = (len(mapObj[0]) - 1) * (TILEHEIGHT - TILEFLOORHEIGHT)
 + TILEHEIGHT
545.     mapSurf = pygame.Surface((mapSurfWidth, mapSurfHeight))
546.     mapSurf.fill(BGCOLOR) # start with a blank color on the surface.
```

drawMap()函数将返回一个Surface对象，整个地图（以及玩家和星星）都绘制于其上。该Surface所需的宽度和高度，都必须通过mapObj来计算（这在第543行和第544行完成）。所有内容都绘制于其上的Surface对象在第545行创建。为了开始这一步，在第546行，整个Surface对象都绘制到背景颜色之上。

```
548.     # Draw the tile sprites onto this surface.
549.     for x in range(len(mapObj)):
550.         for y in range(len(mapObj[x])):
551.             spaceRect = pygame.Rect((x * TILEWIDTH, y * (TILEHEIGHT -
 TILEFLOORHEIGHT), TILEWIDTH, TILEHEIGHT))
```

第549行到第550行的for循环将遍历地图上的每一个可能的XY坐标，并且在该位置绘制相应的贴片图像。

第 9 章 Star Pusher

```
552.              if mapObj[x][y] in TILEMAPPING:
553.                  baseTile = TILEMAPPING[mapObj[x][y]]
554.              elif mapObj[x][y] in OUTSIDEDECOMAPPING:
555.                  baseTile = TILEMAPPING[' ']
556.
557.              # First draw the base ground/wall tile.
558.              mapSurf.blit(baseTile, spaceRect)
559.
```

baseTile 变量设置为在该迭代中在当前 XY 坐标所要绘制的贴片图像的 Surface 对象。如果该单字符字符串位于 OUTSIDEDECOMAPPING 字典中，那么将使用 TILEMAPPING[' ']（基本的室外地板贴片的单字符字符串）。

```
560.              if mapObj[x][y] in OUTSIDEDECOMAPPING:
561.                  # Draw any tree/rock decorations that are on this tile.
562.                  mapSurf.blit(OUTSIDEDECOMAPPING[mapObj[x][y]], spaceRect)
```

此外，如果该贴片列于 OUTSIDEDECOMAPPING 字典之中，相应的树木或岩石图像将会绘制到刚刚绘制于 XY 坐标的贴片之上。

```
563.              elif (x, y) in gameStateObj['stars']:
564.                  if (x, y) in goals:
565.                      # A goal AND star are on this space, draw goal first.
566.                      mapSurf.blit(IMAGESDICT['covered goal'], spaceRect)
567.                  # Then draw the star sprite.
568.                  mapSurf.blit(IMAGESDICT['star'], spaceRect)
```

如果在地图上的 XY 坐标处有一个星星（可以通过检查(x, y)是否位于 gameStateObj['stars']中的列表中来发现），那么，应该在这个 XY 坐标绘制一个星星（这在第 568 行完成）。在绘制星星之前，代码应该首先检查该位置是否也有一个目标，如果有一个目标，应该先绘制"覆盖的目标"贴片。

```
569.              elif (x, y) in goals:
570.                  # Draw a goal without a star on it.
571.                  mapSurf.blit(IMAGESDICT['uncovered goal'], spaceRect)
```

如果在地图的这个 XY 坐标处有一个目标，那么应该在贴片之上绘制"打开的目标"。之所以要绘制这个打开的目标，是因为如果执行到了第 569 行的 elif 语句，我们知道第 563 行的 elif 语句的条件为 False，并且该 XY 坐标上没有星星。

```
573.              # Last draw the player on the board.
574.              if (x, y) == gameStateObj['player']:
```

9.11 检查关卡是否完成

```
575.                    # Note: The value "currentImage" refers
576.                    # to a key in "PLAYERIMAGES" which has the
577.                    # specific player image we want to show.
578.                    mapSurf.blit(PLAYERIMAGES[currentImage], spaceRect)
579.
580.    return mapSurf
```

最后，drawMap()函数检查玩家是否位于该 XY 坐标上，如果是的，玩家的图像将会绘制到贴片之上。第 580 行位于嵌套的 for 循环之外，该循环从第 549 行和第 550 行开始，因此，当 Surface 对象返回的时候，整个地图已经绘制于其上了。

9.11 检查关卡是否完成

```
583. def isLevelFinished(levelObj, gameStateObj):
584.     """Returns True if all the goals have stars in them."""
585.     for goal in levelObj['goals']:
586.         if goal not in gameStateObj['stars']:
587.             # Found a space with a goal but no star on it.
588.             return False
589.     return True
```

如果所有的目标都被覆盖上了星星，isLevelFinished()函数返回 True。有些关卡的星星的数目可能比目标的数目要多，因此，重要的是要检查是否所有的目标都覆盖了星星，而不是是否所有的星星都覆盖在目标之上。

第 585 行的 for 循环遍历了 levelObj['goals']中的目标（这是每个目标的 XY 坐标的一个元组的列表），并且检查 gameStateObj['stars']中是否有一个星星拥有相同的 XY 坐标（这里的 not in 操作符有效，因为 gameStateObj['stars']是那些相同的 XY 坐标的元组的一个列表）。代码第一次找到在相同位置上没有星星的一个目标，该函数返回 False。

如果遍历了所有的目标，并且发现每个目标上都有一个星星，isLevelFinished()返回 True。

```
592. def terminate():
593.     pygame.quit()
594.     sys.exit()
```

terminate()函数和前面的游戏中的该函数相同。

第 9 章 Star Pusher

```
597. if __name__ == '__main__':
598.     main()
```

在所有的函数定义完之后,在第 602 行调用 main() 函数以开始游戏。

9.12 本章小结

在 Squirrel Eat Squirrel 游戏中,游戏世界相当简单,只有一个无限的绿色平地,其中草的图像随机地分布其上。Star Pusher 游戏引入了一些新东西,即拥有使用贴片图形独特设计的关卡。为了将这些关卡存储为计算机能够读取的一种格式,将它们都输入到一个文本文件中,程序中的代码读取这些文件并且为关卡创建数据结构。

实际上,Star Pusher 程序并不只是用单个的地图来制作一款简单的游戏,它更多是基于关卡文件来加载定制地图的一个系统。只需要修改关卡文件,我们就可以修改墙、星星和目标出现在游戏世界中的位置。Star Pusher 程序可以处理关卡文件所设置的任何配置(只要它通过了确保地图有意义的 assert 语句)。

你甚至不必知道如何编写 Python 程序来制作自己的关卡。要拥有自己的 Star Pusher 游戏关卡编辑器,只需要一个能够修改 *starPusherLevels.txt* 文件的文本编辑器程序就够了。

要进行更多的编程练习,可以从 http://invpy.com/buggy/starpusher 下载 Star Pusher 的带有 Bug 的版本,并尝试搞清楚如何修正这些 Bug。

第 10 章 4 款其他游戏

本章包含了 4 款其他的游戏的源代码。遗憾的是，本章中只包含源代码（以及注释），而没有对代码进行任何详细的介绍。但是现在，你应该能够玩这些游戏，并且通过查看源代码和注释来搞清楚代码是如何工作的。

这些游戏包括：
- Flippy——"Othello"游戏的一个翻版，其中玩家试图翻转计算机 AI 玩家的贴片。
- Ink Spill——"Flood It"游戏的一个翻版，利用了漫水填充算法。
- Four in a Row——"Connect Four"游戏的翻版，和计算机 AI 玩家对战。
- Gemgem——"Bejeweled"的翻版，其中玩家交换宝石以试图让 3 个相同的宝石排成一行。

如果你对本书中的源代码有任何的问题，可以发邮件给作者：al@inventwithpython.com。

如果想要练习修正 Bug 的话，可以通过下面的链接下载带有 Bug 的源代码。
- http://invpy.com/buggy/flippy
- http://invpy.com/buggy/inkspill
- http://invpy.com/buggy/fourinarow
- http://invpy.com/buggy/gemgem

10.1 Flippy，Othello 的翻版

Othello 更为常见的名称是 Reversi，它拥有一个 8×8 的游戏板，其中带有一些贴片，一面是黑色的，而另一面是白色的。开始的时候，游戏板如图 10-1 所示。每个玩家在自己的轮次中，都要放下一个自己的颜色的新的贴片。在新的贴片和另一个相同颜色贴片之间，任何对手的贴片都要翻转。游戏的目标是拥有尽可能多的自己的颜色的贴片。

例如，图 10-2 显示了白色玩家在 5, 6 位置放置一个新的白色贴片的样子。

5, 5 的黑色贴片，位于新的白色贴片和 5, 4 处已有的白色贴片之间。这个黑色贴片将翻转，并且变为一个新的白色贴片，这使得游戏板看上去如图 10-3 所示。黑色玩家接下来进行一次类似的移动，在 4, 6 放置一个黑色贴片，这将会翻转位于 4, 5 的白色贴片。最终的游戏板如图 10-3 所示。

第 10 章 4 款其他游戏

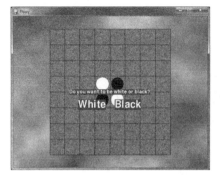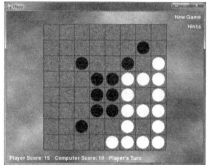

图 10-1

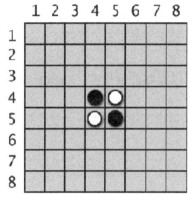

刚开始的Reversi游戏板有两个白色
的贴片和两个黑色贴片

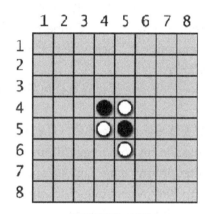

白色玩家放置一个新贴片

图 10-2

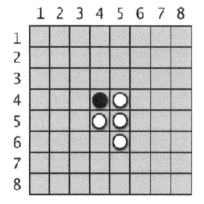

白色玩家的移动将会翻转一个黑色贴片

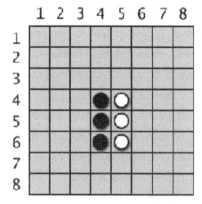

黑色玩家放置一个新的贴片,
它将会翻转一个白色贴片

图 10-3

10.2 Flippy 的源代码

不管在什么方向上，如果贴片位于玩家的新贴片和已有的贴片之间，它都会翻转。在图 10-4 中，白色玩家在 3, 6 放置了一个贴片，这翻转了两个方向上的黑色贴片（用线条画了出来）。结果如图 10-4 所示。

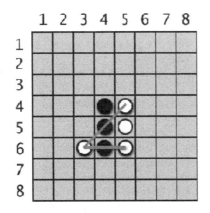

黑色玩家放置一个新的贴片，它将会翻转一个白色贴片　　　白色玩家两次翻转后的游戏板

图 10-4

正如你所看到的，每个玩家都可以移动一次或两次，就在游戏板的贴片上获得优势。玩家必须总是移动以至少捕获一个贴片。当玩家无法移动的时候，或者游戏板已经完全占满的时候，游戏结束。拥有较多的自己的颜色贴片的玩家获胜。

可以从 Wiki 页面 http://en.wikipedia.org/wiki/Reversi 了解 Reversi 的更多信息。

《Invent Your Own Computer Games with Python》一书的第 15 章，给出了该游戏的一个文本版本，它使用 print() 和 input() 而不是 Pygame。你可以阅读该章以详细了解如何将计算机 AI 算法整合到其中。

对该游戏使用计算机 AI 是很不错的，因为计算机较为容易模仿每一次可能的移动并通过走一步来翻转尽可能多的贴片。我和计算机玩的时候，它通常会赢我。

10.2 Flippy 的源代码

可以从 http://invpy.com/flippy.py 下载其源代码。

从 http://invpy.com/flippyimages.zip 下载 Flippy 所用到的图文件。

```
1. # Flippy (an Othello or Reversi clone)
2. # By Al Sweigart al@inventwithpython.com
3. # http://inventwithpython.com/pygame
4. # Released under a "Simplified BSD" license
5.
```

```
 6. # Based on the "reversi.py" code that originally appeared in "Invent
 7. # Your Own Computer Games with Python", chapter 15:
 8. #    http://inventwithpython.com/chapter15.html
 9.
10. import random, sys, pygame, time, copy
11. from pygame.locals import *
12.
13. FPS = 10 # frames per second to update the screen
14. WINDOWWIDTH = 640 # width of the program's window, in pixels
15. WINDOWHEIGHT = 480 # height in pixels
16. SPACESIZE = 50 # width & height of each space on the board, in pixels
17. BOARDWIDTH = 8 # how many columns of spaces on the game board
18. BOARDHEIGHT = 8 # how many rows of spaces on the game board
19. WHITE_TILE = 'WHITE_TILE' # an arbitrary but unique value
20. BLACK_TILE = 'BLACK_TILE' # an arbitrary but unique value
21. EMPTY_SPACE = 'EMPTY_SPACE' # an arbitrary but unique value
22. HINT_TILE = 'HINT_TILE' # an arbitrary but unique value
23. ANIMATIONSPEED = 25 # integer from 1 to 100, higher is faster animation
24.
25. # Amount of space on the left & right side (XMARGIN) or above and below
26. # (YMARGIN) the game board, in pixels.
27. XMARGIN = int((WINDOWWIDTH - (BOARDWIDTH * SPACESIZE)) / 2)
28. YMARGIN = int((WINDOWHEIGHT - (BOARDHEIGHT * SPACESIZE)) / 2)
29.
30. #              R    G    B
31. WHITE      = (255, 255, 255)
32. BLACK      = (  0,   0,   0)
33. GREEN      = (  0, 155,   0)
34. BRIGHTBLUE = (  0,  50, 255)
35. BROWN      = (174,  94,   0)
36.
37. TEXTBGCOLOR1 = BRIGHTBLUE
38. TEXTBGCOLOR2 = GREEN
39. GRIDLINECOLOR = BLACK
40. TEXTCOLOR = WHITE
41. HINTCOLOR = BROWN
42.
43.
44. def main():
45.     global MAINCLOCK, DISPLAYSURF, FONT, BIGFONT, BGIMAGE
46.
47.     pygame.init()
48.     MAINCLOCK = pygame.time.Clock()
49.     DISPLAYSURF = pygame.display.set_mode((WINDOWWIDTH, WINDOWHEIGHT))
50.     pygame.display.set_caption('Flippy')
51.     FONT = pygame.font.Font('freesansbold.ttf', 16)
```

```
52.         BIGFONT = pygame.font.Font('freesansbold.ttf', 32)
53.
54.         # Set up the background image.
55.         boardImage = pygame.image.load('flippyboard.png')
56.         # Use smoothscale() to stretch the board image to fit the entire board:
57.         boardImage = pygame.transform.smoothscale(boardImage, (BOARDWIDTH * SPACESIZE, BOARDHEIGHT * SPACESIZE))
58.         boardImageRect = boardImage.get_rect()
59.         boardImageRect.topleft = (XMARGIN, YMARGIN)
60.         BGIMAGE = pygame.image.load('flippybackground.png')
61.         # Use smoothscale() to stretch the background image to fit the entire window:
62.         BGIMAGE = pygame.transform.smoothscale(BGIMAGE, (WINDOWWIDTH, WINDOWHEIGHT))
63.         BGIMAGE.blit(boardImage, boardImageRect)
64.
65.         # Run the main game.
66.         while True:
67.             if runGame() == False:
68.                 break
69.
70.
71.     def runGame():
72.         # Plays a single game of reversi each time this function is called.
73.
74.         # Reset the board and game.
75.         mainBoard = getNewBoard()
76.         resetBoard(mainBoard)
77.         showHints = False
78.         turn = random.choice(['computer', 'player'])
79.
80.         # Draw the starting board and ask the player what color they want.
81.         drawBoard(mainBoard)
82.         playerTile, computerTile = enterPlayerTile()
83.
84.         # Make the Surface and Rect objects for the "New Game" and "Hints" buttons
85.         newGameSurf = FONT.render('New Game', True, TEXTCOLOR, TEXTBGCOLOR2)
86.         newGameRect = newGameSurf.get_rect()
87.         newGameRect.topright = (WINDOWWIDTH - 8, 10)
88.         hintsSurf = FONT.render('Hints', True, TEXTCOLOR, TEXTBGCOLOR2)
89.         hintsRect = hintsSurf.get_rect()
90.         hintsRect.topright = (WINDOWWIDTH - 8, 40)
91.
92.         while True: # main game loop
```

```
93.            # Keep looping for player and computer's turns.
94.            if turn == 'player':
95.                # Player's turn:
96.                if getValidMoves(mainBoard, playerTile) == []:
97.                    # If it's the player's turn but they
98.                    # can't move, then end the game.
99.                    break
100.               movexy = None
101.               while movexy == None:
102.                   # Keep looping until the player clicks on a valid space.
103.
104.                   # Determine which board data structure to use for display.
105.                   if showHints:
106.                       boardToDraw = getBoardWithValidMoves(mainBoard, playerTile)
107.                   else:
108.                       boardToDraw = mainBoard
109.
110.                   checkForQuit()
111.                   for event in pygame.event.get(): # event handling loop
112.                       if event.type == MOUSEBUTTONUP:
113.                           # Handle mouse click events
114.                           mousex, mousey = event.pos
115.                           if newGameRect.collidepoint( (mousex, mousey) ):
116.                               # Start a new game
117.                               return True
118.                           elif hintsRect.collidepoint( (mousex, mousey) ):
119.                               # Toggle hints mode
120.                               showHints = not showHints
121.                           # movexy is set to a two-item tuple XY coordinate, or None value
122.                           movexy = getSpaceClicked(mousex, mousey)
123.                           if movexy != None and not isValidMove(mainBoard, playerTile, movexy[0], movexy[1]):
124.                               movexy = None
125.
126.                   # Draw the game board.
127.                   drawBoard(boardToDraw)
128.                   drawInfo(boardToDraw, playerTile, computerTile, turn)
129.
130.                   # Draw the "New Game" and "Hints" buttons.
131.                   DISPLAYSURF.blit(newGameSurf, newGameRect)
132.                   DISPLAYSURF.blit(hintsSurf, hintsRect)
133.
134.                   MAINCLOCK.tick(FPS)
135.                   pygame.display.update()
```

```
136.
137.                # Make the move and end the turn.
138.                makeMove(mainBoard, playerTile, movexy[0], movexy[1], True)
139.                if getValidMoves(mainBoard, computerTile) != []:
140.                    # Only set for the computer's turn if it can make a move.
141.                    turn = 'computer'
142.
143.            else:
144.                # Computer's turn:
145.                if getValidMoves(mainBoard, computerTile) == []:
146.                    # If it was set to be the computer's turn but
147.                    # they can't move, then end the game.
148.                    break
149.
150.                # Draw the board.
151.                drawBoard(mainBoard)
152.                drawInfo(mainBoard, playerTile, computerTile, turn)
153.
154.                # Draw the "New Game" and "Hints" buttons.
155.                DISPLAYSURF.blit(newGameSurf, newGameRect)
156.                DISPLAYSURF.blit(hintsSurf, hintsRect)
157.
158.                # Make it look like the computer is thinking by pausing a bit.
159.                pauseUntil = time.time() + random.randint(5, 15) * 0.1
160.                while time.time() < pauseUntil:
161.                    pygame.display.update()
162.
163.                # Make the move and end the turn.
164.                x, y = getComputerMove(mainBoard, computerTile)
165.                makeMove(mainBoard, computerTile, x, y, True)
166.                if getValidMoves(mainBoard, playerTile) != []:
167.                    # Only set for the player's turn if they can make a move.
168.                    turn = 'player'
169.
170.        # Display the final score.
171.        drawBoard(mainBoard)
172.        scores = getScoreOfBoard(mainBoard)
173.
174.        # Determine the text of the message to display.
175.        if scores[playerTile] > scores[computerTile]:
176.            text = 'You beat the computer by %s points! Congratulations!' % \
177.                    (scores[playerTile] - scores[computerTile])
178.        elif scores[playerTile] < scores[computerTile]:
179.            text = 'You lost. The computer beat you by %s points.' % \
180.                    (scores[computerTile] - scores[playerTile])
181.        else:
```

```
182.            text = 'The game was a tie!'
183.
184.        textSurf = FONT.render(text, True, TEXTCOLOR, TEXTBGCOLOR1)
185.        textRect = textSurf.get_rect()
186.        textRect.center = (int(WINDOWWIDTH / 2), int(WINDOWHEIGHT / 2))
187.        DISPLAYSURF.blit(textSurf, textRect)
188.
189.        # Display the "Play again?" text with Yes and No buttons.
190.        text2Surf = BIGFONT.render('Play again?', True, TEXTCOLOR,
TEXTBGCOLOR1)
191.        text2Rect = text2Surf.get_rect()
192.        text2Rect.center = (int(WINDOWWIDTH / 2), int(WINDOWHEIGHT / 2) + 50)
193.
194.        # Make "Yes" button.
195.        yesSurf = BIGFONT.render('Yes', True, TEXTCOLOR, TEXTBGCOLOR1)
196.        yesRect = yesSurf.get_rect()
197.        yesRect.center = (int(WINDOWWIDTH / 2) - 60, int(WINDOWHEIGHT / 2) +
90)
198.
199.        # Make "No" button.
200.        noSurf = BIGFONT.render('No', True, TEXTCOLOR, TEXTBGCOLOR1)
201.        noRect = noSurf.get_rect()
202.        noRect.center = (int(WINDOWWIDTH / 2) + 60, int(WINDOWHEIGHT / 2) +
90)
203.
204.        while True:
205.            # Process events until the user clicks on Yes or No.
206.            checkForQuit()
207.            for event in pygame.event.get(): # event handling loop
208.                if event.type == MOUSEBUTTONUP:
209.                    mousex, mousey = event.pos
210.                    if yesRect.collidepoint( (mousex, mousey) ):
211.                        return True
212.                    elif noRect.collidepoint( (mousex, mousey) ):
213.                        return False
214.            DISPLAYSURF.blit(textSurf, textRect)
215.            DISPLAYSURF.blit(text2Surf, text2Rect)
216.            DISPLAYSURF.blit(yesSurf, yesRect)
217.            DISPLAYSURF.blit(noSurf, noRect)
218.            pygame.display.update()
219.            MAINCLOCK.tick(FPS)
220.
221.
222.    def translateBoardToPixelCoord(x, y):
223.        return XMARGIN + x * SPACESIZE + int(SPACESIZE / 2), YMARGIN + y *
SPACESIZE + int(SPACESIZE / 2)
```

```
224.
225.
226. def animateTileChange(tilesToFlip, tileColor, additionalTile):
227.     # Draw the additional tile that was just laid down. (Otherwise we'd
228.     # have to completely redraw the board & the board info.)
229.     if tileColor == WHITE_TILE:
230.         additionalTileColor = WHITE
231.     else:
232.         additionalTileColor = BLACK
233.     additionalTileX, additionalTileY = translateBoardToPixelCoord(additionalTile[0], additionalTile[1])
234.     pygame.draw.circle(DISPLAYSURF, additionalTileColor, (additionalTileX, additionalTileY), int(SPACESIZE / 2) - 4)
235.     pygame.display.update()
236.
237.     for rgbValues in range(0, 255, int(ANIMATIONSPEED * 2.55)):
238.         if rgbValues > 255:
239.             rgbValues = 255
240.         elif rgbValues < 0:
241.             rgbValues = 0
242.
243.         if tileColor == WHITE_TILE:
244.             color = tuple([rgbValues] * 3) # rgbValues goes from 0 to 255
245.         elif tileColor == BLACK_TILE:
246.             color = tuple([255 - rgbValues] * 3) # rgbValues goes from 255 to 0
247.
248.         for x, y in tilesToFlip:
249.             centerx, centery = translateBoardToPixelCoord(x, y)
250.             pygame.draw.circle(DISPLAYSURF, color, (centerx, centery), int(SPACESIZE / 2) - 4)
251.         pygame.display.update()
252.         MAINCLOCK.tick(FPS)
253.         checkForQuit()
254.
255.
256. def drawBoard(board):
257.     # Draw background of board.
258.     DISPLAYSURF.blit(BGIMAGE, BGIMAGE.get_rect())
259.
260.     # Draw grid lines of the board.
261.     for x in range(BOARDWIDTH + 1):
262.         # Draw the horizontal lines.
263.         startx = (x * SPACESIZE) + XMARGIN
264.         starty = YMARGIN
265.         endx = (x * SPACESIZE) + XMARGIN
```

```
266.            endy = YMARGIN + (BOARDHEIGHT * SPACESIZE)
267.            pygame.draw.line(DISPLAYSURF, GRIDLINECOLOR, (startx, starty),
(endx, endy))
268.        for y in range(BOARDHEIGHT + 1):
269.            # Draw the vertical lines.
270.            startx = XMARGIN
271.            starty = (y * SPACESIZE) + YMARGIN
272.            endx = XMARGIN + (BOARDWIDTH * SPACESIZE)
273.            endy = (y * SPACESIZE) + YMARGIN
274.            pygame.draw.line(DISPLAYSURF, GRIDLINECOLOR, (startx, starty),
(endx, endy))
275.
276.        # Draw the black & white tiles or hint spots.
277.        for x in range(BOARDWIDTH):
278.            for y in range(BOARDHEIGHT):
279.                centerx, centery = translateBoardToPixelCoord(x, y)
280.                if board[x][y] == WHITE_TILE or board[x][y] == BLACK_TILE:
281.                    if board[x][y] == WHITE_TILE:
282.                        tileColor = WHITE
283.                    else:
284.                        tileColor = BLACK
285.                    pygame.draw.circle(DISPLAYSURF, tileColor, (centerx,
centery), int(SPACESIZE / 2) - 4)
286.                if board[x][y] == HINT_TILE:
287.                    pygame.draw.rect(DISPLAYSURF, HINTCOLOR, (centerx - 4,
centery - 4, 8, 8))
288.
289.
290. def getSpaceClicked(mousex, mousey):
291.     # Return a tuple of two integers of the board space coordinates where
292.     # the mouse was clicked. (Or returns None not in any space.)
293.     for x in range(BOARDWIDTH):
294.         for y in range(BOARDHEIGHT):
295.             if mousex > x * SPACESIZE + XMARGIN and \
296.                mousex < (x + 1) * SPACESIZE + XMARGIN and \
297.                mousey > y * SPACESIZE + YMARGIN and \
298.                mousey < (y + 1) * SPACESIZE + YMARGIN:
299.                 return (x, y)
300.     return None
301.
302.
303. def drawInfo(board, playerTile, computerTile, turn):
304.     # Draws scores and whose turn it is at the bottom of the screen.
305.     scores = getScoreOfBoard(board)
306.     scoreSurf = FONT.render("Player Score: %s    Computer Score: %s
%s's Turn" % (str(scores[playerTile]), str(scores[computerTile]),
```

10.2 Flippy 的源代码

```
    turn.title()), True, TEXTCOLOR)
307.        scoreRect = scoreSurf.get_rect()
308.        scoreRect.bottomleft = (10, WINDOWHEIGHT - 5)
309.        DISPLAYSURF.blit(scoreSurf, scoreRect)
310.
311.
312. def resetBoard(board):
313.        # Blanks out the board it is passed, and sets up starting tiles.
314.        for x in range(BOARDWIDTH):
315.            for y in range(BOARDHEIGHT):
316.                board[x][y] = EMPTY_SPACE
317.
318.        # Add starting pieces to the center
319.        board[3][3] = WHITE_TILE
320.        board[3][4] = BLACK_TILE
321.        board[4][3] = BLACK_TILE
322.        board[4][4] = WHITE_TILE
323.
324.
325. def getNewBoard():
326.        # Creates a brand new, empty board data structure.
327.        board = []
328.        for i in range(BOARDWIDTH):
329.            board.append([EMPTY_SPACE] * BOARDHEIGHT)
330.
331.        return board
332.
333.
334. def isValidMove(board, tile, xstart, ystart):
335.        # Returns False if the player's move is invalid. If it is a valid
336.        # move, returns a list of spaces of the captured pieces.
337.        if board[xstart][ystart] != EMPTY_SPACE or not isOnBoard(xstart,
ystart):
338.            return False
339.
340.        board[xstart][ystart] = tile # temporarily set the tile on the board.
341.
342.        if tile == WHITE_TILE:
343.            otherTile = BLACK_TILE
344.        else:
345.            otherTile = WHITE_TILE
346.
347.        tilesToFlip = []
348.        # check each of the eight directions:
349.        for xdirection, ydirection in [[0, 1], [1, 1], [1, 0], [1, -1], [0, -
1], [-1, -1], [-1, 0], [-1, 1]]:
```

```
350.            x, y = xstart, ystart
351.            x += xdirection
352.            y += ydirection
353.            if isOnBoard(x, y) and board[x][y] == otherTile:
354.                # The piece belongs to the other player next to our piece.
355.                x += xdirection
356.                y += ydirection
357.                if not isOnBoard(x, y):
358.                    continue
359.                while board[x][y] == otherTile:
360.                    x += xdirection
361.                    y += ydirection
362.                    if not isOnBoard(x, y):
363.                        break # break out of while loop, continue in for loop
364.                if not isOnBoard(x, y):
365.                    continue
366.                if board[x][y] == tile:
367.                    # There are pieces to flip over. Go in the reverse
368.                    # direction until we reach the original space, noting all
369.                    # the tiles along the way.
370.                    while True:
371.                        x -= xdirection
372.                        y -= ydirection
373.                        if x == xstart and y == ystart:
374.                            break
375.                        tilesToFlip.append([x, y])
376.
377.        board[xstart][ystart] = EMPTY_SPACE # make space empty
378.        if len(tilesToFlip) == 0: # If no tiles flipped, this move is invalid
379.            return False
380.        return tilesToFlip
381.
382.
383.    def isOnBoard(x, y):
384.        # Returns True if the coordinates are located on the board.
385.        return x >= 0 and x < BOARDWIDTH and y >= 0 and y < BOARDHEIGHT
386.
387.
388.    def getBoardWithValidMoves(board, tile):
389.        # Returns a new board with hint markings.
390.        dupeBoard = copy.deepcopy(board)
391.
392.        for x, y in getValidMoves(dupeBoard, tile):
393.            dupeBoard[x][y] = HINT_TILE
394.        return dupeBoard
395.
```

```
396.
397. def getValidMoves(board, tile):
398.     # Returns a list of (x,y) tuples of all valid moves.
399.     validMoves = []
400.
401.     for x in range(BOARDWIDTH):
402.         for y in range(BOARDHEIGHT):
403.             if isValidMove(board, tile, x, y) != False:
404.                 validMoves.append((x, y))
405.     return validMoves
406.
407.
408. def getScoreOfBoard(board):
409.     # Determine the score by counting the tiles.
410.     xscore = 0
411.     oscore = 0
412.     for x in range(BOARDWIDTH):
413.         for y in range(BOARDHEIGHT):
414.             if board[x][y] == WHITE_TILE:
415.                 xscore += 1
416.             if board[x][y] == BLACK_TILE:
417.                 oscore += 1
418.     return {WHITE_TILE:xscore, BLACK_TILE:oscore}
419.
420.
421. def enterPlayerTile():
422.     # Draws the text and handles the mouse click events for letting
423.     # the player choose which color they want to be.  Returns
424.     # [WHITE_TILE, BLACK_TILE] if the player chooses to be White,
425.     # [BLACK_TILE, WHITE_TILE] if Black.
426.
427.     # Create the text.
428.     textSurf = FONT.render('Do you want to be white or black?', True, TEXTCOLOR, TEXTBGCOLOR1)
429.     textRect = textSurf.get_rect()
430.     textRect.center = (int(WINDOWWIDTH / 2), int(WINDOWHEIGHT / 2))
431.
432.     xSurf = BIGFONT.render('White', True, TEXTCOLOR, TEXTBGCOLOR1)
433.     xRect = xSurf.get_rect()
434.     xRect.center = (int(WINDOWWIDTH / 2) - 60, int(WINDOWHEIGHT / 2) + 40)
435.
436.     oSurf = BIGFONT.render('Black', True, TEXTCOLOR, TEXTBGCOLOR1)
437.     oRect = oSurf.get_rect()
438.     oRect.center = (int(WINDOWWIDTH / 2) + 60, int(WINDOWHEIGHT / 2) + 40)
439.
440.     while True:
```

```
441.            # Keep looping until the player has clicked on a color.
442.            checkForQuit()
443.            for event in pygame.event.get(): # event handling loop
444.                if event.type == MOUSEBUTTONUP:
445.                    mousex, mousey = event.pos
446.                    if xRect.collidepoint( (mousex, mousey) ):
447.                        return [WHITE_TILE, BLACK_TILE]
448.                    elif oRect.collidepoint( (mousex, mousey) ):
449.                        return [BLACK_TILE, WHITE_TILE]
450.
451.            # Draw the screen.
452.            DISPLAYSURF.blit(textSurf, textRect)
453.            DISPLAYSURF.blit(xSurf, xRect)
454.            DISPLAYSURF.blit(oSurf, oRect)
455.            pygame.display.update()
456.            MAINCLOCK.tick(FPS)
457.
458.
459.    def makeMove(board, tile, xstart, ystart, realMove=False):
460.        # Place the tile on the board at xstart, ystart, and flip tiles
461.        # Returns False if this is an invalid move, True if it is valid.
462.        tilesToFlip = isValidMove(board, tile, xstart, ystart)
463.
464.        if tilesToFlip == False:
465.            return False
466.
467.        board[xstart][ystart] = tile
468.
469.        if realMove:
470.            animateTileChange(tilesToFlip, tile, (xstart, ystart))
471.
472.        for x, y in tilesToFlip:
473.            board[x][y] = tile
474.        return True
475.
476.
477.    def isOnCorner(x, y):
478.        # Returns True if the position is in one of the four corners.
479.        return (x == 0 and y == 0) or \
480.               (x == BOARDWIDTH and y == 0) or \
481.               (x == 0 and y == BOARDHEIGHT) or \
482.               (x == BOARDWIDTH and y == BOARDHEIGHT)
483.
484.
485.    def getComputerMove(board, computerTile):
486.        # Given a board and the computer's tile, determine where to
```

```
487.        # move and return that move as a [x, y] list.
488.        possibleMoves = getValidMoves(board, computerTile)
489.
490.        # randomize the order of the possible moves
491.        random.shuffle(possibleMoves)
492.
493.        # always go for a corner if available.
494.        for x, y in possibleMoves:
495.            if isOnCorner(x, y):
496.                return [x, y]
497.
498.        # Go through all possible moves and remember the best scoring move
499.        bestScore = -1
500.        for x, y in possibleMoves:
501.            dupeBoard = copy.deepcopy(board)
502.            makeMove(dupeBoard, computerTile, x, y)
503.            score = getScoreOfBoard(dupeBoard)[computerTile]
504.            if score > bestScore:
505.                bestMove = [x, y]
506.                bestScore = score
507.        return bestMove
508.
509.
510.    def checkForQuit():
511.        for event in pygame.event.get((QUIT, KEYUP)): # event handling loop
512.            if event.type == QUIT or (event.type == KEYUP and event.key == K_ESCAPE):
513.                pygame.quit()
514.                sys.exit()
515.
516.
517.    if __name__ == '__main__':
518.        main()
```

10.3 Ink Spill，Flood It 游戏的翻版

  Flood It 游戏开始的时候，游戏板填满了彩色的贴片，如图 10-5 所示。在每一轮中，玩家选择一种新的颜色来绘制左上方的贴片以及和它相邻的具有相同颜色的贴片。该游戏利用了漫水填充算法（在第 9 章中介绍过）。游戏的目标是在用尽轮次之前，将整个游戏板变成一种单一的颜色。

  该游戏还有一个 Setting 屏幕，玩家可以在其中修改游戏板的大小和难度。如果玩家对原来的颜色感到厌烦，还有几种其他的颜色方案可供选择。

第 10 章　4 款其他游戏

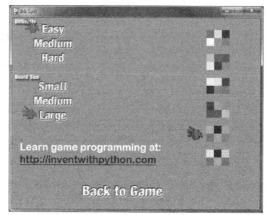

图 10-5

10.4　Ink Spill 的源代码

可以从 http://invpy.com/inkspill.py 下载 Ink Spill 的源代码。

还可以从 http://invpy.com/inkspillimages.zip 下载 Ink Spill 所用到的图像文件。

```
1.  # Ink Spill (a Flood It clone)
2.  # http://inventwithpython.com/pygame
3.  # By Al Sweigart al@inventwithpython.com
4.  # Released under a "Simplified BSD" license
5.
6.  import random, sys, webbrowser, copy, pygame
7.  from pygame.locals import *
8.
9.  # There are different box sizes, number of boxes, and
10. # life depending on the "board size" setting selected.
11. SMALLBOXSIZE  = 60 # size is in pixels
12. MEDIUMBOXSIZE = 20
13. LARGEBOXSIZE  = 11
14.
15. SMALLBOARDSIZE  = 6 # size is in boxes
16. MEDIUMBOARDSIZE = 17
17. LARGEBOARDSIZE  = 30
18.
19. SMALLMAXLIFE  = 10 # number of turns
20. MEDIUMMAXLIFE = 30
21. LARGEMAXLIFE  = 64
22.
23. FPS = 30
```

```
 24. WINDOWWIDTH = 640
 25. WINDOWHEIGHT = 480
 26. boxSize = MEDIUMBOXSIZE
 27. PALETTEGAPSIZE = 10
 28. PALETTESIZE = 45
 29. EASY = 0   # arbitrary but unique value
 30. MEDIUM = 1 # arbitrary but unique value
 31. HARD = 2   # arbitrary but unique value
 32.
 33. difficulty = MEDIUM # game starts in "medium" mode
 34. maxLife = MEDIUMMAXLIFE
 35. boardWidth = MEDIUMBOARDSIZE
 36. boardHeight = MEDIUMBOARDSIZE
 37.
 38.
 39. #              R    G    B
 40. WHITE    = (255, 255, 255)
 41. DARKGRAY = ( 70,  70,  70)
 42. BLACK    = (  0,   0,   0)
 43. RED      = (255,   0,   0)
 44. GREEN    = (  0, 255,   0)
 45. BLUE     = (  0,   0, 255)
 46. YELLOW   = (255, 255,   0)
 47. ORANGE   = (255, 128,   0)
 48. PURPLE   = (255,   0, 255)
 49.
 50. # The first color in each scheme is the background color, the next six are the palette colors.
 51. COLORSCHEMES = (((150, 200, 255), RED, GREEN, BLUE, YELLOW, ORANGE, PURPLE),
 52.                 ((0, 155, 104),   (97, 215, 164),  (228, 0, 69),   (0, 125, 50),   (204, 246, 0),   (148, 0, 45),    (241, 109, 149)),
 53.                 ((195, 179, 0),   (255, 239, 115), (255, 226, 0),  (147, 3, 167),  (24, 38, 176),   (166, 147, 0),   (197, 97, 211)),
 54.                 ((85, 0, 0),      (155, 39, 102),  (0, 201, 13),   (255, 118, 0),  (206, 0, 113),   (0, 130, 9),     (255, 180, 115)),
 55.                 ((191, 159, 64), (183, 182, 208), (4, 31, 183),    (167, 184, 45), (122, 128, 212), (37, 204, 7),    (88, 155, 213)),
 56.                 ((200, 33, 205), (116, 252, 185), (68, 56, 56),    (52, 238, 83),  (23, 149, 195),  (222, 157, 227), (212, 86, 185)))
 57. for i in range(len(COLORSCHEMES)):
 58.     assert len(COLORSCHEMES[i]) == 7, 'Color scheme %s does not have exactly 7 colors.' % (i)
 59. bgColor = COLORSCHEMES[0][0]
 60. paletteColors =  COLORSCHEMES[0][1:]
 61.
```

第 10 章　4 款其他游戏

```
62. def main():
63.     global FPSCLOCK, DISPLAYSURF, LOGOIMAGE, SPOTIMAGE, SETTINGSIMAGE,
SETTINGSBUTTONIMAGE, RESETBUTTONIMAGE
64.
65.     pygame.init()
66.     FPSCLOCK = pygame.time.Clock()
67.     DISPLAYSURF = pygame.display.set_mode((WINDOWWIDTH, WINDOWHEIGHT))
68.
69.     # Load images
70.     LOGOIMAGE = pygame.image.load('inkspilllogo.png')
71.     SPOTIMAGE = pygame.image.load('inkspillspot.png')
72.     SETTINGSIMAGE = pygame.image.load('inkspillsettings.png')
73.     SETTINGSBUTTONIMAGE = pygame.image.load('inkspillsettingsbutton.png')
74.     RESETBUTTONIMAGE = pygame.image.load('inkspillresetbutton.png')
75.
76.     pygame.display.set_caption('Ink Spill')
77.     mousex = 0
78.     mousey = 0
79.     mainBoard = generateRandomBoard(boardWidth, boardHeight, difficulty)
80.     life = maxLife
81.     lastPaletteClicked = None
82.
83.     while True: # main game loop
84.         paletteClicked = None
85.         resetGame = False
86.
87.         # Draw the screen.
88.         DISPLAYSURF.fill(bgColor)
89.         drawLogoAndButtons()
90.         drawBoard(mainBoard)
91.         drawLifeMeter(life)
92.         drawPalettes()
93.
94.         checkForQuit()
95.         for event in pygame.event.get(): # event handling loop
96.             if event.type == MOUSEBUTTONUP:
97.                 mousex, mousey = event.pos
98.                 if pygame.Rect(WINDOWWIDTH -
SETTINGSBUTTONIMAGE.get_width(),
99.                                 WINDOWHEIGHT -
SETTINGSBUTTONIMAGE.get_height(),
100.                                 SETTINGSBUTTONIMAGE.get_width(),
101.
SETTINGSBUTTONIMAGE.get_height()).collidepoint(mousex, mousey):
102.                     resetGame = showSettingsScreen() # clicked on Settings
```

```
button
103.                elif pygame.Rect(WINDOWWIDTH -
RESETBUTTONIMAGE.get_width(),
104.                       WINDOWHEIGHT -
SETTINGSBUTTONIMAGE.get_height() - RESETBUTTONIMAGE.get_height(),
105.                       RESETBUTTONIMAGE.get_width(),
106.
RESETBUTTONIMAGE.get_height()).collidepoint(mousex, mousey):
107.                    resetGame = True # clicked on Reset button
108.                else:
109.                    # check if a palette button was clicked
110.                    paletteClicked = getColorOfPaletteAt(mousex, mousey)
111.
112.        if paletteClicked != None and paletteClicked !=
lastPaletteClicked:
113.            # a palette button was clicked that is different from the
114.            # last palette button clicked (this check prevents the player
115.            # from accidentally clicking the same palette twice)
116.            lastPaletteClicked = paletteClicked
117.            floodAnimation(mainBoard, paletteClicked)
118.            life -= 1
119.
120.            resetGame = False
121.            if hasWon(mainBoard):
122.                for i in range(4): # flash border 4 times
123.                    flashBorderAnimation(WHITE, mainBoard)
124.                resetGame = True
125.                pygame.time.wait(2000) # pause so the player can bask in
victory
126.            elif life == 0:
127.                # life is zero, so player has lost
128.                drawLifeMeter(0)
129.                pygame.display.update()
130.                pygame.time.wait(400)
131.                for i in range(4):
132.                    flashBorderAnimation(BLACK, mainBoard)
133.                resetGame = True
134.                pygame.time.wait(2000) # pause so the player can suffer in
their defeat
135.
136.        if resetGame:
137.            # start a new game
138.            mainBoard = generateRandomBoard(boardWidth, boardHeight,
difficulty)
139.            life = maxLife
```

```
140.              lastPaletteClicked = None
141.
142.          pygame.display.update()
143.          FPSCLOCK.tick(FPS)
144.
145.
146. def checkForQuit():
147.     # Terminates the program if there are any QUIT or escape key events.
148.     for event in pygame.event.get(QUIT): # get all the QUIT events
149.         pygame.quit() # terminate if any QUIT events are present
150.         sys.exit()
151.     for event in pygame.event.get(KEYUP): # get all the KEYUP events
152.         if event.key == K_ESCAPE:
153.             pygame.quit() # terminate if the KEYUP event was for the Esc key
154.             sys.exit()
155.         pygame.event.post(event) # put the other KEYUP event objects back
156.
157.
158. def hasWon(board):
159.     # if the entire board is the same color, player has won
160.     for x in range(boardWidth):
161.         for y in range(boardHeight):
162.             if board[x][y] != board[0][0]:
163.                 return False # found a different color, player has not won
164.     return True
165.
166.
167. def showSettingsScreen():
168.     global difficulty, boxSize, boardWidth, boardHeight, maxLife, paletteColors, bgColor
169.
170.     # The pixel coordinates in this function were obtained by loading
171.     # the inkspillsettings.png image into a graphics editor and reading
172.     # the pixel coordinates from there. Handy trick.
173.
174.     origDifficulty = difficulty
175.     origBoxSize = boxSize
176.     screenNeedsRedraw = True
177.
178.     while True:
179.         if screenNeedsRedraw:
180.             DISPLAYSURF.fill(bgColor)
181.             DISPLAYSURF.blit(SETTINGSIMAGE, (0,0))
182.
183.             # place the ink spot marker next to the selected difficulty
```

```
184.            if difficulty == EASY:
185.                DISPLAYSURF.blit(SPOTIMAGE, (30, 4))
186.            if difficulty == MEDIUM:
187.                DISPLAYSURF.blit(SPOTIMAGE, (8, 41))
188.            if difficulty == HARD:
189.                DISPLAYSURF.blit(SPOTIMAGE, (30, 76))
190.
191.            # place the ink spot marker next to the selected size
192.            if boxSize == SMALLBOXSIZE:
193.                DISPLAYSURF.blit(SPOTIMAGE, (22, 150))
194.            if boxSize == MEDIUMBOXSIZE:
195.                DISPLAYSURF.blit(SPOTIMAGE, (11, 185))
196.            if boxSize == LARGEBOXSIZE:
197.                DISPLAYSURF.blit(SPOTIMAGE, (24, 220))
198.
199.            for i in range(len(COLORSCHEMES)):
200.                drawColorSchemeBoxes(500, i * 60 + 30, i)
201.
202.            pygame.display.update()
203.
204.        screenNeedsRedraw = False # by default, don't redraw the screen
205.        for event in pygame.event.get(): # event handling loop
206.            if event.type == QUIT:
207.                pygame.quit()
208.                sys.exit()
209.            elif event.type == KEYUP:
210.                if event.key == K_ESCAPE:
211.                    # Esc key on settings screen goes back to game
212.                    return not (origDifficulty == difficulty and origBoxSize == boxSize)
213.            elif event.type == MOUSEBUTTONUP:
214.                screenNeedsRedraw = True # screen should be redrawn
215.                mousex, mousey = event.pos # syntactic sugar
216.
217.                # check for clicks on the difficulty buttons
218.                if pygame.Rect(74, 16, 111, 30).collidepoint(mousex, mousey):
219.                    difficulty = EASY
220.                elif pygame.Rect(53, 50, 104, 29).collidepoint(mousex, mousey):
221.                    difficulty = MEDIUM
222.                elif pygame.Rect(72, 85, 65, 31).collidepoint(mousex, mousey):
223.                    difficulty = HARD
224.
225.                # check for clicks on the size buttons
```

```
226.                elif pygame.Rect(63, 156, 84, 31).collidepoint(mousex,
mousey):
227.                    # small board size setting:
228.                    boxSize = SMALLBOXSIZE
229.                    boardWidth = SMALLBOARDSIZE
230.                    boardHeight = SMALLBOARDSIZE
231.                    maxLife = SMALLMAXLIFE
232.                elif pygame.Rect(52, 192, 106,32).collidepoint(mousex,
mousey):
233.                    # medium board size setting:
234.                    boxSize = MEDIUMBOXSIZE
235.                    boardWidth = MEDIUMBOARDSIZE
236.                    boardHeight = MEDIUMBOARDSIZE
237.                    maxLife = MEDIUMMAXLIFE
238.                elif pygame.Rect(67, 228, 58, 37).collidepoint(mousex,
mousey):
239.                    # large board size setting:
240.                    boxSize = LARGEBOXSIZE
241.                    boardWidth = LARGEBOARDSIZE
242.                    boardHeight = LARGEBOARDSIZE
243.                    maxLife = LARGEMAXLIFE
244.                elif pygame.Rect(14, 299, 371, 97).collidepoint(mousex,
mousey):
245.                    # clicked on the "learn programming" ad
246.                    webbrowser.open('http://inventwithpython.com') # opens a web browser
247.                elif pygame.Rect(178, 418, 215, 34).collidepoint(mousex,
mousey):
248.                    # clicked on the "back to game" button
249.                    return not (origDifficulty == difficulty and origBoxSize == boxSize)
250.
251.                for i in range(len(COLORSCHEMES)):
252.                    # clicked on a color scheme button
253.                    if pygame.Rect(500, 30 + i * 60, MEDIUMBOXSIZE * 3, MEDIUMBOXSIZE * 2).collidepoint(mousex, mousey):
254.                        bgColor = COLORSCHEMES[i][0]
255.                        paletteColors = COLORSCHEMES[i][1:]
256.
257.
258. def drawColorSchemeBoxes(x, y, schemeNum):
259.     # Draws the color scheme boxes that appear on the "Settings" screen.
260.     for boxy in range(2):
261.         for boxx in range(3):
262.             pygame.draw.rect(DISPLAYSURF, COLORSCHEMES[schemeNum][3 * boxy + boxx + 1], (x + MEDIUMBOXSIZE * boxx, y + MEDIUMBOXSIZE * boxy,
```

```
                    MEDIUMBOXSIZE, MEDIUMBOXSIZE))
263.              if paletteColors == COLORSCHEMES[schemeNum][1:]:
264.                  # put the ink spot next to the selected color scheme
265.                  DISPLAYSURF.blit(SPOTIMAGE, (x - 50, y))
266.
267.
268. def flashBorderAnimation(color, board, animationSpeed=30):
269.     origSurf = DISPLAYSURF.copy()
270.     flashSurf = pygame.Surface(DISPLAYSURF.get_size())
271.     flashSurf = flashSurf.convert_alpha()
272.     for start, end, step in ((0, 256, 1), (255, 0, -1)):
273.         # the first iteration on the outer loop will set the inner loop
274.         # to have transparency go from 0 to 255, the second iteration will
275.         # have it go from 255 to 0. This is the "flash".
276.         for transparency in range(start, end, animationSpeed * step):
277.             DISPLAYSURF.blit(origSurf, (0, 0))
278.             r, g, b = color
279.             flashSurf.fill((r, g, b, transparency))
280.             DISPLAYSURF.blit(flashSurf, (0, 0))
281.             drawBoard(board) # draw board ON TOP OF the transparency layer
282.             pygame.display.update()
283.             FPSCLOCK.tick(FPS)
284.     DISPLAYSURF.blit(origSurf, (0, 0)) # redraw the original surface
285.
286.
287. def floodAnimation(board, paletteClicked, animationSpeed=25):
288.     origBoard = copy.deepcopy(board)
289.     floodFill(board, board[0][0], paletteClicked, 0, 0)
290.
291.     for transparency in range(0, 255, animationSpeed):
292.         # The "new" board slowly become opaque over the original board.
293.         drawBoard(origBoard)
294.         drawBoard(board, transparency)
295.         pygame.display.update()
296.         FPSCLOCK.tick(FPS)
297.
298.
299. def generateRandomBoard(width, height, difficulty=MEDIUM):
300.     # Creates a board data structure with random colors for each box.
301.     board = []
302.     for x in range(width):
303.         column = []
304.         for y in range(height):
305.             column.append(random.randint(0, len(paletteColors) - 1))
306.         board.append(column)
307.
```

```
308.        # Make board easier by setting some boxes to same color as a neighbor.
309.
310.        # Determine how many boxes to change.
311.        if difficulty == EASY:
312.            if boxSize == SMALLBOXSIZE:
313.                boxesToChange = 100
314.            else:
315.                boxesToChange = 1500
316.        elif difficulty == MEDIUM:
317.            if boxSize == SMALLBOXSIZE:
318.                boxesToChange = 5
319.            else:
320.                boxesToChange = 200
321.        else:
322.            boxesToChange = 0
323.
324.        # Change neighbor's colors:
325.        for i in range(boxesToChange):
326.            # Randomly choose a box whose color to copy
327.            x = random.randint(1, width-2)
328.            y = random.randint(1, height-2)
329.
330.            # Randomly choose neighbors to change.
331.            direction = random.randint(0, 3)
332.            if direction == 0: # change left and up neighbor
333.                board[x-1][y] == board[x][y]
334.                board[x][y-1] == board[x][y]
335.            elif direction == 1: # change right and down neighbor
336.                board[x+1][y] == board[x][y]
337.                board[x][y+1] == board[x][y]
338.            elif direction == 2: # change right and up neighbor
339.                board[x][y-1] == board[x][y]
340.                board[x+1][y] == board[x][y]
341.            else: # change left and down neighbor
342.                board[x][y+1] == board[x][y]
343.                board[x-1][y] == board[x][y]
344.    return board
345.
346.
347.    def drawLogoAndButtons():
348.        # draw the Ink Spill logo and Settings and Reset buttons.
349.        DISPLAYSURF.blit(LOGOIMAGE, (WINDOWWIDTH - LOGOIMAGE.get_width(), 0))
350.        DISPLAYSURF.blit(SETTINGSBUTTONIMAGE, (WINDOWWIDTH -
SETTINGSBUTTONIMAGE.get_width(), WINDOWHEIGHT -
SETTINGSBUTTONIMAGE.get_height()))
351.        DISPLAYSURF.blit(RESETBUTTONIMAGE, (WINDOWWIDTH -
```

10.4　Ink Spill 的源代码

```
RESETBUTTONIMAGE.get_width(), WINDOWHEIGHT - SETTINGSBUTTONIMAGE.get_height() -
RESETBUTTONIMAGE.get_height()))
352.
353.
354. def drawBoard(board, transparency=255):
355.     # The colored squares are drawn to a temporary surface which is then
356.     # drawn to the DISPLAYSURF surface. This is done so we can draw the
357.     # squares with transparency on top of DISPLAYSURF as it currently is.
358.     tempSurf = pygame.Surface(DISPLAYSURF.get_size())
359.     tempSurf = tempSurf.convert_alpha()
360.     tempSurf.fill((0, 0, 0, 0))
361.
362.     for x in range(boardWidth):
363.         for y in range(boardHeight):
364.             left, top = leftTopPixelCoordOfBox(x, y)
365.             r, g, b = paletteColors[board[x][y]]
366.             pygame.draw.rect(tempSurf, (r, g, b, transparency), (left,
top, boxSize, boxSize))
367.     left, top = leftTopPixelCoordOfBox(0, 0)
368.     pygame.draw.rect(tempSurf, BLACK, (left-1, top-1, boxSize * boardWidth
+ 1, boxSize * boardHeight + 1), 1)
369.     DISPLAYSURF.blit(tempSurf, (0, 0))
370.
371.
372. def drawPalettes():
373.     # Draws the six color palettes at the bottom of the screen.
374.     numColors = len(paletteColors)
375.     xmargin = int((WINDOWWIDTH - ((PALETTESIZE * numColors) +
(PALETTEGAPSIZE * (numColors - 1)))) / 2)
376.     for i in range(numColors):
377.         left = xmargin + (i * PALETTESIZE) + (i * PALETTEGAPSIZE)
378.         top = WINDOWHEIGHT - PALETTESIZE - 10
379.         pygame.draw.rect(DISPLAYSURF, paletteColors[i], (left, top,
PALETTESIZE, PALETTESIZE))
380.         pygame.draw.rect(DISPLAYSURF, bgColor,  (left + 2, top + 2,
PALETTESIZE - 4, PALETTESIZE - 4), 2)
381.
382.
383. def drawLifeMeter(currentLife):
384.     lifeBoxSize = int((WINDOWHEIGHT - 40) / maxLife)
385.
386.     # Draw background color of life meter.
387.     pygame.draw.rect(DISPLAYSURF, bgColor, (20, 20, 20, 20 + (maxLife *
lifeBoxSize)))
388.
389.     for i in range(maxLife):
```

第 10 章　4 款其他游戏

```
390.            if currentLife >= (maxLife - i): # draw a solid red box
391.                pygame.draw.rect(DISPLAYSURF, RED, (20, 20 + (i * lifeBoxSize), 20, lifeBoxSize))
392.                pygame.draw.rect(DISPLAYSURF, WHITE, (20, 20 + (i * lifeBoxSize), 20, lifeBoxSize), 1) # draw white outline
393.
394.
395. def getColorOfPaletteAt(x, y):
396.     # Returns the index of the color in paletteColors that the x and y parameters
397.     # are over. Returns None if x and y are not over any palette.
398.     numColors = len(paletteColors)
399.     xmargin = int((WINDOWWIDTH - ((PALETTESIZE * numColors) + (PALETTEGAPSIZE * (numColors - 1)))) / 2)
400.     top = WINDOWHEIGHT - PALETTESIZE - 10
401.     for i in range(numColors):
402.         # Find out if the mouse click is inside any of the palettes.
403.         left = xmargin + (i * PALETTESIZE) + (i * PALETTEGAPSIZE)
404.         r = pygame.Rect(left, top, PALETTESIZE, PALETTESIZE)
405.         if r.collidepoint(x, y):
406.             return i
407.     return None # no palette exists at these x, y coordinates
408.
409.
410. def floodFill(board, oldColor, newColor, x, y):
411.     # This is the flood fill algorithm.
412.     if oldColor == newColor or board[x][y] != oldColor:
413.         return
414.
415.     board[x][y] = newColor # change the color of the current box
416.
417.     # Make the recursive call for any neighboring boxes:
418.     if x > 0:
419.         floodFill(board, oldColor, newColor, x - 1, y) # on box to the left
420.     if x < boardWidth - 1:
421.         floodFill(board, oldColor, newColor, x + 1, y) # on box to the right
422.     if y > 0:
423.         floodFill(board, oldColor, newColor, x, y - 1) # on box to up
424.     if y < boardHeight - 1:
425.         floodFill(board, oldColor, newColor, x, y + 1) # on box to down
426.
427.
428. def leftTopPixelCoordOfBox(boxx, boxy):
429.     # Returns the x and y of the left-topmost pixel of the xth & yth box.
```

```
430.        xmargin = int((WINDOWWIDTH - (boardWidth * boxSize)) / 2)
431.        ymargin = int((WINDOWHEIGHT - (boardHeight * boxSize)) / 2)
432.        return (boxx * boxSize + xmargin, boxy * boxSize + ymargin)
433.
434.
435. if __name__ == '__main__':
436.     main()
```

10.5 Four-In-A-Row，Connect Four 的翻版

游戏 Connect Four 拥有一个 7×6 的游戏板，玩家可以在自己的轮次中把棋子放置到游戏板的顶部，如图 10-6 所示。棋子将会从每一列的顶部落下，直到到达游戏板的底部，或者落到了该列的最上部的棋子之上。当玩家在水平方向、垂直方向或者对角线方向上将 4 个棋子排成一条线的时候，就获胜了。

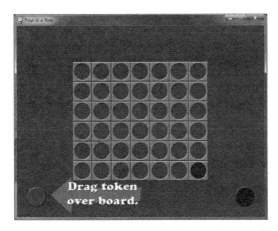
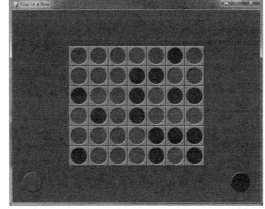

图 10-6

这款游戏的 AI 也相当不错。它模拟了所能做出的移动的每一种可能，然后，模拟了人类玩家可能对这些每次移动做出应对的每一种可能的移动，然后模拟了自己可以"对此做出应对"的每一种可能的移动，并再次模拟了人类玩家的每一次可能的应对。在所有这些思考之后，计算机判定了怎么移动才最有可能获胜。

因此，计算机是很难被打败的。我通常会输给它。

由于在你下棋的时候有 7 种可能的移动（除非某些列被占满了），而计算机对手下棋的时候也有 7 种可能的移动，对此可能有 7 种应对，然后又有 7 种应对，这意味着，在每一轮中，计算机要考虑 7×7×7×7 = 2 401 种可能的移动。你可以通过将 DIFFICULTY 常量设置为一个更高的数字，以便让计算机更深入地思考该游戏，但是，当我将这个值设置为 2 以上的数

字的时候，计算机需要花较长的时间思考走哪一步棋。

可以将 DIFFICULTY 设置为 1，以使得游戏更容易一些。然后，计算机只思考自己的移动和玩家对这些移动做出的应对。如果你将 DIFFICULTY 设置为 0，那么，计算机就失去了所有的智能，并且只是随机地移动。

10.6 Four-In-A-Row 的源代码

可以从 http://invpy.com/fourinarow.py 下载 Four-In-A-Row 的源代码。

可以从 http://invpy.com/fourinarowimages.zip 下载 Four-In-A-Row 所用到的图像文件。

```
 1. # Four-In-A-Row (a Connect Four clone)
 2. # By Al Sweigart al@inventwithpython.com
 3. # http://inventwithpython.com/pygame
 4. # Released under a "Simplified BSD" license
 5.
 6. import random, copy, sys, pygame
 7. from pygame.locals import *
 8.
 9. BOARDWIDTH = 7  # how many spaces wide the board is
10. BOARDHEIGHT = 6 # how many spaces tall the board is
11. assert BOARDWIDTH >= 4 and BOARDHEIGHT >= 4, 'Board must be at least 4x4.'
12.
13. DIFFICULTY = 2 # how many moves to look ahead. (>2 is usually too slow)
14.
15. SPACESIZE = 50 # size of the tokens and individual board spaces in pixels
16.
17. FPS = 30 # frames per second to update the screen
18. WINDOWWIDTH = 640 # width of the program's window, in pixels
19. WINDOWHEIGHT = 480 # height in pixels
20.
21. XMARGIN = int((WINDOWWIDTH - BOARDWIDTH * SPACESIZE) / 2)
22. YMARGIN = int((WINDOWHEIGHT - BOARDHEIGHT * SPACESIZE) / 2)
23.
24. BRIGHTBLUE = (0, 50, 255)
25. WHITE = (255, 255, 255)
26.
27. BGCOLOR = BRIGHTBLUE
28. TEXTCOLOR = WHITE
29.
30. RED = 'red'
31. BLACK = 'black'
32. EMPTY = None
33. HUMAN = 'human'
34. COMPUTER = 'computer'
```

10.6 Four-In-A-Row 的源代码

```
35.
36.
37.  def main():
38.      global FPSCLOCK, DISPLAYSURF, REDPILERECT, BLACKPILERECT, REDTOKENIMG
39.      global BLACKTOKENIMG, BOARDIMG, ARROWIMG, ARROWRECT, HUMANWINNERIMG
40.      global COMPUTERWINNERIMG, WINNERRECT, TIEWINNERIMG
41.
42.      pygame.init()
43.      FPSCLOCK = pygame.time.Clock()
44.      DISPLAYSURF = pygame.display.set_mode((WINDOWWIDTH, WINDOWHEIGHT))
45.      pygame.display.set_caption('Four in a Row')
46.
47.      REDPILERECT = pygame.Rect(int(SPACESIZE / 2), WINDOWHEIGHT - int(3 * SPACESIZE / 2), SPACESIZE, SPACESIZE)
48.      BLACKPILERECT = pygame.Rect(WINDOWWIDTH - int(3 * SPACESIZE / 2), WINDOWHEIGHT - int(3 * SPACESIZE / 2), SPACESIZE, SPACESIZE)
49.      REDTOKENIMG = pygame.image.load('4row_red.png')
50.      REDTOKENIMG = pygame.transform.smoothscale(REDTOKENIMG, (SPACESIZE, SPACESIZE))
51.      BLACKTOKENIMG = pygame.image.load('4row_black.png')
52.      BLACKTOKENIMG = pygame.transform.smoothscale(BLACKTOKENIMG, (SPACESIZE, SPACESIZE))
53.      BOARDIMG = pygame.image.load('4row_board.png')
54.      BOARDIMG = pygame.transform.smoothscale(BOARDIMG, (SPACESIZE, SPACESIZE))
55.
56.      HUMANWINNERIMG = pygame.image.load('4row_humanwinner.png')
57.      COMPUTERWINNERIMG = pygame.image.load('4row_computerwinner.png')
58.      TIEWINNERIMG = pygame.image.load('4row_tie.png')
59.      WINNERRECT = HUMANWINNERIMG.get_rect()
60.      WINNERRECT.center = (int(WINDOWWIDTH / 2), int(WINDOWHEIGHT / 2))
61.
62.      ARROWIMG = pygame.image.load('4row_arrow.png')
63.      ARROWRECT = ARROWIMG.get_rect()
64.      ARROWRECT.left = REDPILERECT.right + 10
65.      ARROWRECT.centery = REDPILERECT.centery
66.
67.      isFirstGame = True
68.
69.      while True:
70.          runGame(isFirstGame)
71.          isFirstGame = False
72.
73.
74.  def runGame(isFirstGame):
75.      if isFirstGame:
```

```
 76.         # Let the computer go first on the first game, so the player
 77.         # can see how the tokens are dragged from the token piles.
 78.         turn = COMPUTER
 79.         showHelp = True
 80.     else:
 81.         # Randomly choose who goes first.
 82.         if random.randint(0, 1) == 0:
 83.             turn = COMPUTER
 84.         else:
 85.             turn = HUMAN
 86.         showHelp = False
 87. 
 88.     # Set up a blank board data structure.
 89.     mainBoard = getNewBoard()
 90. 
 91.     while True: # main game loop
 92.         if turn == HUMAN:
 93.             # Human player's turn.
 94.             getHumanMove(mainBoard, showHelp)
 95.             if showHelp:
 96.                 # turn off help arrow after the first move
 97.                 showHelp = False
 98.             if isWinner(mainBoard, RED):
 99.                 winnerImg = HUMANWINNERIMG
100.                 break
101.             turn = COMPUTER # switch to other player's turn
102.         else:
103.             # Computer player's turn.
104.             column = getComputerMove(mainBoard)
105.             animateComputerMoving(mainBoard, column)
106.             makeMove(mainBoard, BLACK, column)
107.             if isWinner(mainBoard, BLACK):
108.                 winnerImg = COMPUTERWINNERIMG
109.                 break
110.             turn = HUMAN # switch to other player's turn
111. 
112.         if isBoardFull(mainBoard):
113.             # A completely filled board means it's a tie.
114.             winnerImg = TIEWINNERIMG
115.             break
116. 
117.     while True:
118.         # Keep looping until player clicks the mouse or quits.
119.         drawBoard(mainBoard)
120.         DISPLAYSURF.blit(winnerImg, WINNERRECT)
121.         pygame.display.update()
```

```
122.            FPSCLOCK.tick()
123.            for event in pygame.event.get(): # event handling loop
124.                if event.type == QUIT or (event.type == KEYUP and event.key == K_ESCAPE):
125.                    pygame.quit()
126.                    sys.exit()
127.                elif event.type == MOUSEBUTTONUP:
128.                    return
129.
130.
131. def makeMove(board, player, column):
132.     lowest = getLowestEmptySpace(board, column)
133.     if lowest != -1:
134.         board[column][lowest] = player
135.
136.
137. def drawBoard(board, extraToken=None):
138.     DISPLAYSURF.fill(BGCOLOR)
139.
140.     # draw tokens
141.     spaceRect = pygame.Rect(0, 0, SPACESIZE, SPACESIZE)
142.     for x in range(BOARDWIDTH):
143.         for y in range(BOARDHEIGHT):
144.             spaceRect.topleft = (XMARGIN + (x * SPACESIZE), YMARGIN + (y * SPACESIZE))
145.             if board[x][y] == RED:
146.                 DISPLAYSURF.blit(REDTOKENIMG, spaceRect)
147.             elif board[x][y] == BLACK:
148.                 DISPLAYSURF.blit(BLACKTOKENIMG, spaceRect)
149.
150.     # draw the extra token
151.     if extraToken != None:
152.         if extraToken['color'] == RED:
153.             DISPLAYSURF.blit(REDTOKENIMG, (extraToken['x'], extraToken['y'], SPACESIZE, SPACESIZE))
154.         elif extraToken['color'] == BLACK:
155.             DISPLAYSURF.blit(BLACKTOKENIMG, (extraToken['x'], extraToken['y'], SPACESIZE, SPACESIZE))
156.
157.     # draw board over the tokens
158.     for x in range(BOARDWIDTH):
159.         for y in range(BOARDHEIGHT):
160.             spaceRect.topleft = (XMARGIN + (x * SPACESIZE), YMARGIN + (y * SPACESIZE))
161.             DISPLAYSURF.blit(BOARDIMG, spaceRect)
162.
```

```
163.        # draw the red and black tokens off to the side
164.        DISPLAYSURF.blit(REDTOKENIMG, REDPILERECT) # red on the left
165.        DISPLAYSURF.blit(BLACKTOKENIMG, BLACKPILERECT) # black on the right
166.
167.
168. def getNewBoard():
169.        board = []
170.        for x in range(BOARDWIDTH):
171.            board.append([EMPTY] * BOARDHEIGHT)
172.        return board
173.
174.
175. def getHumanMove(board, isFirstMove):
176.        draggingToken = False
177.        tokenx, tokeny = None, None
178.        while True:
179.            for event in pygame.event.get(): # event handling loop
180.                if event.type == QUIT:
181.                    pygame.quit()
182.                    sys.exit()
183.                elif event.type == MOUSEBUTTONDOWN and not draggingToken and REDPILERECT.collidepoint(event.pos):
184.                    # start of dragging on red token pile.
185.                    draggingToken = True
186.                    tokenx, tokeny = event.pos
187.                elif event.type == MOUSEMOTION and draggingToken:
188.                    # update the position of the red token being dragged
189.                    tokenx, tokeny = event.pos
190.                elif event.type == MOUSEBUTTONUP and draggingToken:
191.                    # let go of the token being dragged
192.                    if tokeny < YMARGIN and tokenx > XMARGIN and tokenx < WINDOWWIDTH - XMARGIN:
193.                        # let go at the top of the screen.
194.                        column = int((tokenx - XMARGIN) / SPACESIZE)
195.                        if isValidMove(board, column):
196.                            animateDroppingToken(board, column, RED)
197.                            board[column][getLowestEmptySpace(board, column)] = RED
198.                            drawBoard(board)
199.                            pygame.display.update()
200.                            return
201.                    tokenx, tokeny = None, None
202.                    draggingToken = False
203.            if tokenx != None and tokeny != None:
204.                drawBoard(board, {'x':tokenx - int(SPACESIZE / 2), 'y':tokeny - int(SPACESIZE / 2), 'color':RED})
```

10.6 Four-In-A-Row 的源代码

```
205.        else:
206.            drawBoard(board)
207.
208.        if isFirstMove:
209.            # Show the help arrow for the player's first move.
210.            DISPLAYSURF.blit(ARROWIMG, ARROWRECT)
211.
212.        pygame.display.update()
213.        FPSCLOCK.tick()
214.
215.
216. def animateDroppingToken(board, column, color):
217.     x = XMARGIN + column * SPACESIZE
218.     y = YMARGIN - SPACESIZE
219.     dropSpeed = 1.0
220.
221.     lowestEmptySpace = getLowestEmptySpace(board, column)
222.
223.     while True:
224.         y += int(dropSpeed)
225.         dropSpeed += 0.5
226.         if int((y - YMARGIN) / SPACESIZE) >= lowestEmptySpace:
227.             return
228.         drawBoard(board, {'x':x, 'y':y, 'color':color})
229.         pygame.display.update()
230.         FPSCLOCK.tick()
231.
232.
233. def animateComputerMoving(board, column):
234.     x = BLACKPILERECT.left
235.     y = BLACKPILERECT.top
236.     speed = 1.0
237.     # moving the black tile up
238.     while y > (YMARGIN - SPACESIZE):
239.         y -= int(speed)
240.         speed += 0.5
241.         drawBoard(board, {'x':x, 'y':y, 'color':BLACK})
242.         pygame.display.update()
243.         FPSCLOCK.tick()
244.     # moving the black tile over
245.     y = YMARGIN - SPACESIZE
246.     speed = 1.0
247.     while x > (XMARGIN + column * SPACESIZE):
248.         x -= int(speed)
249.         speed += 0.5
250.         drawBoard(board, {'x':x, 'y':y, 'color':BLACK})
```

```
251.          pygame.display.update()
252.          FPSCLOCK.tick()
253.      # dropping the black tile
254.      animateDroppingToken(board, column, BLACK)
255.
256.
257. def getComputerMove(board):
258.     potentialMoves = getPotentialMoves(board, BLACK, DIFFICULTY)
259.     # get the best fitness from the potential moves
260.     bestMoveFitness = -1
261.     for i in range(BOARDWIDTH):
262.         if potentialMoves[i] > bestMoveFitness and isValidMove(board, i):
263.             bestMoveFitness = potentialMoves[i]
264.     # find all potential moves that have this best fitness
265.     bestMoves = []
266.     for i in range(len(potentialMoves)):
267.         if potentialMoves[i] == bestMoveFitness and isValidMove(board, i):
268.             bestMoves.append(i)
269.     return random.choice(bestMoves)
270.
271.
272. def getPotentialMoves(board, tile, lookAhead):
273.     if lookAhead == 0 or isBoardFull(board):
274.         return [0] * BOARDWIDTH
275.
276.     if tile == RED:
277.         enemyTile = BLACK
278.     else:
279.         enemyTile = RED
280.
281.     # Figure out the best move to make.
282.     potentialMoves = [0] * BOARDWIDTH
283.     for firstMove in range(BOARDWIDTH):
284.         dupeBoard = copy.deepcopy(board)
285.         if not isValidMove(dupeBoard, firstMove):
286.             continue
287.         makeMove(dupeBoard, tile, firstMove)
288.         if isWinner(dupeBoard, tile):
289.             # a winning move automatically gets a perfect fitness
290.             potentialMoves[firstMove] = 1
291.             break # don't bother calculating other moves
292.         else:
293.             # do other player's counter moves and determine best one
294.             if isBoardFull(dupeBoard):
295.                 potentialMoves[firstMove] = 0
296.             else:
```

10.6 Four-In-A-Row 的源代码

```
297.                for counterMove in range(BOARDWIDTH):
298.                    dupeBoard2 = copy.deepcopy(dupeBoard)
299.                    if not isValidMove(dupeBoard2, counterMove):
300.                        continue
301.                    makeMove(dupeBoard2, enemyTile, counterMove)
302.                    if isWinner(dupeBoard2, enemyTile):
303.                        # a losing move automatically gets the worst fitness
304.                        potentialMoves[firstMove] = -1
305.                        break
306.                    else:
307.                        # do the recursive call to getPotentialMoves()
308.                        results = getPotentialMoves(dupeBoard2, tile, lookAhead - 1)
309.                        potentialMoves[firstMove] += (sum(results) / BOARDWIDTH) / BOARDWIDTH
310.    return potentialMoves
311.
312.
313. def getLowestEmptySpace(board, column):
314.     # Return the row number of the lowest empty row in the given column.
315.     for y in range(BOARDHEIGHT-1, -1, -1):
316.         if board[column][y] == EMPTY:
317.             return y
318.     return -1
319.
320.
321. def isValidMove(board, column):
322.     # Returns True if there is an empty space in the given column.
323.     # Otherwise returns False.
324.     if column < 0 or column >= (BOARDWIDTH) or board[column][0] != EMPTY:
325.         return False
326.     return True
327.
328.
329. def isBoardFull(board):
330.     # Returns True if there are no empty spaces anywhere on the board.
331.     for x in range(BOARDWIDTH):
332.         for y in range(BOARDHEIGHT):
333.             if board[x][y] == EMPTY:
334.                 return False
335.     return True
336.
337.
338. def isWinner(board, tile):
339.     # check horizontal spaces
```

```
340.        for x in range(BOARDWIDTH - 3):
341.            for y in range(BOARDHEIGHT):
342.                if board[x][y] == tile and board[x+1][y] == tile and board[x+2][y] == tile and board[x+3][y] == tile:
343.                    return True
344.        # check vertical spaces
345.        for x in range(BOARDWIDTH):
346.            for y in range(BOARDHEIGHT - 3):
347.                if board[x][y] == tile and board[x][y+1] == tile and board[x][y+2] == tile and board[x][y+3] == tile:
348.                    return True
349.        # check / diagonal spaces
350.        for x in range(BOARDWIDTH - 3):
351.            for y in range(3, BOARDHEIGHT):
352.                if board[x][y] == tile and board[x+1][y-1] == tile and board[x+2][y-2] == tile and board[x+3][y-3] == tile:
353.                    return True
354.        # check \ diagonal spaces
355.        for x in range(BOARDWIDTH - 3):
356.            for y in range(BOARDHEIGHT - 3):
357.                if board[x][y] == tile and board[x+1][y+1] == tile and board[x+2][y+2] == tile and board[x+3][y+3] == tile:
358.                    return True
359.    return False
360.
361.
362. if __name__ == '__main__':
363.    main()
```

10.7 Gemgem，Bejeweled 的翻版

在 Bejeweled 游戏中，宝石落下填满了游戏板，如图 10-7 所示。玩家可以交换任何两个相邻的宝石，以试图将 3 个宝石排成一行（垂直方向上或者水平方向上，但是对角线不可以）。排成行的宝石就会消失，为从顶部落下的新的宝石腾出空间。排出 3 个以上的宝石，或者导致宝石排成行的连锁反应，都将获得更多的分数。玩家的分数随着时间而缓慢下降，因此，玩家必须不断地进行宝石匹配。当不可能在游戏板上进行任何匹配的时候，游戏结束。

10.8　Gemgem 的源代码

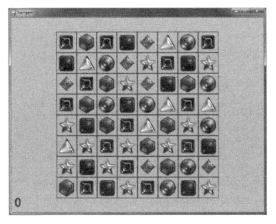 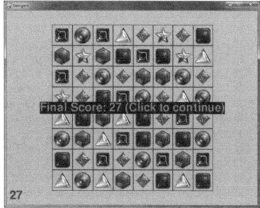

图 10-7

10.8　Gemgem 的源代码

可以从 http://invpy.com/gemgem.py 下载 Gemgem 的源代码。

可以从 http://invpy.com/gemgemimages.zip 下载 Gemgem 所用到的图像文件。

```
 1. # Gemgem (a Bejeweled clone)
 2. # By Al Sweigart al@inventwithpython.com
 3. # http://inventwithpython.com/pygame
 4. # Released under a "Simplified BSD" license
 5.
 6. """
 7. This program has "gem data structures", which are basically dictionaries
 8. with the following keys:
 9.   'x' and 'y' - The location of the gem on the board. 0,0 is the top left.
10.                 There is also a ROWABOVEBOARD row that 'y' can be set to,
11.                 to indicate that it is above the board.
12.   'direction' - one of the four constant variables UP, DOWN, LEFT, RIGHT.
13.                 This is the direction the gem is moving.
14.   'imageNum'  - The integer index into GEMIMAGES to denote which image
15.                 this gem uses.
16. """
17.
18. import random, time, pygame, sys, copy
19. from pygame.locals import *
20.
21. FPS = 30 # frames per second to update the screen
22. WINDOWWIDTH = 600  # width of the program's window, in pixels
```

```
23. WINDOWHEIGHT = 600 # height in pixels
24.
25. BOARDWIDTH = 8 # how many columns in the board
26. BOARDHEIGHT = 8 # how many rows in the board
27. GEMIMAGESIZE = 64 # width & height of each space in pixels
28.
29. # NUMGEMIMAGES is the number of gem types. You will need .png image
30. # files named gem0.png, gem1.png, etc. up to gem(N-1).png.
31. NUMGEMIMAGES = 7
32. assert NUMGEMIMAGES >= 5 # game needs at least 5 types of gems to work
33.
34. # NUMMATCHSOUNDS is the number of different sounds to choose from when
35. # a match is made. The .wav files are named match0.wav, match1.wav, etc.
36. NUMMATCHSOUNDS = 6
37.
38. MOVERATE = 25 # 1 to 100, larger num means faster animations
39. DEDUCTSPEED = 0.8 # reduces score by 1 point every DEDUCTSPEED seconds.
40.
41. #              R    G    B
42. PURPLE    = (255,   0, 255)
43. LIGHTBLUE = (170, 190, 255)
44. BLUE      = (  0,   0, 255)
45. RED       = (255, 100, 100)
46. BLACK     = (  0,   0,   0)
47. BROWN     = ( 85,  65,   0)
48. HIGHLIGHTCOLOR = PURPLE # color of the selected gem's border
49. BGCOLOR = LIGHTBLUE # background color on the screen
50. GRIDCOLOR = BLUE # color of the game board
51. GAMEOVERCOLOR = RED # color of the "Game over" text.
52. GAMEOVERBGCOLOR = BLACK # background color of the "Game over" text.
53. SCORECOLOR = BROWN # color of the text for the player's score
54.
55. # The amount of space to the sides of the board to the edge of the window
56. # is used several times, so calculate it once here and store in variables.
57. XMARGIN = int((WINDOWWIDTH - GEMIMAGESIZE * BOARDWIDTH) / 2)
58. YMARGIN = int((WINDOWHEIGHT - GEMIMAGESIZE * BOARDHEIGHT) / 2)
59.
60. # constants for direction values
61. UP = 'up'
62. DOWN = 'down'
63. LEFT = 'left'
64. RIGHT = 'right'
65.
66. EMPTY_SPACE = -1 # an arbitrary, nonpositive value
67. ROWABOVEBOARD = 'row above board' # an arbitrary, noninteger value
68.
```

10.8 Gemgem 的源代码

```
 69. def main():
 70.     global FPSCLOCK, DISPLAYSURF, GEMIMAGES, GAMESOUNDS, BASICFONT,
BOARDRECTS
 71.
 72.     # Initial set up.
 73.     pygame.init()
 74.     FPSCLOCK = pygame.time.Clock()
 75.     DISPLAYSURF = pygame.display.set_mode((WINDOWWIDTH, WINDOWHEIGHT))
 76.     pygame.display.set_caption('Gemgem')
 77.     BASICFONT = pygame.font.Font('freesansbold.ttf', 36)
 78.
 79.     # Load the images
 80.     GEMIMAGES = []
 81.     for i in range(1, NUMGEMIMAGES+1):
 82.         gemImage = pygame.image.load('gem%s.png' % i)
 83.         if gemImage.get_size() != (GEMIMAGESIZE, GEMIMAGESIZE):
 84.             gemImage = pygame.transform.smoothscale(gemImage,
(GEMIMAGESIZE, GEMIMAGESIZE))
 85.         GEMIMAGES.append(gemImage)
 86.
 87.     # Load the sounds.
 88.     GAMESOUNDS = {}
 89.     GAMESOUNDS['bad swap'] = pygame.mixer.Sound('badswap.wav')
 90.     GAMESOUNDS['match'] = []
 91.     for i in range(NUMMATCHSOUNDS):
 92.         GAMESOUNDS['match'].append(pygame.mixer.Sound('match%s.wav' % i))
 93.
 94.     # Create pygame.Rect objects for each board space to
 95.     # do board-coordinate-to-pixel-coordinate conversions.
 96.     BOARDRECTS = []
 97.     for x in range(BOARDWIDTH):
 98.         BOARDRECTS.append([])
 99.         for y in range(BOARDHEIGHT):
100.             r = pygame.Rect((XMARGIN + (x * GEMIMAGESIZE),
101.                              YMARGIN + (y * GEMIMAGESIZE),
102.                              GEMIMAGESIZE,
103.                              GEMIMAGESIZE))
104.             BOARDRECTS[x].append(r)
105.
106.     while True:
107.         runGame()
108.
109.
110. def runGame():
111.     # Plays through a single game. When the game is over, this function
returns.
```

```
112.
113.        # initialize the board
114.        gameBoard = getBlankBoard()
115.        score = 0
116.        fillBoardAndAnimate(gameBoard, [], score) # Drop the initial gems.
117.
118.        # initialize variables for the start of a new game
119.        firstSelectedGem = None
120.        lastMouseDownX = None
121.        lastMouseDownY = None
122.        gameIsOver = False
123.        lastScoreDeduction = time.time()
124.        clickContinueTextSurf = None
125.
126.        while True: # main game loop
127.            clickedSpace = None
128.            for event in pygame.event.get(): # event handling loop
129.                if event.type == QUIT or (event.type == KEYUP and event.key == K_ESCAPE):
130.                    pygame.quit()
131.                    sys.exit()
132.                elif event.type == KEYUP and event.key == K_BACKSPACE:
133.                    return # start a new game
134.
135.                elif event.type == MOUSEBUTTONUP:
136.                    if gameIsOver:
137.                        return # after games ends, click to start a new game
138.
139.                    if event.pos == (lastMouseDownX, lastMouseDownY):
140.                        # This event is a mouse click, not the end of a mouse drag.
141.                        clickedSpace = checkForGemClick(event.pos)
142.                    else:
143.                        # this is the end of a mouse drag
144.                        firstSelectedGem = checkForGemClick((lastMouseDownX, lastMouseDownY))
145.                        clickedSpace = checkForGemClick(event.pos)
146.                        if not firstSelectedGem or not clickedSpace:
147.                            # if not part of a valid drag, deselect both
148.                            firstSelectedGem = None
149.                            clickedSpace = None
150.                elif event.type == MOUSEBUTTONDOWN:
151.                    # this is the start of a mouse click or mouse drag
152.                    lastMouseDownX, lastMouseDownY = event.pos
153.
154.            if clickedSpace and not firstSelectedGem:
```

```
155.            # This was the first gem clicked on.
156.            firstSelectedGem = clickedSpace
157.        elif clickedSpace and firstSelectedGem:
158.            # Two gems have been clicked on and selected. Swap the gems.
159.            firstSwappingGem, secondSwappingGem =
getSwappingGems(gameBoard, firstSelectedGem, clickedSpace)
160.            if firstSwappingGem == None and secondSwappingGem == None:
161.                # If both are None, then the gems were not adjacent
162.                firstSelectedGem = None # deselect the first gem
163.                continue
164.
165.            # Show the swap animation on the screen.
166.            boardCopy = getBoardCopyMinusGems(gameBoard,
(firstSwappingGem, secondSwappingGem))
167.            animateMovingGems(boardCopy, [firstSwappingGem,
secondSwappingGem], [], score)
168.
169.            # Swap the gems in the board data structure.
170.            gameBoard[firstSwappingGem['x']][firstSwappingGem['y']] =
secondSwappingGem['imageNum']
171.            gameBoard[secondSwappingGem['x']][secondSwappingGem['y']] =
firstSwappingGem['imageNum']
172.
173.            # See if this is a matching move.
174.            matchedGems = findMatchingGems(gameBoard)
175.            if matchedGems == []:
176.                # Was not a matching move; swap the gems back
177.                GAMESOUNDS['bad swap'].play()
178.                animateMovingGems(boardCopy, [firstSwappingGem,
secondSwappingGem], [], score)
179.                gameBoard[firstSwappingGem['x']][firstSwappingGem['y']] =
firstSwappingGem['imageNum']
180.                gameBoard[secondSwappingGem['x']][secondSwappingGem['y']]
= secondSwappingGem['imageNum']
181.            else:
182.                # This was a matching move.
183.                scoreAdd = 0
184.                while matchedGems != []:
185.                    # Remove matched gems, then pull down the board.
186.
187.                    # points is a list of dicts that tells
fillBoardAndAnimate()
188.                    # where on the screen to display text to show how many
189.                    # points the player got. points is a list because if
190.                    # the player gets multiple matches, then multiple
```

```
points text should appear.
191.                    points = []
192.                    for gemSet in matchedGems:
193.                        scoreAdd += (10 + (len(gemSet) - 3) * 10)
194.                        for gem in gemSet:
195.                            gameBoard[gem[0]][gem[1]] = EMPTY_SPACE
196.                            points.append({'points': scoreAdd,
197.                                            'x': gem[0] * GEMIMAGESIZE + XMARGIN,
198.                                            'y': gem[1] * GEMIMAGESIZE + YMARGIN})
199.                        random.choice(GAMESOUNDS['match']).play()
200.                        score += scoreAdd
201.
202.                        # Drop the new gems.
203.                        fillBoardAndAnimate(gameBoard, points, score)
204.
205.                        # Check if there are any new matches.
206.                        matchedGems = findMatchingGems(gameBoard)
207.                    firstSelectedGem = None
208.
209.                    if not canMakeMove(gameBoard):
210.                        gameIsOver = True
211.
212.        # Draw the board.
213.        DISPLAYSURF.fill(BGCOLOR)
214.        drawBoard(gameBoard)
215.        if firstSelectedGem != None:
216.            highlightSpace(firstSelectedGem['x'], firstSelectedGem['y'])
217.        if gameIsOver:
218.            if clickContinueTextSurf == None:
219.                # Only render the text once. In future iterations, just
220.                # use the Surface object already in clickContinueTextSurf
221.                clickContinueTextSurf = BASICFONT.render('Final Score: %s (Click to continue)' % (score), 1, GAMEOVERCOLOR, GAMEOVERBGCOLOR)
222.                clickContinueTextRect = clickContinueTextSurf.get_rect()
223.                clickContinueTextRect.center = int(WINDOWWIDTH / 2), int(WINDOWHEIGHT / 2)
224.            DISPLAYSURF.blit(clickContinueTextSurf, clickContinueTextRect)
225.        elif score > 0 and time.time() - lastScoreDeduction > DEDUCTSPEED:
226.            # score drops over time
227.            score -= 1
228.            lastScoreDeduction = time.time()
229.        drawScore(score)
230.        pygame.display.update()
231.        FPSCLOCK.tick(FPS)
```

10.8 Gemgem 的源代码

```
232.
233.
234. def getSwappingGems(board, firstXY, secondXY):
235.     # If the gems at the (X, Y) coordinates of the two gems are adjacent,
236.     # then their 'direction' keys are set to the appropriate direction
237.     # value to be swapped with each other.
238.     # Otherwise, (None, None) is returned.
239.     firstGem = {'imageNum': board[firstXY['x']][firstXY['y']],
240.                 'x': firstXY['x'],
241.                 'y': firstXY['y']}
242.     secondGem = {'imageNum': board[secondXY['x']][secondXY['y']],
243.                  'x': secondXY['x'],
244.                  'y': secondXY['y']}
245.     highlightedGem = None
246.     if firstGem['x'] == secondGem['x'] + 1 and firstGem['y'] == secondGem['y']:
247.         firstGem['direction'] = LEFT
248.         secondGem['direction'] = RIGHT
249.     elif firstGem['x'] == secondGem['x'] - 1 and firstGem['y'] == secondGem['y']:
250.         firstGem['direction'] = RIGHT
251.         secondGem['direction'] = LEFT
252.     elif firstGem['y'] == secondGem['y'] + 1 and firstGem['x'] == secondGem['x']:
253.         firstGem['direction'] = UP
254.         secondGem['direction'] = DOWN
255.     elif firstGem['y'] == secondGem['y'] - 1 and firstGem['x'] == secondGem['x']:
256.         firstGem['direction'] = DOWN
257.         secondGem['direction'] = UP
258.     else:
259.         # These gems are not adjacent and can't be swapped.
260.         return None, None
261.     return firstGem, secondGem
262.
263.
264. def getBlankBoard():
265.     # Create and return a blank board data structure.
266.     board = []
267.     for x in range(BOARDWIDTH):
268.         board.append([EMPTY_SPACE] * BOARDHEIGHT)
269.     return board
270.
271.
272. def canMakeMove(board):
273.     # Return True if the board is in a state where a matching
```

```
274.        # move can be made on it. Otherwise return False.
275.
276.        # The patterns in oneOffPatterns represent gems that are configured
277.        # in a way where it only takes one move to make a triplet.
278.        oneOffPatterns = (((0,1), (1,0), (2,0)),
279.                          ((0,1), (1,1), (2,0)),
280.                          ((0,0), (1,1), (2,0)),
281.                          ((0,1), (1,0), (2,1)),
282.                          ((0,0), (1,0), (2,1)),
283.                          ((0,0), (1,1), (2,1)),
284.                          ((0,0), (0,2), (0,3)),
285.                          ((0,0), (0,1), (0,3)))
286.
287.        # The x and y variables iterate over each space on the board.
288.        # If we use + to represent the currently iterated space on the
289.        # board, then this pattern: ((0,1), (1,0), (2,0))refers to identical
290.        # gems being set up like this:
291.        #
292.        #     +A
293.        #     B
294.        #     C
295.        #
296.        # That is, gem A is offset from the + by (0,1), gem B is offset
297.        # by (1,0), and gem C is offset by (2,0). In this case, gem A can
298.        # be swapped to the left to form a vertical three-in-a-row triplet.
299.        #
300.        # There are eight possible ways for the gems to be one move
301.        # away from forming a triple, hence oneOffPattern has 8 patterns.
302.
303.        for x in range(BOARDWIDTH):
304.            for y in range(BOARDHEIGHT):
305.                for pat in oneOffPatterns:
306.                    # check each possible pattern of "match in next move" to
307.                    # see if a possible move can be made.
308.                    if (getGemAt(board, x+pat[0][0], y+pat[0][1]) == \
309.                        getGemAt(board, x+pat[1][0], y+pat[1][1]) == \
310.                        getGemAt(board, x+pat[2][0], y+pat[2][1]) != None) or \
311.                       (getGemAt(board, x+pat[0][1], y+pat[0][0]) == \
312.                        getGemAt(board, x+pat[1][1], y+pat[1][0]) == \
313.                        getGemAt(board, x+pat[2][1], y+pat[2][0]) != None):
314.                            return True # return True the first time you find a pattern
315.        return False
316.
317.
```

```
318. def drawMovingGem(gem, progress):
319.     # Draw a gem sliding in the direction that its 'direction' key
320.     # indicates. The progress parameter is a number from 0 (just
321.     # starting) to 100 (slide complete).
322.     movex = 0
323.     movey = 0
324.     progress *= 0.01
325.
326.     if gem['direction'] == UP:
327.         movey = -int(progress * GEMIMAGESIZE)
328.     elif gem['direction'] == DOWN:
329.         movey = int(progress * GEMIMAGESIZE)
330.     elif gem['direction'] == RIGHT:
331.         movex = int(progress * GEMIMAGESIZE)
332.     elif gem['direction'] == LEFT:
333.         movex = -int(progress * GEMIMAGESIZE)
334.
335.     basex = gem['x']
336.     basey = gem['y']
337.     if basey == ROWABOVEBOARD:
338.         basey = -1
339.
340.     pixelx = XMARGIN + (basex * GEMIMAGESIZE)
341.     pixely = YMARGIN + (basey * GEMIMAGESIZE)
342.     r = pygame.Rect( (pixelx + movex, pixely + movey, GEMIMAGESIZE, GEMIMAGESIZE) )
343.     DISPLAYSURF.blit(GEMIMAGES[gem['imageNum']], r)
344.
345.
346. def pullDownAllGems(board):
347.     # pulls down gems on the board to the bottom to fill in any gaps
348.     for x in range(BOARDWIDTH):
349.         gemsInColumn = []
350.         for y in range(BOARDHEIGHT):
351.             if board[x][y] != EMPTY_SPACE:
352.                 gemsInColumn.append(board[x][y])
353.         board[x] = ([EMPTY_SPACE] * (BOARDHEIGHT - len(gemsInColumn))) + gemsInColumn
354.
355.
356. def getGemAt(board, x, y):
357.     if x < 0 or y < 0 or x >= BOARDWIDTH or y >= BOARDHEIGHT:
358.         return None
359.     else:
360.         return board[x][y]
361.
```

```
362.
363. def getDropSlots(board):
364.     # Creates a "drop slot" for each column and fills the slot with a
365.     # number of gems that that column is lacking. This function assumes
366.     # that the gems have been gravity dropped already.
367.     boardCopy = copy.deepcopy(board)
368.     pullDownAllGems(boardCopy)
369.
370.     dropSlots = []
371.     for i in range(BOARDWIDTH):
372.         dropSlots.append([])
373.
374.     # count the number of empty spaces in each column on the board
375.     for x in range(BOARDWIDTH):
376.         for y in range(BOARDHEIGHT-1, -1, -1): # start from bottom, going up
377.             if boardCopy[x][y] == EMPTY_SPACE:
378.                 possibleGems = list(range(len(GEMIMAGES)))
379.                 for offsetX, offsetY in ((0, -1), (1, 0), (0, 1), (-1, 0)):
380.                     # Narrow down the possible gems we should put in the
381.                     # blank space so we don't end up putting an two of
382.                     # the same gems next to each other when they drop.
383.                     neighborGem = getGemAt(boardCopy, x + offsetX, y + offsetY)
384.                     if neighborGem != None and neighborGem in possibleGems:
385.                         possibleGems.remove(neighborGem)
386.
387.                 newGem = random.choice(possibleGems)
388.                 boardCopy[x][y] = newGem
389.                 dropSlots[x].append(newGem)
390.     return dropSlots
391.
392.
393. def findMatchingGems(board):
394.     gemsToRemove = [] # a list of lists of gems in matching triplets that should be removed
395.     boardCopy = copy.deepcopy(board)
396.
397.     # loop through each space, checking for 3 adjacent identical gems
398.     for x in range(BOARDWIDTH):
399.         for y in range(BOARDHEIGHT):
400.             # look for horizontal matches
401.             if getGemAt(boardCopy, x, y) == getGemAt(boardCopy, x + 1, y) == getGemAt(boardCopy, x + 2, y) and getGemAt(boardCopy, x, y) != EMPTY_SPACE:
```

```
402.                targetGem = boardCopy[x][y]
403.                offset = 0
404.                removeSet = []
405.                while getGemAt(boardCopy, x + offset, y) == targetGem:
406.                    # keep checking, in case there's more than 3 gems in a row
407.                    removeSet.append((x + offset, y))
408.                    boardCopy[x + offset][y] = EMPTY_SPACE
409.                    offset += 1
410.                gemsToRemove.append(removeSet)
411.
412.            # look for vertical matches
413.            if getGemAt(boardCopy, x, y) == getGemAt(boardCopy, x, y + 1) == getGemAt(boardCopy, x, y + 2) and getGemAt(boardCopy, x, y) != EMPTY_SPACE:
414.                targetGem = boardCopy[x][y]
415.                offset = 0
416.                removeSet = []
417.                while getGemAt(boardCopy, x, y + offset) == targetGem:
418.                    # keep checking if there's more than 3 gems in a row
419.                    removeSet.append((x, y + offset))
420.                    boardCopy[x][y + offset] = EMPTY_SPACE
421.                    offset += 1
422.                gemsToRemove.append(removeSet)
423.
424.    return gemsToRemove
425.
426.
427. def highlightSpace(x, y):
428.     pygame.draw.rect(DISPLAYSURF, HIGHLIGHTCOLOR, BOARDRECTS[x][y], 4)
429.
430.
431. def getDroppingGems(board):
432.     # Find all the gems that have an empty space below them
433.     boardCopy = copy.deepcopy(board)
434.     droppingGems = []
435.     for x in range(BOARDWIDTH):
436.         for y in range(BOARDHEIGHT - 2, -1, -1):
437.             if boardCopy[x][y + 1] == EMPTY_SPACE and boardCopy[x][y] != EMPTY_SPACE:
438.                 # This space drops if not empty but the space below it is
439.                 droppingGems.append( {'imageNum': boardCopy[x][y], 'x': x, 'y': y, 'direction': DOWN} )
440.                 boardCopy[x][y] = EMPTY_SPACE
441.     return droppingGems
442.
```

```
443.
444. def animateMovingGems(board, gems, pointsText, score):
445.     # pointsText is a dictionary with keys 'x', 'y', and 'points'
446.     progress = 0 # progress at 0 represents beginning, 100 means finished.
447.     while progress < 100: # animation loop
448.         DISPLAYSURF.fill(BGCOLOR)
449.         drawBoard(board)
450.         for gem in gems: # Draw each gem.
451.             drawMovingGem(gem, progress)
452.         drawScore(score)
453.         for pointText in pointsText:
454.             pointsSurf = BASICFONT.render(str(pointText['points']), 1, SCORECOLOR)
455.             pointsRect = pointsSurf.get_rect()
456.             pointsRect.center = (pointText['x'], pointText['y'])
457.             DISPLAYSURF.blit(pointsSurf, pointsRect)
458.
459.         pygame.display.update()
460.         FPSCLOCK.tick(FPS)
461.         progress += MOVERATE # progress the animation a little bit more for the next frame
462.
463.
464. def moveGems(board, movingGems):
465.     # movingGems is a list of dicts with keys x, y, direction, imageNum
466.     for gem in movingGems:
467.         if gem['y'] != ROWABOVEBOARD:
468.             board[gem['x']][gem['y']] = EMPTY_SPACE
469.             movex = 0
470.             movey = 0
471.             if gem['direction'] == LEFT:
472.                 movex = -1
473.             elif gem['direction'] == RIGHT:
474.                 movex = 1
475.             elif gem['direction'] == DOWN:
476.                 movey = 1
477.             elif gem['direction'] == UP:
478.                 movey = -1
479.             board[gem['x'] + movex][gem['y'] + movey] = gem['imageNum']
480.         else:
481.             # gem is located above the board (where new gems come from)
482.             board[gem['x']][0] = gem['imageNum'] # move to top row
483.
484.
485. def fillBoardAndAnimate(board, points, score):
486.     dropSlots = getDropSlots(board)
```

```
487.     while dropSlots != [[]] * BOARDWIDTH:
488.         # do the dropping animation as long as there are more gems to drop
489.         movingGems = getDroppingGems(board)
490.         for x in range(len(dropSlots)):
491.             if len(dropSlots[x]) != 0:
492.                 # cause the lowest gem in each slot to begin moving in the
DOWN direction
493.                 movingGems.append({'imageNum': dropSlots[x][0], 'x': x,
'y': ROWABOVEBOARD, 'direction': DOWN})
494.
495.         boardCopy = getBoardCopyMinusGems(board, movingGems)
496.         animateMovingGems(boardCopy, movingGems, points, score)
497.         moveGems(board, movingGems)
498.
499.         # Make the next row of gems from the drop slots
500.         # the lowest by deleting the previous lowest gems.
501.         for x in range(len(dropSlots)):
502.             if len(dropSlots[x]) == 0:
503.                 continue
504.             board[x][0] = dropSlots[x][0]
505.             del dropSlots[x][0]
506.
507.
508. def checkForGemClick(pos):
509.     # See if the mouse click was on the board
510.     for x in range(BOARDWIDTH):
511.         for y in range(BOARDHEIGHT):
512.             if BOARDRECTS[x][y].collidepoint(pos[0], pos[1]):
513.                 return {'x': x, 'y': y}
514.     return None # Click was not on the board.
515.
516.
517. def drawBoard(board):
518.     for x in range(BOARDWIDTH):
519.         for y in range(BOARDHEIGHT):
520.             pygame.draw.rect(DISPLAYSURF, GRIDCOLOR, BOARDRECTS[x][y], 1)
521.             gemToDraw = board[x][y]
522.             if gemToDraw != EMPTY_SPACE:
523.                 DISPLAYSURF.blit(GEMIMAGES[gemToDraw], BOARDRECTS[x][y])
524.
525.
526. def getBoardCopyMinusGems(board, gems):
527.     # Creates and returns a copy of the passed board data structure,
528.     # with the gems in the "gems" list removed from it.
529.     #
530.     # Gems is a list of dicts, with keys x, y, direction, imageNum
```

```
531.
532.        boardCopy = copy.deepcopy(board)
533.
534.        # Remove some of the gems from this board data structure copy.
535.        for gem in gems:
536.            if gem['y'] != ROWABOVEBOARD:
537.                boardCopy[gem['x']][gem['y']] = EMPTY_SPACE
538.        return boardCopy
539.
540.
541.    def drawScore(score):
542.        scoreImg = BASICFONT.render(str(score), 1, SCORECOLOR)
543.        scoreRect = scoreImg.get_rect()
544.        scoreRect.bottomleft = (10, WINDOWHEIGHT - 6)
545.        DISPLAYSURF.blit(scoreImg, scoreRect)
546.
547.
548.    if __name__ == '__main__':
549.        main()
```

10.9 本章小结

我希望这些游戏程序能够让你对自己想要制作什么样的游戏以及如何编写其代码产生一些想法。即便你自己没有任何想法，尝试编写你玩过的游戏的一个翻版，这也是很不错的。

如下是教授更多 Python 编程知识的几个 Web 站点。

- http://pygame.org——Pygame 官方网站，拥有人们使用 Pygame 库所编写的数以百计的游戏的源代码。你可以通过下载和阅读别人的源代码而学到很多。
- http://python.org/doc——更多的 Python 教程，并且包括所有的 Python 模块和函数的文档。
- http://pygame.org/docs/—— 关于 Pygame 的模块和函数的完整文档。
- http://reddit.com/r/learnpython 和 http://reddit.com/r/learnprogramming——有几个用户，能够帮助你找到学习编程的资源。
- http://inventwithpython.com/pygame—— 本书的 Web 站点，包括了这些程序的所有源代码以及额外信息。这个站点还有 Pygame 程序中用到的图像和声音文件。
- http://inventwithpython.com——《Invent Your Own Computer Games with Python》一书的 Web 站点，这本书介绍了 Python 编程的基础知识。
- http://invpy.com/wiki——介绍各种 Python 编程概念的 wiki 页面，如果你想要学习一些特定内容的话，可以在上面查找。
- http://invpy.com/traces—— 这是一个 Web 应用程序，能够帮助你一步一步地跟踪本书中的程序的运行。
- http://invpy.com/videos——和本书中的程序配套的视频。
- http://gamedevlessons.com——关于设计和编写视频游戏的一个有用的 Web 站点。
- al@inventwithpython.com——我的 Email 地址。关于本书或 Python 编程的问题，可以写信给我。

也许你通过上网搜索可以找到更多有关 Python 的信息。请访问 Goolge 并搜索"Python programming"或"Python tutorials"以找到教授你有关 Python 编程的更多知识的 Web 站点。

现在，开始去编写自己的游戏吧！祝你好运！

术语表

Alpha 值（Alpha Value） 表示颜色透明度的量。在 Pygame 中，alpha 值的范围从 0（完全透明）到 255（完全不透明）。

抗锯齿（Anti-Aliasing） 通过给形状的边界添加模糊色，使其看上去更加平滑而减少块状化效果。抗锯齿的绘制看上去更加平滑。而带有锯齿的绘制看上去块状化。

属性（Attributes） 作为对象的一部分的一个变量。例如，Rect 对象拥有诸如 top 和 lef 这样的成员，其中保存了用于该 Rect 对象的整数值。

向后兼容（Backwards Compatibility） 编写能够与软件的旧版本兼容的代码。Python 3 有一些向后兼容 Python 2 的功能，但是，编写向后兼容 Python 2 的 Python 3 程序也是可能的。

基本条件（Base Case） 在递归中，基本条件是停止进一步的递归函数调用的条件。基本条件是防止栈溢出错误所必需的。

复制（Blitting） Blitting 这个单词表示将一个 Surface 对象上的内容复制到另一个 Surface 对象上。在编程中，通常这意味着将一幅图像复制到另一幅图像。

边界矩形（Bounding Rectangle） 能够围绕另一个形状绘制的最小的矩形。

相机（Camera） 游戏世界的一个特定部分的视图。当游戏世界太大了，无法放到玩家的屏幕之上的时候，就需要使用相机。

标题（Caption） 在编程中，标题是窗口的标题栏上的文本。在 Pygame 中，可以使用 pygame.display.set_caption()函数来设置标题。

CLI 参见命令行界面（Command Line Interface）

命令行界面（Command Line Interface） 这是一个程序，用户可以用来查看屏幕上的文本并通过键盘输入文本。旧的计算机只能通过运行 CLI 程序来使用，但是，新的计算机有了图形化用户界面（Graphical User Interfaces）。

构造函数（Constructor Function） 创建一个新的对象的函数。在 Python 中，这些函数与它们所产生类具有相同的名称。例如，pygame.Rect()创建 Rect 对象。

显示 Surface（Display Surface） 调用 pygame.display.set_mode()所返回的 Surface 对象。这个 Surface 对象很特别，因为使用 Pygame 的绘制函数或复制函数绘制于其上的任何内容，在调用 pygame.display.update()的时候都会显示于屏幕上。

绘制图元（Drawing Primitives） Pygame 中基本的图形绘制函数的名称。绘制图源包括矩形、线条和椭圆形。绘制图元并不包括诸如.png 或.jpg 等文件中的图像。

事件处理（Event Handling） 执行动作以响应用户所生成的 Event 对象（如按键按下

或点击鼠标）的代码。

事件处理循环（Event Handling Loop）　事件处理代码通常在一个循环中，这个循环处理从上一次执行事件处理循环开始所生成的每一个事件。

事件队列（Event Queue）　当鼠标点击或按键按下这样的事件发生的时候，Pygame 将其存储到一个内部的队列数据结构中。可以删除事件，并且通过调用 pygame.event.get() 从事件队列获取事件。

FPS　参见帧每秒

帧（Frame）　作为动画的一部分在屏幕上显示的单个图像。动画图形是由很多的帧组成的，每一帧都在 1 秒的某一个部分中显示。

帧速率（Frame Rate）　参见刷新速率。

帧每秒（Frames Per Second）　度量动画每秒钟显示多少帧的单位。对于游戏来说，通常是每秒显示 30 帧或更多。

游戏循环（Game Loop）　游戏循环包括执行事件处理、更新游戏世界状态和将游戏世界的状态绘制到屏幕的代码。这会在一秒钟内完成很多次。

游戏状态（Game State）　构成游戏世界的值的完整的集合。这可能包括有关玩家角色、游戏板上的每一个砖块，或者得分和关卡编号等信息。

图形化用户界面（Graphical User Interface）　向用户显示图形作为输出，并且能够接受键盘按键和鼠标点击作为输入的一个程序。

GUI　参见图形化用户界面

不可变的（Immutable）　不能够改变或修改的。在 Python 中，列表值是可变的，元组值是不可变的。

交互式 shell（Interactive Shell）　一个程序（是 IDLE 的一部分），每次执行一条 Python 指令。交互式 shell 是体验一行代码做什么的一种好办法。

解释器（Interpreter）　执行用 Python 编程语言编写的指令的软件。在 Windows 中，这是 python.exe。当某人说"Python 运行这个程序"的时候，他的意思是说："Python 解释器软件运行该程序"。

幻数（Magic Numbers）　在程序中使用而没有说明的整数或浮点数。幻数应该使用具有描述性很强的名称的常量变量来替代，从而增加程序的可读性。

主循环（Main Loop）　参见游戏循环

成员变量（Member Variable）　参见属性。

模除操作符（Modulus Operator）　在 Python 中，模除操作符是 % 符号。它执行"求余数"的数学运算。例如，22 / 7 得 3 余 1，因此，22 % 7 的结果为 1。

多维（Multidimensional）　拥有 1 个以上的维度。在 Python 中，这通常表示一个列表包含另一个列表，或者一个字典包含一个元组（该元组反过来可以包含其他的列表、元组或字典）。

术语表

可变的（Mutable） 可以改变或可以修改。在 Python 中，列表值是可变的，元组值是不可变的。

Pi 是表示一个圆的周长是其直径长度的多少倍的一个数值。不管圆有多大，Pi 值是相同的。这个值可以通过 math 模块的 math.pi 来使用，它就是浮点值 3.1415926535897931。

像素（Pixels） 表示"picture element"。像素是计算机屏幕上单个的彩色方块。屏幕由成千上百的像素组成，它们可以针对一幅图像而设置为不同的颜色。

点（Point） Python 中的点表示为两个整数（或浮点值）的一个元组，表示一个 2D Surface 上的一个位置的坐标。

属性（Properties） 参见属性（Attributes）。

实时（Real-time） 持续运行而不会等待玩家做某些事情的程序，称之为实时运行。

递归调用（Recursive Call） 递归函数中，对相同的函数的调用。

刷新速率（Refresh Rate） 计算机屏幕更新其图像的频率。较高或较快的刷新速率会使得动画显得很平滑，而较低或较慢的刷新速率，会使得动画看上去很卡顿。刷新速率用 FPS 或赫兹来度量（其含义是相同的）。

RGB 值（RGB Values） RGB 值是一种特定颜色的具体的值。RGB 表示 red、green 和 blue。在 Pygame 中，RGB 值是 3 个整数的一个元组（这 3 个整数都在 0 到 255 之间），分别表示颜色中的红色、绿色和蓝色的量。

Shell 参见交互式 shell。

精灵（Sprite） 给图像的一个名称。游戏通常针对其中每一种对象都有一个精灵。

栈溢出（Stack Overflow） 当一个递归函数没有一个基本条件的时候会引发的一种错误。

语法糖（Syntactic Sugar） 编写来使得程序更加可读的一些代码，即便对于程序的工作来说它们不是必须的。

贴片精灵（Tile Sprites） 贴片是设计来在 2D 栅格上绘制的一种精灵。它们通常是背景图像，如地板或墙壁。

标题栏（Title Bar） 总是位于程序顶部的一栏，通常包含了程序的标题和关闭按钮。不同的操作系统之中，标题栏的样式也各不相同。

X 轴（X-axis） 笛卡尔坐标系中用于水平布局的数字。X 坐标越小，位置越偏左，越大越偏右。

Y 轴（Y-axis） 笛卡尔坐标系中用于垂直布局的数字。Y 坐标越小，位置越偏上，越大越偏下（这和 Y 轴在数学中的作用相反）。